Cut-off Grades and Optimising the Strategic Mine Plan

by Brian Hall

The Australasian Institute of Mining and Metallurgy
Spectrum Series 20

AusIMM

Published by:
THE AUSTRALASIAN INSTITUTE OF MINING AND METALLURGY
Ground Floor, 204 Lygon Street, Carlton Victoria 3053 Australia

ISBN 978 1 925100 21 1

Desktop published by Stephanie Ashworth and Kelly Steele for
The Australasian Institute of Mining and Metallurgy

Front cover image:
A risk officer looks out over the Mogalakwena Platinum Mine, South Africa.
Image supplied and reproduced by the kind permission of Anglo American.

Foreword

At a time when the mining industry is reviewing its performance and approach in delivering value to shareholders, this book provides the intellectual tools needed by mine managers in order to turn what their directors are promising into reality. Too often, good intentions at a board level do not translate into appropriate actions by technical staff in mine planning and mine optimisation. Inappropriately calculated cut-offs cause many operations to mine and process large quantities of material that does not add value to the operation or its owners.

This book is a worthy successor to Ken Lane's 1988 book *The Economic Definition of Ore*, which is now out of print. Mr Hall has built upon the theory in a very readable manner that will be readily understood by mining engineers and geologists, but this book goes well beyond Lane's mathematics and explains how mine planning and optimisation are undertaken using the latest tools and techniques. All explanations are in plain language and are obviously the result of many years of practice in this field, and of training others in the techniques.

The gold industry, in particular, will welcome this book as it seeks to move from a focus on ounces in reserves or ounces of annual production to a focus on cash flow and the payment of regular dividends. With the emergence of gold contracts for difference (CFDs) in recent years, gold stocks no longer provide leverage to the gold price. It is essential that operators understand that many of the ounces they have previously claimed in resources and reserves are break-even ounces or even loss-making ounces and should be left in the ground, at least until later generations with new technologies can turn them to a profit.

While the primary appeal of this book will be to mining engineers and geologists, metallurgists and other specialists must contribute to the mine optimisation process and will benefit from the concepts explained. Corporate managers, company directors, mineral economists and in particular mining analysts should take the time to work through the concepts, if only to understand just how complex mining and mine optimisation really are.

A book that has the potential to transform the performance of an industry comes along only once in a generation or so. This is one of those books. Readers who take the trouble to understand the concepts and processes and put them into practice will gain a valuable advantage over their competitors, both for their employers and for their own careers.

Geoff Sharrock FAusIMM(CP)
AusIMM President 2013–2014

Preface

In *De Re Metallica*, first published in 1556, Georgius Agricola wrote, in Book I:

> *… it may be asked whether the art of mining is really profitable or not to those who are engaged in it … a learned and experienced miner differs from one ignorant and unskilled in the art. The latter digs out the ore without any careful discrimination, while the former first assays and proves it, and when he finds the veins either too narrow and hard, or too wide and soft, he infers therefrom that these cannot be mined profitably, and so works only the approved ones. What wonder then if we find the incompetent miner suffers loss, while the competent one is rewarded by an abundant return from his mining?*

And again in Book VII, dealing with 'methods of assaying ores':

> *… miners can determine with certainty whether ores contain any metals in them or not; [then] ascertain by such tests the method by which the metal can be separated from that part of the ore devoid of it …*

We see in this early mining industry text the necessity of identifying what is to be mined, what is to be treated and what is not. Discrimination is required to ensure that the mine owners do not suffer 'great loss'. The underlying aim of mining is clearly profitability or economic gain. People go into mining to make money, not to simply generate ore or metal.

Jump forward some 400 years. In 1964, Ken Lane published his seminal paper on cut-off theory, and later, his text book on the topic (1988). In the introduction of the second edition, Lane (1997) wrote:

> *… ores in general are defined operationally by a cut-off grade; material with a mineral content above the cut-off is scheduled for treatment, other material is left or dumped as waste.*

Upon learning this, Lane asked the question: why work to a certain cut-off value grade rather than some other value?

Some of the answers he received were:

- We have always worked to 0.3%.
- Head office decided on 5% combined metals way back.
- It's a technical matter that we leave to the people on site.

Such responses stimulated Lane's interest in defining ore. He claims that the subject had received minimal attention in text books and papers, and there was no authoritative text on the topic:

> *It seemed ironic, in an industry devoted to mining ore that its definition of ore should be so taken for granted.*

Major mining engineering handbooks over the years, such as Peele's (1941) and the SME's (Cummins and Given, 1973), gloss over the cut-off as if it needs no explanation. The latter work contains one sample break-even cut-off grade calculation for an open pit with an assumed ratio of waste to ore, which implies that a cut-off has been specified before the cut-off has been determined! The concept is apparently not important enough for the word 'cut-off' to appear in the index, and even the sample calculation does not appear in the expanded second edition published in 1992.

A number of mining companies (especially the majors) are now applying cut-off and strategy optimisation principles to some extent in their strategic mine planning. Otherwise, little seems to have changed. There is still a widespread belief in the industry that cut-off and break-even are the same thing; that any cut-off not derived by a break-even calculation is somehow wrong. This is despite Lane identifying some 50 years ago that the optimum cut-off to maximise net present value – a commonly expressed goal of many mining companies – may not be a break-even grade, and even if it is, the costs included in the calculation may significantly differ from what is commonly used. Either way, the optimum cut-off will be higher than traditional break-evens.

Like Lane, I have become interested in the closely related topics of cut-offs and strategy optimisation because of the apparent lack of understanding in what they mean. It has become apparent over many years that there are a number of ways that cut-offs can be determined, and no one method is necessarily wrong. What is clear is that these represent distinct models of cut-off theory and derivation. I see the understanding of cut-off theory as analogous to how our understanding of the cosmos has grown – from the insights of Newton and Einstein to the latest theories advanced by physicists – and I expand on this theme in the earlier chapters of this volume. Simple models of the universe, such as Newton's laws of motion, turn out to be approximations of the more comprehensive theory and are useful for many purposes but inappropriate for others. Similarly, basic cut-off models may be adequate for some purposes but are inadequate for others.

Unfortunately, many involved in the industry do not recognise the validity of higher-dimension cut-off models. Industry analysts are heard to say at conferences that cash is king, and chastise mining company managers for failing to deliver cash benefits to shareholders. Yet the measures they focus on are uncorrelated – or, worse, negatively correlated – with cash generation. Since many decision-makers in the industry have no technical background – and those who do have minimal exposure to cut-off theory and strategy optimisation – there is no one in a position to tell mining investors that strategies they believe are correlated with cash generation actually reduce it, and vice versa.

Consider the ounces in reserve for a gold operation. Anecdotal evidence suggests that this is a key metric for industry analysts – indeed, some gold mining companies have stated in their public documents that their aim is to maximise reserve ounces. This might sound good in principle, but if this maximises cash generation the best cut-off policy is obvious: set it to zero – that is, mine and treat every bit of mineralisation with gold as ore. Depending on the grade distribution in the mineralised zone, this may generate a positive net cash flow for the operation, so it is not necessarily a silly suggestion. But there will be howls of protest that some of the low-grade ounces are losing money and not covering their costs of production, with which I concur. Clearly there are two types of ounces: those that make money and those that do not.

So if cash is king and not all ounces generate cash, how many of the ounces in the published reserves of gold mining companies are losing more money than they are making? Based on years of real studies for mining companies, the answer is: a lot. This is perhaps not surprising. Some years ago, a colleague was undertaking a course by the Securities Institute of Australia. The course notes for the mining industry unit taught specifically that cut-off is determined by a break-even calculation. If that's what industry analysts are being taught about cut-offs by an authoritative institution, it is little wonder that that's what they believe.

While presenting short courses on the material in this book for over ten years, I've had several students say that they thought I was just going to clarify what numbers to put into their break-even formulas – which I do, as does this book. They then admit that they had no idea there was so much more to it.

When discussing Lane's theory, I usually ask four questions:

- Who has heard of it?
- Who has started reading Lane's book?
- Who finished it?
- Who understood it?

I've probably asked several hundred people over the years these questions. Of those who ought to have heard of it (say, mining engineers and geologists in planning or technical roles), perhaps 80 per cent have. Of those, about 80 per cent have started to read the book. Unfortunately, it's now out of print, so those numbers are dropping. The real problem comes with the latter two questions. About 20 per have finished it, and I can count on one hand those who claim to have understood it. Most who start but don't finish admit to giving up when they stumble on the rigorous mathematical derivations of Lane's formulas in Chapter 3. This is a pity because, while it's good to know that they have been proven, you don't need to understand the proof to understand the implications of the formulas, and you then miss everything that follows.

My aim with this book has been to develop a framework of understanding, so that everything fits together, building from initial concepts and limited examples to more comprehensive models of cut-off theory and practice. Each chapter concludes with a chapter summary, so that key issues discussed are highlighted and reinforced. My hope is that this book will add to the understanding of cut-offs and strategy optimisation in the industry worldwide and help others to see further than they have in the past.

Brian Hall MAusIMM(CP)
Brisbane, 2014

About the author

Brian's primary expertise is in planning for underground metalliferous mining operations including detailed mine designs, technical assessments, feasibility studies and strategic planning. He began his career as a graduate trainee with Consolidated Gold Fields Australia Ltd in 1973, and after a stint at Mt Lyell in Tasmania, moved to Gold Fields of South Africa in 1977 with posts at West Driefontein Gold Mine and Black Mountain base metals operation.

In 1982 Brian joined Mount Isa Mines in Queensland where the next 12 years saw him fill the roles of Underground Manager (lead and copper), Planning Manager and Chief Mining Engineer. Here he was responsible for the development of an in-house computerised mine planning system, and was involved in studies across a range of mining, mineral processing and pyrometallurgical systems.

Brian joined AMC Consultants in 1994 where his experience covers planning, design and scheduling, technical audits, strategic reviews, feasibility studies, cut-off grade studies, mine strategy optimizations and computer simulations for operations in Australasia, Asia, Europe and North America. Over the last decade, his main areas of work have involved cut-off and strategy optimisation studies for underground and open pit mines across a range of commodities.

Holding a Bachelor of Engineering (Mining) (Honours) and Bachelor of Commerce, Brian has developed and presented courses at undergraduate and postgraduate level at universities and provided technical training for clients. Since 2001 he has presented workshops on project evaluation and the theory and practical application of cut-off grade and mine and business strategy optimization, and has presented a number of technical papers on these subjects. In addition to being a Chartered Professional of the Minerals Institute, Brian is a Chartered Engineer with the Institution of Materials, Minerals and Mining and a Registered Professional Engineer of Queensland.

Acknowledgements

As with any such work, this book is the culmination of the work and input of many people. My sincere thanks are due to the many who have had a hand in its preparation in some way. First, I must thank the management of AMC Consultants Pty Ltd, who have very generously given me the intellectual property rights to the material in the book – most of my understanding has been developed while working as an AMC employee, and any intellectual property rights would therefore normally belong to AMC. Numerous colleagues at AMC have also been involved in the projects on which I have worked, and their assistance and insights have contributed to the development of my own understanding of principles and useful practices.

Thanks are due to the many clients of AMC on whose projects I have worked over the years, both the companies in general and, more particularly, the various staff members with whom I have had the privilege of working and forming friendships over the years. Although there are many similarities between projects, no two operations are the same: each has its own specific issues and much of my understanding has evolved by addressing the intellectual challenges associated with each case. This has only been possible because of the explanations given by the clients' staff members, resulting from their detailed understanding of their operations, and always given ungrudgingly despite their own heavy workloads. Also, understanding has developed by addressing challenging questions thrown at me by various members of each client's team. I mightn't have always had the answer on the spot, particularly in the earlier days, but the questions got me thinking more widely to be able to address more real-world issues adequately. More-recent clients have benefited from insights gained while working on projects for earlier clients; future clients will benefit from current projects; and where a number of investigations have been conducted for the same client over the years, they have also benefited from insights I have gained at their own sites.

Thanks also to the many attendees at my courses over the years. I have finally bowed to the pressure of those who have said it needed to be put into a book. Your questions and responses have in a very real way shaped the overall development of the discussion in this book, and indeed were continuing to do so even as the manuscript was approaching completion.

Thanks to the reviewers of the text. Anthony Allman and Craig Stewart have been most helpful with their comments and criticisms of the early drafts, as have Brett King, Peter McCarthy and Jeff Whittle as official reviewers for the AusIMM. Glen Williamson has contributed details of open pit planning procedures in Chapter 7 that are outside my own areas of expertise and Michael Samis has provided advice on the discussion of real options valuation in Chapter 11. Any remaining errors and controversial opinions are of course my sole responsibility.

Thanks also to the AusIMM Publications Committee, particularly Kristy Burt, who have been remarkably patient during the extended preparation of this book. In mitigation, I plead the pressures of working in an exciting industry at a time when there is a significant shortage of professional staff worldwide. I hope the wait has been worthwhile.

Finally, I would be most remiss not to acknowledge the support of my family. My son Ben has generated all the figures, drawn from a variety of sources, in a consistent format in both colour and monochrome versions for the electronic and paper versions of the book. Most especially, thanks to my wife Jennifer, who has been my support living and working in a number of mining towns and major cities for some 40 years, as you do in this industry—or at least as those of us did who grew up in the days of remote mining towns, before there was such a thing as fly-in, fly-out. She's kept the family operating smoothly while I've been on the road visiting clients in many interesting parts of the world. During the writing of this book, she has spent many evenings and weekends effectively home alone, when we could have been out doing things together, while I have worked away out the back in my office on the text. Couldn't have done it without you, Jen.

Brian Hall MAusIMM(CP)
Brisbane, 2014

Sponsors

The AusIMM would like to thank the following sponsors for their generous support of this volume.

PRINCIPAL SPONSOR

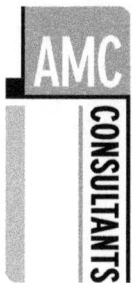

AMC CONSULTANTS

MAJOR SPONSORS

ALFORD MINING SYSTEMS   AngloAmerican   BARRICK   VALE

GENERAL SPONSORS

GOLDFIELDS   Mount Gibson Iron

Major sponsor profile
Alford Mining Systems

Alford Mining Systems (AMS), based in Melbourne (Australia) and established in 1999, is a specialist supplier of research capability, software and services to the mining industry. The principal, Chris Alford, trained as a mining engineer at the University of Melbourne, and has broad experience working in industry, research and development, and academia.

AMS has made significant contributions to the development and application of optimisation techniques for mine design. Over the past ten years the focus has been primarily on new techniques required to solve design tasks specific to underground mine design.

AMS is currently project manager and lead researcher for two large international research projects coordinated by AMIRA International (P1037 and P1043) and funded by seven major international mining companies (BHPB, Newmont, Xstrata, Vale, Barrick, MMG and Boliden) and four mining software suppliers (CAE, Maptek, MineRP and Deswik) to develop tools for stope design and strategic life-of-mine optimisation for underground mine design.

The AMS business model is to develop optimisation engines that can be integrated into the software platforms of the mining software suppliers, utilise their proprietary database APIs, and provide the look and feel of the host platform. AMS has released two engines for automated stope design and optimisation of decline designs. Further products will be developed from the AMIRA project research.

AMS has partnered with AMC Consultants to conduct the mining research and test and evaluate new optimisation approaches and software prototypes.

The philosophy behind the strategic underground optimisation approach developed on the AMIRA research projects follows techniques outlined by Brian Hall in this book, to produce hills of value to allow the impact of key drivers in project evaluation to be analysed and visualised – not just to produce the single optimal result but to evaluate the tradeoffs and sensitivities in mine design choices.

Major sponsor profile
Anglo American

Anglo American is one of the world's largest mining companies, is headquartered in the UK and listed on the London and Johannesburg stock exchanges. Our portfolio of mining businesses meets our customers' changing needs and spans bulk commodities – iron ore and manganese, metallurgical coal and thermal coal; base metals and minerals – copper, nickel, niobium and phosphates; and precious metals and minerals – in which we are a global leader in both platinum and diamonds. At Anglo American, we are committed to working together with our stakeholders – our investors, our partners and our employees – to create sustainable value that makes a real difference, while upholding the highest standards of safety and responsibility across all our businesses and geographies. The company's mining operations, pipeline of growth projects and exploration activities span southern Africa, South America, Australia, North America, Asia and Europe.

Major sponsor profile
Barrick

Barrick operates mines and advanced exploration and development projects on five continents, and holds large land positions on some of the most prolific and prospective mineral trends. Our vision is to be the world's best gold mining company by operating in a safe, profitable and responsible manner. Our strategic priorities are:

Capital Discipline. Disciplined capital allocation drives every decision we make. All investment alternatives compete for capital based on their ability to generate attractive risk-adjusted returns and free cash flow.

Operational Excellence. Execution is a key component of investor confidence. Barrick has an excellent track record in this area and has met its gold production guidance for 11 years in a row.

Corporate Responsibility. Our success depends on our ability to develop our resources responsibly and share the benefits of our business with local communities, governments and other stakeholders.

Shareholder Returns. Our efforts have a common goal – ultimately we are focused on value creation for our shareholders through higher returns.

Major sponsor profile
Vale

 VALE

Vale is one of the world's leading mining companies, with a presence in over 38 countries across five continents worldwide. Vale is also the world's second largest producer of nickel, with its Base Metals business headquartered in Toronto. Operating in Canada for more than 100 years, Vale's Canadian operations employ 6000 people and produce nickel, copper, platinum group metals, precious metals and cobalt. The company's mission is to transform resources into prosperity and sustainable development. For its corporate social responsibility efforts, in 2013 Vale was named one of the 50 most sustainable corporations in the world and one of the 50 top foreign corporate citizens in Canada, according to Corporate Knights.

Contents

CHAPTER 1

Introduction

INTRODUCTORY COMMENTS

Cut-off grade is perhaps the least understood driver of value for a mining operation. Many believe that it is determined relatively easily by knowing the mining and extraction costs, metallurgical recovery and an estimate of the price for the metal/s under consideration. Determining the values for these parameters may not necessarily be easy but, having determined them, calculating the cut-off as a break-even is straightforward. For others, the cut-off is what it is because that's what it has always been.

Nothing could be further from the truth. Unfortunately, it would seem that many of the key executives and decision-makers in the industry are unaware of this, as are many of the industry analysts who commentate on the actions and decisions of company executives. The result would appear to be that strategies and policies that maximise returns for stakeholders and reduce risk are given the thumbs down, while those that are guaranteed to minimise value and increase the risk of the operation are applauded.

Cut-off specification is an integral part of optimising the mining strategy. It should be an outcome of the strategic planning and optimisation process, not an input into it. That statement alone may raise questions in the minds of some readers. Typical practice, particularly for an underground mine, is first to derive a cut-off that defines the size and shape of the orebody and then to develop mining and extraction plans for the one orebody thus defined. This commonly used model for cut-off derivation is inadequate in many cases.

What is meant by a 'cut-off model'? Simply, it is the mental picture we have of the what, why and how of cut-off specification – what a cut-off is, why we have one and how we determine it.

By way of analogy consider classical or Newtonian physics, the physics of forces and motion described by Sir Isaac Newton in the late 17th century. It works in three-dimensional space, and is satisfactory for most day-to-day activities. The physical formulas relating mass, length, time, force, velocity and acceleration are useful because they work consistently at the scales of time, distance and accuracy that most of us are dealing with for most if not all of the time.

Early in the 20th century, Albert Einstein turned all this upside down with his special and general theories of relativity. Suddenly we no longer existed in three-dimensional space, but in four-dimensional space–time. Whereas we'd always thought such quantities as mass, length and time were constant, it transpired that all of these are relative, and the speed of light relative to any observer is what is constant. The concepts of relativity are counter-intuitive based on our common experience, but they have been proven to be correct. They work at extremely large scales and high velocities, while the Newtonian model does not. Yet at the scales and velocities with which we are generally accustomed, the Newtonian formulas are accurate approximations of the more complex theory.

At the other extreme, classical physics also breaks down. Quantum mechanics, initially developed in the early decades of the 20th century, describes behaviour at this scale but the concepts are perhaps even more difficult to comprehend than those of relativity. 'Quantum weirdness' is so far beyond our day-to-day experience that even physicists who are experts in the field admit to having difficulty comprehending it; however, they understand the mathematics and can draw appropriate inferences, even though it is all beyond normal understanding.

The current 'holy grail' of physical research is 'a grand unified theory of everything' that effectively combines relativity and quantum mechanics and explains all four of the fundamental forces of nature in one overarching theory. Classical physics, relativity and quantum mechanics are all seen to be approximations of the single unified theory that work satisfactorily within different sets of limits. At the time of writing, super-string and M-theory working in ten- or 11-dimensional space–time appears to have the best chance of leading to this goal,[1] but as in all fundamental scientific research, nothing is assured.

How does this relate to cut-off? Just as we see an increasing number of dimensions in our model of the cosmos and how it works, so too are there an increasing number of dimensions in progressively more complex (but also more useful and generally applicable) models of cut-off and strategy optimisation. The more dimensions in the cut-off and strategy optimisation model, the more parameters that can be accounted for and the better the potential outcome. As in physics, simpler cut-off models might nevertheless still be useful under certain circumstances.

THE EVOLUTION OF CUT-OFF THEORY – INCREASING NUMBERS OF DIMENSIONS

The first part of this book deals with the principles underlying progressively more complex models of approaches to cut-off and strategy optimisation.

The development of these concepts is supported by a number of important principles and definitions, which are established in Chapter 2. In particular, the discussion deals with definitions of cut-off and the classification of costs and materials dealt with.

---

1. For example, see Hawking and Mlodinow (2010). Even M-theory now seems to be not the desired single unified theory but rather several models that work in different circumstances. These models appear to overlap and, in the areas of overlap, give similar results, thereby generating a continuum of models that potentially cover the whole range of physical reality.

Many of these are counter to common industry practice and much of the confusion surrounding cut-off derivation stems from the general misunderstanding of how the distinctions apply.

Cut-off has always been an important concept in mining. Old texts talk of pay limits, and pay dirt is a common term in popular culture when dealing with old-time prospectors and miners. The term *cut-off* is used in a number of these texts but without any formal definition or explanation. It appears that the writers have assumed that cut-off is a fundamental concept that is so well understood that it needs no description. The author is not aware of any formalised definitions but suggests that the terminology and the way it appears to be applied implies some sort of break-even concept.

Break-even analysis is essentially what we might call a one-dimensional process. It is based on financial parameters only. Prices, adjusted to account for metallurgical recovery, are compared with costs. Simplistically, if the grade of mineralised material is sufficient for the revenue obtained to pay the costs associated with extracting, processing and selling, it is classed as ore. Break-evens are discussed in more detail in Chapter 3.

Although widely used in the industry, break-even is a very limited cut-off model. In particular, it does not account for the geology, especially the grade distribution of the mineralisation, nor the capacities of the mining and processing plant. One could be excused for asking how a cut-off model that ignores geology and production system capacities could be useful. The answer is that the simple break-even model can be useful so long as it is used within the constraints of the outcomes of a more comprehensive model; however, the reality is that in practice a simple break-even model is all that is used by many operations. As we shall see, this can lead to mine plans that are almost guaranteed to not deliver the company's stated goals. Indeed there is no guarantee that an orebody delineated by a break-even cut-off – derived according to typical industry practices – will even deliver a profit. Use of break-even cut-offs has probably been the primary reason why mining companies have not been generating returns for shareholders in line with increases in metal prices in recent years.[2]

Cut-off theory advanced somewhat with the publication of a little-known paper on South African gold mining grade control practices in the late 1940s (Mortimer, 1950), a section of which gives rise to what the author refers to as *Mortimer's Definition*. This can be thought of as a two-dimensional model of cut-offs. As well as the financial aspects of break-evens, it brings in the need to consider the geology and nature of the mineralisation as expressed by the tonnage and grade versus cut-off relationships. This ensures not only that break-even conditions are satisfied but also that the orebody delineated will deliver a specified profit target. Mortimer's Definition is discussed in more detail in Chapter 4.

Ken Lane published what has, for many years, been state-of-the-art in cut-off optimisation, initially in a technical paper (Lane, 1964) and later in a text book (Lane, 1988, 1997). Continuing the dimensional metaphor, Lane's methodology could be described as a three-dimensional process. As well as the financial and geological dimensions of the break-even and Mortimer approaches, Lane accounts for the production system's capacities to handle three classes of material: rock, ore and product. The rationale for this classification of materials is fundamental for all cut-off determinations, and is therefore described in more detail in Chapter 2.

---

2. The author has not carried out any research to justify this comment but is recounting here complaints by analysts at several conferences in recent years, as reported by various industry-related online news sites.

Lane developed his theory using rigorous mathematical processes. From discussions with mining industry professionals from a number of companies and countries, this author perceives that the vast majority of mining engineers and geologists who attempt to read and understand Lane give up when they get to the mathematical derivations of Lane's formulas. However, these formulas may be derived logically, if not completely rigorously, without resorting to complex mathematics. This process and other important aspects of Lane's methodology are described in Chapter 5, which is this author's attempt to 'demystify Lane'.

Lane and other workers in the field in the 1970s and 1980s introduced a number of concepts to the development of cut-off theory, which are also described in Chapter 5. Worth noting is the concept of a balancing cut-off, which ensures that two of the three production system components (for rock, ore and product) are operating at their capacities. A balancing cut-off is a function of the geology and plant capacities only and is not at all related to costs and prices. It will often be the optimum cut-off to apply. In other words, the optimum cut-off may actually not depend in any way on prices and costs, a revelation for many who encounter it for the first time after believing all their working lives that cut-off and break-even are synonymous. Also worthy of note is the concept of opportunity cost, which has always been implicitly understood in the industry. It is, for example, understood that one can mine down to the marginal break-even to 'fill the mill' if there is a shortfall in ore supply. It is also recognised that this is only valid so long as lower-grade material is not displacing higher-grade feed. The opportunity cost quantifies the extent to which such displacement can occur economically. Lane's methodology automatically accounts for it in the cut-off optimisation process.

With the development of powerful digital computers from the 1990s on, full mine strategy optimisation has increasingly become the goal of cutting-edge mine planners and mining software developers.[3] This may be thought of as a multidimensional analysis taking account of, notionally, everything. The values of the strategic decision variables can all be optimised while taking account of the uncontrollable projected economic, financial, social and geological parameters in the future. The decision variables include (but are not limited to) such items as cut-offs; production capacities in various parts of the overall production process, perhaps including multiple plants; sequencing and timing of mining from stopes, mining blocks, pits and separate mines; products and product mixes; and stockpiling policies. All of these can vary over time or by location, or both.

In principle, if something can be described, it can be modelled and included in analyses and evaluations. The only limits are computing power and the evaluators' imagination and ingenuity in identifying important parameters, the relationships between them and suitably simple, accurate and efficient ways to model these within the limitations of the evaluation methodology and, where appropriate, the software being employed.

In all the models of cut-off prior to strategy optimisation, all parameters other than cut-off are effectively assumed to be pre-specified and the cut-off is derived for that case. In strategy optimisation, cut-off is but one of many decisions to be made, and it is optimised concurrently with all other decisions. Chapter 6 describes the principles underlying a full strategy optimisation evaluation and the processes typically involved.

---

3. Complete strategy optimisation, however, is still too large a problem for current and reasonably foreseeable computer power. Although significant advances have been made, we are still only looking at subproblems of the whole problem, or we model the whole problem crudely or work on small, simple problems.

It can be seen that there are a number of cut-off models available to the mine planner. Depending on the cut-off derivation methodology being used, it might at times be necessary to make complex assumptions regarding some inputs for the methodology to work. Or perhaps some important parameters are not adequately accounted for. This will usually be an indication that the cut-off model used is inadequate and a higher-dimensional model is required to deal with the real complexities of the situation. Chapter 7 suggests when particular cut-off models may be appropriate and how they fit into the overall mine planning process.

CONDUCTING CUT-OFF EVALUATIONS AND STRATEGY OPTIMISATIONS

The first part of this book deals with the principles underlying progressively more complex models of approaches to cut-off and strategy optimisation. The second part deals with specific aspects of the associated activities.

Chapter 8 describes in some detail the inputs required for a full strategy optimisation evaluation. A number of the practicalities involved in developing and using an evaluation and optimisation model are discussed.

An early consideration is deciding upon the appropriate grade descriptor. This is the value given to each block of rock to describe its value relative to other blocks – the number that tells us that 'this bit' is more valuable than 'that bit'. Actual metal grades, metal equivalents and money equivalents are commonly used; however, the derivation of equivalents is not necessarily done as well as it could be and other measures may be more appropriate. Chapter 9 describes how alternative grade descriptors might be derived and how the best one is selected.

Chapter 10 discusses some of the measures of value that might be used to enable us to say that Strategy A is better than Strategy B. If cash is indeed king, then value measures that quantify cash generation, such as net present value (NPV) and real options value (ROV), must be the key measures for any strategy optimisation. Cut-offs and other components of operating strategies must be focused on maximising these. Many companies have other value measures that interest them, and are prepared to accept a lower NPV for improvements in other measures. Chapter 10 also discusses some typical alternative measures. Unfortunately, it would seem that in many cases these other measures are treated as surrogates for cash generation; however, many of these are not correlated with cash generation, and in some cases are actually negatively correlated. The popular 'ounces in reserve' measure is one such parameter, which receives a lot of attention from senior managers and industry analysts, but can be simply demonstrated to be negatively correlated with cash generation – in other words, the more ounces the less cash, and vice versa. Chapter 10, as well as discussing various measures also describes how they each correlate with cash generation.

Chapter 11 describes alternative methodologies to derive some of these value measures. Chapter 12 then considers how optimisation techniques can be applied to deliver more complex corporate goals than simply, for example, maximising NPV or some similar single measure of value. In particular, the trade-off between maximising the reward from making correct assumptions about future conditions, such as price, and minimising the downside risk of making wrong predictions is discussed.

Chapter 13 describes briefly some of the techniques available to optimise strategies. This is not intended to replace a detailed text on operations research nor detail particular algorithms, but rather give the flavour of several commonly used techniques, touching on their advantages and disadvantages. All of these are useful and none can necessarily be relied upon to give the best answer. The project evaluator might well be advised to use a combination of techniques to allow the strengths of one to complement the weaknesses of another.

Chapter 14 addresses some of the misconceptions and objections that apply to various aspects of cut-off and strategy optimisation. Chapter 15 summarises the major themes established in the preceding chapters and poses some philosophical challenges for the industry into the future.

To conclude, the appendix describes several case studies, indicating how many principles and practices described have been applied to identify optimum strategies and how the value obtained from each has been increased.

GENERAL TERMINOLOGY AND UNITS USED

The strategic mine plan

Throughout this book, unless the context implies otherwise,[4] *mining, the mine plan* and similar terms relate to the total extraction process of valuable material from the earth to the point of sale of product. These mining terms are assumed to include:

- the drill, blast, load, haul and processes typically associated with the mine to extract ore and related waste
- the metallurgical treatment processes performed on ore to extract valuable product/s in the plant or the mill
- integrated smelting and refining or similar downstream processes to add value to an already saleable product (such as the smelting of a base metal concentrate) if the operation or the company has such facilities
- transport of all products to the point of sale, accounting for any physical or market limitations on the amount of product that can be handled or sold into the market.

The reason for this usage is that optimisation of the corporate business plan for an operation or multiple operations must take account of the complete process stream, from having the mineralisation identified and described by geologists through to final sale and receipt of revenue. While it is possible in practice to limit an evaluation to one part of the process, such as the mine, and assume that all other processes are not going to change, the reality is that suboptimisation of components of a complex process cannot be guaranteed to find the best overall plan. This can only be done if the whole process is considered together in one integrated set of evaluations.

Much of Part 1 of this book, however, will be dealing purely with selecting a cut-off grade, and this is largely a mine-focused activity. By implication, parameters relating to processes outside the mine will be taken as given; nevertheless, they will enter into the cut-off derivation in a number of ways, and the need for concurrent optimisation of the total production process will become evident.

Cut-offs

The term *cut-off* may be used as both an adjective and a noun. It is defined more fully in Chapter 2.

---

4. The primary exception will be the use of the term 'mining' to refer to a clearly defined part of the overall production process when discussing Lane's cut-off methodology in Chapter 5.

Cut-off policy refers to a planned sequence of cut-offs for an operation over time. The term recognises that it is rare for the best cut-off to be the same unchanged value over the life of the mine, even if such issues as costs and product prices were to remain unchanged.

Grade is used generically to refer to any parameter that is used to describe the relative value of a block of rock. It could be the metal grade in a single-metal deposit, or a metal equivalent grade or monetary value measure in a polymetallic deposit. Therefore, there is generally no distinction in this book between such terms as *cut-off grade* and *cut-off value*, or indeed *cut-off* on its own. In any discussion where the distinction between actual metal grades and other grade descriptors or measures of value is important, it will be explicitly explained. In practice, of course, a cut-off should be expressed in unambiguous units so that it is clear which grade or other measure its value is to be compared with to distinguish ore from waste.

Ore and reserves

The terms *orebody*, *reserve* and similar are used in their colloquial sense and not according to the strict definitions specified in the JORC (Joint Ore Reserves Committee) Code, NI 43-101 (National Instrument), SAMREC (South African Mineral Resource Committee) and similar codes for public reporting. The cut-off derivation and strategy optimisation processes described in this book are internal to the company performing the evaluation, where the use of such terminology is not required by law or other external codes of practice. This should be conducted before a reserve is publicly reported in accordance with the code of the country where publication occurs. These processes will typically involve identifying a number of sets of potential reserves at different cut-offs, one of which will ultimately form the basis for the mine plan that supports the publicly reported reserve.

As this book discusses processes that will ultimately lead to specifying a cut-off that will potentially lead to a publicly reported reserve, *ore* has a specific meaning that is quite different from the definitions in the reserves reporting codes. It is defined more fully in Chapter 2, along with *rock* and *product*, but for now may be thought of as the mineralised material above a particular cut-off or that which is intended to be sent through the mine's ore-handling system for processing through the treatment plant for the case being considered. It should be noted, however, that specific resource categories – Measured, Indicated and Inferred – are commonly used in many companies' planning procedures, and nothing here should imply that this is wrong or be discouraged. Indeed, the opposite is the case, as the definitions for these terms convey commonly understood information regarding the certainty of the underlying geological information and other factors.

It will always be appropriate to comply with company policies that impose restrictions on the use of, for example, Measured and Indicated Resources for final plans; however, it is *not* appropriate to restrict investigations to these categories while developing the mine plans. All available information should be used, regardless of its resource category or other descriptor of certainty, while obviously taking account of the different levels of accuracy and certainty associated with each item of data.[5] This, of course, allows for

---

5. As an aside, the author strongly recommends that when resource categories are recorded in geological block models, these should represent geological confidence only. For all mine planning purposes, it is just as important to know how certain the characteristics of low-grade material are as it is for high-grade material; however, it is not uncommon to find that material with a grade below the resource cut-off is given a resource category code that identifies it as waste, and only material above the resource cut-off is identified as Measured, Indicated, Inferred, etc. This practice effectively destroys any record of the geological confidence of low-grade material

the possibility that some plans might be developed using only Measured and Indicated Resources and others using all known mineralisation. This would help to identify whether strategies would differ if what is currently classified as lower confidence material was subsequently found to actually exist or not, and therefore how important it might be to conduct exploration sooner rather than later.

Units

The International System of Units (SI) and related units are generally used throughout this book, although some non-SI units that are commonly used in the industry are also employed. Key units and the corresponding abbreviations used in this book are:

- Mass – tonnes (t), typically for rock, ore and waste quantities but also for product quantities for base metals; ounces (oz), typically for product quantities for precious metals.
- Length – metres (m).
- Volume – cubic metres (m³).
- Density – tonnes per cubic metre (t/m³). Note that specific gravity (SG) is the ratio of the density of the material to that of water, and is therefore dimensionless; however, in SI units, density and SG are numerically the same.
- Grade – grams per tonne (g/t), typically for precious metals; per cent (%), typically for base metals; parts per million (ppm), typically for trace elements, contaminants and penalty elements. When discussing general principles, the term *grade unit*, or *gu* is used to refer to the quantity in the numerator of the grade symbol, with the generalised grade being reported in *grade units per tonne* or *gu/t*.
- Recovery – per cent (%).

Monetary amounts are generally referred to as dollars.

When expressing grades as monetary units, the term dollar value ($ value) is used, and the unit is typically dollars per tonne ($/t).

Value drivers, options and scenarios

The term *value driver* is used to refer to a parameter that has a major influence on the value of the operation. Value drivers will be described in more detail in Chapter 6 and later, but for now may be thought of as such items as 'production rate' and 'metal prices'. The two examples noted here also exemplify the different classes of value drivers:

1. Those for which the company can specify a value or setting, such as specifying a production rate or the type of metallurgical processes to be used. These are referred to as *options*.

---

without any corresponding gain to offset that loss of information. Whether it is resource or not can be assessed by reference to the resource cut-off grade: a separate waste flag is redundant. Extraction of resource quantities of various categories for reporting is therefore not compromised by having the resource category code defining only geological confidence. It is simply a matter of applying simultaneously the two essential characteristics that define whether material falls into one of the various categories of resource: its geological confidence and its grade relative to the resource cut-off grade. Indeed, if the resource cut-off is reduced, the recommended procedure simplifies matters; it is only necessary to rerun the data extraction process with the new cut-off. It is unnecessary to repeat whatever was done to specify the geological confidence unless there is additional geological information that changes the level of confidence.

2. Those which the company cannot control or for which it does not have adequate knowledge. Metal prices, for instance, are typically predictions of future conditions and are intrinsically unknowable at the time an evaluation is being performed. Certain costs or metallurgical recoveries, on the other hand, are not selectable as an option, but may be uncertain at the time an evaluation is being undertaken. In both cases, alternative possible values may need to be considered in the evaluation. These are referred to as *scenario parameters*.

The term *case* is used to refer to a complete set of specifications of all the values used for all value drivers, both options and scenarios, in one particular evaluation.

The terms *cost drivers* and *revenue drivers* will be used for such things as ore and product quantities, which will typically be the production outcomes of the strategic decisions made regarding various value drivers classed as options.

PART 1

DEVELOPING AN
UNDERSTANDING OF
CUT-OFF THEORY –
FROM SIMPLE
BREAK-EVENS TO FULL
STRATEGY OPTIMISATION

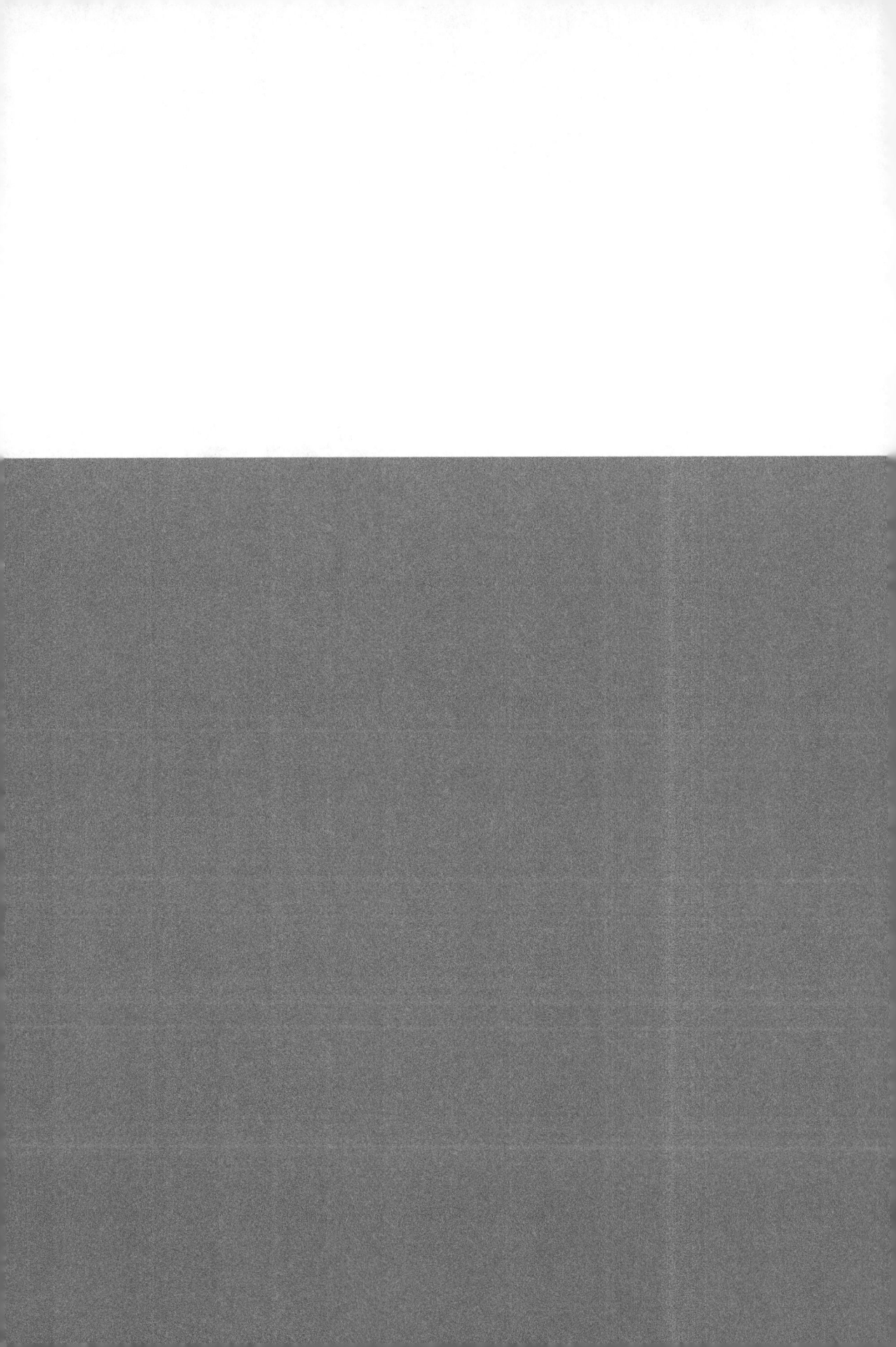

CHAPTER 2

Establishing Underlying Principles

INTRODUCTORY COMMENTS

Before proceeding to discuss cut-offs and mine strategy optimisation, it is essential first to establish some underlying principles and vocabulary. Every specialised field has its unique set of terms with specific meaning within that context. Often these will be common terms with a wider, colloquial meaning for laypersons, and the difference between the informal and formal meanings often being a source of confusion; this is compounded by specialists alternating between meanings depending on the context. Cut-off specification is no exception. Lane (1988, 1997) generated cut-off-specific meanings for several common mining industry terms. From discussions with technical staff from various countries and companies over the years, the distinction is reasonably well understood for some terms but not for others. This chapter therefore sets out a number of terms with special meanings for cut-off and strategy optimisation as they are used in this book. Many of the issues discussed here will be developed in more detail in later chapters.

WHAT IS CUT-OFF GRADE?

Much of the misunderstanding surrounding cut-offs would appear to come from a failure to distinguish what a cut-off actually is, and why and how it is derived. In particular, there seems to be confusion relating the 'what' with the 'how'. It is therefore important to distinguish between cut-off definition, purpose and derivation.

Definition

A cut-off is simply a number that indicates the point between two alternative courses of action. Material with a grade above the cut-off is dealt with in one way, while material with a grade below the cut-off is dealt with in another way. In particular, a cut-off is used to distinguish between ore and waste – material with a grade above the cut-off is ore and below the cut-off is waste.[1]

There are, of course, numerous cut-offs that can be specified. Other common cut-offs are for stockpiles of ore that is to be treated at some time after it has been mined, rather than immediately. This is often separated into a number of categories, such as high, medium and low-grade, based on the primary metal or value measure, and/or on the basis of grades of deleterious constituents. Other cut-offs, commonly referred to as *cut-over* grades, separate the ore into components to be treated by different processes; typically, higher-grade material will go to a higher-recovery but higher-cost process, while lower grades will go to a lower-recovery but lower-cost process.

Purpose

The purpose of a cut-off is to achieve some economic or financial goal, such as to ensure profitability and maximise value, etc. Depending on the location of the project, there may be various social goals such as maximising the life of the operations or employment in the region and taxes paid to the government. And wherever the project is located, minimising environmental damage and greenhouse gas emissions is becoming increasingly important and the standard when doing business in the mining industry. In many jurisdictions, companies are required by law to operate in the best interests of shareholders. In most cases, this will mean boosting cash returns. Achieving various societal goals may nevertheless be important for obtaining the so-called 'social licence to operate', which is an essential precursor to being able to operate at all, and hence work in the best interests of shareholders. Other goals set by companies or analysts are perhaps thought to be proxies for cash generation, but many in reality are not. Common goals are discussed in more detail in Chapter 10.

Naturally, a cut-off that delivers one goal will not necessarily deliver other goals. It is therefore imperative that the goals of the cut-off (and other strategic decisions) are stated before the derivation commences; however, in practice this is often not the case.

Method of determination

As is already evident, there are a number of ways in which the value used as a cut-off can be derived. Break-even calculations are common and are discussed in detail in Chapter 3. Lane's methodology is in use at some open pit operations, and has also been applied at a small number of underground mines though its application is often not easy. This is discussed in more detail in Chapter 5. Lane's methodology can be applied using modern spreadsheet software. It is also embedded in some commercially available strategy-optimisation software packages.

---

1. The question of whether material with a grade exactly equal to the cut-off should be classed as ore or waste sometimes arises, but this will depend on how both rock grade values and cut-offs have been derived and expressed, and what the cut-off is actually intended to achieve. This is explored further in subsequent chapters that deal with different methods of deriving the cut-off.

Overall mine strategy optimisation is perhaps the other major process by which cut-offs are specified, and this is discussed in more detail in Chapter 6. As previously noted in Chapter 1, the cut-off policy should be an outcome of the planning process, rather than a cut-off being derived (perhaps as a simple break-even) as a fixed input into the planning process.

Warning regarding terminology

In this book, *cut-off* is merely a number while *break-even* is a generic term describing a method of calculation. This is in contrast to a number of companies, where *cut-off* is used both for the number and as the generic term for the method by which it is derived, which in this book is called a *break-even calculation*. In some companies, the term *break-even* is used to define a specific set of costs included in the calculation of a break-even cut-off. This usage effectively implies that break-even calculations are the only valid way to determine the number to use as the cut-off, since the calculation method is included in the meaning of the term as defined by these companies' cut-off policy manuals.

This book uses the term *break-even cut-off* to describe a cut-off whose value has been derived by a break-even calculation method. Certain break-even cut-offs include *marginal break-even cut-off* and *full-cost break-even cut-off*, in which the included costs are indicated. This is discussed in more detail in Chapter 3.

CLASSIFICATION OF MATERIALS FOR CUT-OFFS AND STRATEGY OPTIMISATION

In order to accurately determine cut-offs, we must correctly allocate costs and identify production system capacities. First, it is necessary to identify the key cost drivers. Lane (1988) introduces these concepts in the development of his cut-off theory, described in more detail in Chapter 5; but the concepts are essential for a full understanding of all types of cut-offs, so they are introduced here.

The term *cut-off grade* itself points to the necessary classification. Grade in its simplest form is the ratio of the amount of product to the amount of ore or rock in which it is contained. The primary purpose of the cut-off is to distinguish ore from the total rock mass – to differentiate what is ore and what is not. Therefore, cut-off theory and practice needs to deal with three classes of material: rock, ore and product.

Rock

In this publication, *rock* refers to all material mined or accessed before it has been distinguished as ore or waste, or alternatively it is the total of ore and waste.

In an open pit mine, rock is typically all the material mined in and transported from the pit. Simplistically, once the material arrives at the rim of the pit, it is then separated into ore and waste by specifying the destination of the truck carrying a load of rock – waste is sent to the waste dump while ore is sent to the treatment plant or to stockpiles for treatment, either immediately or at some time in the future.

In an underground operation, rock is typically the total mineralised material. It may be delimited by, for example, a geological boundary such as the boundary of the lithological unit that contains the mineralisation. Alternatively, it might be by a grade-defined boundary, where the delimiting grade would be at most the lowest possible cut-off to be applied to distinguish ore from waste under very optimistic economic conditions.

Underground, rock is separated into ore and waste by, for example, planners creating designs for stopes where the rock within the design limits is extracted as ore and the remainder remains *in situ* as waste. The common classification of development as being in ore or waste, or as being capital or operating expenditure, is not relevant for our understanding of cut-off theory. The important issue for development is not the grade of the material being mined through, nor how it will be recorded in the company's accounts. Rather it is whether it is providing access to rock that can be then be stoped or not, or if it is directly required for and associated with ore extraction. The classification of development for cut-off derivation is discussed more in Chapter 3.

The author is aware of some companies that use the term *rock* to refer to what is in this book referred to as *waste*. Clarifying the terminology is essential to avoid confusion. In this book, for open pits:

Rock = ore (treated or stockpiled) + waste = total material moved

Underground rock, as indicated earlier, is the total contents of the mineralised envelope, however that might be defined. What is excavated underground and usually classified as waste or rock, depending on local terminology, is typically from major access development, often outside the mineralised envelope. This material, however, is not waste in the specialised sense applied in cut-off theory – it is simply an outcome of the activities undertaken to expose the rock so that the ore it contains can be stoped.

Note that rock being classified as such 'before distinguishing between ore and waste' is a logical concept for the purposes of cut-off specification and strategy optimisation. In practice, the distinction may well have been made before mining occurs so that different mining processes can be applied if appropriate and long-term plans can be developed, but this implies that a cut-off has already been applied. To determine the cut-off in the first place, the total rock mass must be viewed as having a separate identity before the cut-off is imposed.

Ore

Ore is the material intended to be processed. It will typically be the mineralised material to be sent through the mine's ore-handling system and for processing through the treatment plant. It may be treated at the time of mining or stockpiled for treatment at a later date. There may be a sequence of processes dealing with the ore.

As noted in Chapter 1, the term *ore* is used in this book as just described and not according to the strict definitions specified in the JORC Code, NI 43-101, SAMREC and similar codes for public reporting. The cut-off derivation and strategy optimisation processes described in this book are internal to the company performing the evaluations, where the use of such terminology is not required. This typically involves identifying a number of sets of potential reserves at different cut-offs, one of which will ultimately form the basis for the mine plan that supports the publicly reported reserve.

Product

Product is the valuable material separated from the ore. It will typically be the material in the grade unit – for example, gold or copper metal for grades expressed as ounces or grams per tonne, or as a metal percentage; however, this does not necessarily imply that the product for cut-off derivation purposes is the actual product produced and sold by the operation. For instance, a gold mine might produce and sell gold bullion, so the mine's product is also the product for cut-off derivation purposes. However, for a base metal operation, the product for cut-off purposes is the metal, but the product produced and sold by the mine will typically be a concentrate. Some operations may have multiple process stages producing progressively more refined forms of product such as copper concentrate, blister, anode and cathode.

For base and precious metal operations, the product should be a relatively small proportion of the ore but for bulk commodities, such as iron ore and coal, there may be no separation of product from ore (the ore is the product). If there is some beneficiation, the product may be a large proportion of the ore.

In most of the discussions in this book, the terms *product* and *metal* are used interchangeably. Where *metal* is used, it does not preclude the consideration of coal or other non-metallic products. The term *metal grade* will often distinguish a grade that is expressed in terms of units of product per unit of ore – such as gold grams per tonne or copper per cent – from any generic grade measure of value attributed to a block of rock.

Material types in open pit and underground mines

It should be noted that while the operations in an open pit mine deal solely with rock that is subsequently separated into ore and waste, most of the operations in an underground mine typically deal, or are associated, with ore. The exception is usually access to and infrastructure development for a new mining block: this can be thought of as exposing a new block of rock, which is subsequently separated into ore to be stoped and waste to be left *in situ*. Major access development is therefore generally the only rock-related activity in an underground mine for cut-off derivation purposes.

All mines separate ore from waste in some way. There may, however, be operations where all the rock mined in an open pit is ore. This would commonly occur where the nature of the mineralisation is such that the pit is only excavated in material above a specified cut-off, with all lower-grade material left *in situ*. In this case, the ore–waste boundary within the overall rock mass has been defined by the mine planning processes and the operations are, for cut-off derivation and strategy optimisation purposes, more akin to an underground operation, though mined by open pit methods.

CLASSIFICATION OF COSTS FOR CUT-OFFS AND STRATEGY OPTIMISATION

As previously noted, in order to accurately determine cut-offs, it is necessary that we classify and allocate costs. The three material types with which we are dealing – rock, ore and product – have been identified. We'll now identify how costs behave with respect to each of these. Rock, ore and product are also the key physical cost drivers for cut-off derivation and optimisation.

Cost drivers

Although it is essential to separate costs into those driven by rock, by ore and by product, the operation's cost accounting system may not easily facilitate this. For example, in the treatment plant all costs will often be quoted as a cost per tonne of ore, regardless that some are related to the amount of product generated and some incurred by the passing of time, irrespective of quantities of both ore and product handled. It may be necessary to go back to more detailed records to reallocate costs according to the physical parameters that drive them. More detail on applying this cost classification is provided in Chapter 3 in the description of break-evens.

Variable costs

Variable costs are those where the total dollar cost varies in direct proportion to the physical activity or quantity. Unit costs expressed as, for example, dollars per tonne or dollars per ounce are an accurate representation of cost relationships and behaviour. Each additional tonne or ounce of material dealt with requires spending the specified number of dollars to complete the action.

Fixed costs

Fixed costs are identified as costs that do not change relative to physical activity: they are the same for any specified time at all levels of activity. Total dollars of fixed costs over the life of the operation may also be thought of as being variable with respect to elapsed time, rather than to some physical quantity such as tonnes. Fixed costs are commonly expressed as dollars per year or per month. Again, the correct expression of a fixed cost describes its behaviour: each additional month or year of operation will require spending the stated number of dollars, regardless of how much or how little production activity occurs.

Lump sum costs

A lump sum cost may be defined as a cost for which the quantum does not change with either physical activity or time. Usually it is expended to provide some capability and will be a capital item, such as the construction or expansion of a part of the plant. It may also be the cost of development to expose and make available new rock. In an underground mine it will be the cost to access a zone of mineralisation remote from the current mining operation or, for example, to extend the decline to make the next sublevel block of the resource available. Although the quantum of the lump sum may derive from what appears to be a variable cost relationship (for example, the required metres of development multiplied by the cost per metre), it is nevertheless a cost that is *not* driven by the tonnage of rock, ore or product that may be exposed, mined or treated, nor by the time taken. The cost of exposing a new sublevel block is not dependent on how much rock (or mineralised lithology) is exposed, nor on how much ore is delineated within that rock (it could be all the mineralisation, none of it, or any amount in between), nor on whether the development is done quickly or slowly.

Certain lump sum costs will, of course, be related to rock, ore or product quantities. The cost to construct a concentrator with a specific ore treatment rate will obviously be less than that for a plant of double the capacity; but this is the cost of acquiring a capability or alternative capabilities, not the cost of actually treating the material. A discussion of capital costs follows in this chapter, and the handling of different production capabilities is covered in Chapters 5 and 8.

Other cost behaviours

Not all costs behave in a purely linear fashion that can be separated into fixed and variable components. There may be a continuous non-linear relationship between dollars spent and physical activity, or step changes at certain levels. These types of behaviours are discussed in more detail in Chapter 8. At this point in our understanding, it is sufficient to recognise that the various cut-off models we will discuss effectively require costs to be either fixed or variable. Whatever the true relationship of cost versus activity may be, that relationship will have a gradient that can be resolved into fixed and variable components.

Project capital and sustaining capital costs

The financial accounting distinction between capital and operating costs is largely irrelevant for cut-off derivation purposes. These classifications have been developed by accountants for public accounting reports for the same sorts of reasons that categories of Mineral Resources and Ore Reserves have been defined by codes for public reporting. This classification of costs is, however, misleading for cut-off derivation. What is important is the way the costs behave and what they achieve. It should be noted though that if a cut-off evaluation is being conducted on an after-tax basis, the accounting classification of capital and operating costs must be retained to ensure that tax deductibility and tax costs are applied and derived appropriately.

For cut-off derivation and strategy optimisation we do, however, need to identify two types of capital expenditure: sustaining capital and project capital.

Sustaining capital maintains existing capabilities. It is a regular, ongoing cost incurred for such things as fleet and plant replacement, and may be thought of as maintenance costs spent irregularly. For maintaining existing capital items, there is no conceptual difference between replacing an oil filter every 50 hours (an operating cost) and replacing a whole truck every 50 000 hours (a capital cost). Sustaining capital for an individual item or category of plant or fleet can therefore be thought of as 'lumpy' or irregular maintenance expenditure that, by virtue of its spending pattern rather than its behaviour, is classed by accountants as a capital rather than an operating cost.

Sustaining capital costs should normally be included in the cut-off derivation in the same way as similarly behaving operating costs – as variable or fixed costs. Because of the irregular nature of replacement capital expenditure for any type of asset, sustaining capital will be used to acquire different items in successive years.Companies will often manage their replacement programs so that the aggregate amount spent will be relatively constant over time. This sometimes results in sustaining capital costs being viewed as fixed costs, but in reality sustaining capital is incurred because of the wearing out of plant due to the level of mining and processing activities. These activities are often planned to be similar from year to year, so sustaining capital will therefore appear to be a regular fixed cost per year. However, if the production rate were to double or halve, the reality is that annual sustaining capital spending would do the same since equipment would be wearing out faster or slower. Since cut-off and strategy optimisation often considers changing production rates, it is necessary to evaluate carefully what is driving sustaining capital costs. Often all or most of what is quoted as a fixed cost needs to be converted to a variable cost related to the current level of production activities, so that variations in production in each time period can be accounted for.

Project capital buys additional capability such as mill expansion, an additional truck, etc as well as improved product quality. It is typically a lump sum cost, though the lump sum dollars could be determined like a variable cost as recently described. The resulting lump sum cost does not depend on the amount of ore mined but rather 'buys' the capability to do something different.

Under this definition, project capital is not for the 'projects' that are the current focus of the existing capital plan; many of these will in fact be sustaining capital. Whereas sustaining capital is a normal, ongoing cost of operating, spending of project capital is discretionary, justified by the difference in values of mine plans with and without the capital expenditure.

There are other ways to consider this issue. The argument is that replacing the truck is an ongoing cost of continuing to produce at the planned rate. That presupposes that the mine will continue to operate at a particular production rate, but it need not do so.

One could alternatively say that the base case is to not spend the replacement capital but instead allow the production rate to decline as the availability and performance of the ageing piece of equipment deteriorates, accepting that the production rate will drop to zero at some time when there is a catastrophic failure. In that scenario, replacing a truck is classed as a project capital item, where the benefit is in avoiding the decline or sudden stoppage of production, justifying the expenditure. But one could also argue the same logic for replacing the oil filter, which is normally classed as an operating cost. The same thinking will apply: if the oil filter is not replaced the engine will deteriorate and fail, and production will cease.

It can again be seen that there is no practical distinction between capital and operating costs, as long as cost behaviours and other assumptions are identified and consistently applied.

APPLYING CUT-OFFS TO DISTINGUISH ORE FROM WASTE

There are a number of ways that cut-offs can be applied to delineate the orebody to be mined and processed, taking account of mining practicalities. Often the grade variability in the mineralisation is such that it is not practical to mine and treat as ore only the material above cut-off while classifying as waste all the material below cut-off. It is common for the practical application of the cut-off to result in excluding some ore-grade material above cut-off from the mining inventory and including some sub-cut-off waste. There is no correct way to do this – each case must be considered on its own merits. It will be part of the cut-off evaluator's task to identify the rules by which cut-offs are to be applied in order to best achieve corporate goals. The following sections describe the more common ways in which cut-offs can be applied, but do not purport to be exhaustive. These descriptions are to establish underlying principles that are important in the discussions in later chapters.

Boundary cut-offs

A boundary cut-off rule is applied to delineate the ore-waste boundary of an orebody, stope, etc. In essence it takes account only of the location of the specified cut-off grade in the rock mass. If the geological ore–waste boundary at the specified cut-off is regular, the boundary cut-off rule may be trivial – a practical mining boundary between ore and waste may be simply defined by following the geological boundary or grade contour. If, however, the geological boundary in the rock at the specified cut-off is irregular, the boundary cut-off rule will indicate how the practical boundary is to be placed relative to the actual boundary. At one extreme, the rule might specify that the mining boundary is to be located entirely inside the geological boundary so that no sub-cut-off material is included in the mining inventory. Conversely, the mining boundary could be entirely outside the geological boundary so that no above-cut-off material is excluded. Within these extremes could be a number of rules specifying how the trade-off between ore loss and waste inclusion is to be handled.

When a boundary cut-off rule is applied in isolation, no cognisance is taken of grades of the resulting stopes, orebodies, etc. Depending on the grade distribution in the rock, the average ore grade defined could be significantly or only marginally above the specified cut-off. Applying a boundary cut-off alone implies nothing about the grade of the resulting ore. There is no guarantee that the ore defined will deliver any particular corporate goal, and depending on the grade distribution and cut-off specified, this may or may not be a problem. At low cut-offs, there will potentially be sufficient higher grade material above cut-off to deliver an average grade higher than the cut-off, whereas at higher cut-offs, the remaining material above cut-off may only be marginally above it. If the boundary cut-off has been applied to the *in situ* grades of the rock and allowance is subsequently made for overbreak and dilution, the final grade of the volume may even be less than the cut-off applied.

Unless otherwise described, cut-offs in this book are assumed to be boundary cut-offs.

Volume cut-offs

Applying a volume cut-off rule addresses the potential shortcomings of the boundary cut-off by accounting for the resulting average grade. A boundary rule would first be applied to generate an initial mining inventory. Volumes, such as individual stopes or subregions within the orebody defined, with average grades below the specified volume cut-off would then be deleted from the inventory.

The volume cut-off need not be the same as the boundary cut-off; however, in the author's experience operations that apply a volume cut-off typically apply the same value as the boundary cut-off.

The volume cut-off rule may allow for factors not accounted for in the boundary cut-off. In particular, the effect of ore recovery and dilution on tonnages and grades to be dealt with if the material is classed as ore may be applied to change the average grade of the material or the unit costs included in the cut-off derivation.

Other rules

Depending on the nature of the mineralisation, other rules may be applied to cull material above other cut-offs from the mining inventory or to include sub-cut-off material in the inventory. For example, an outlier rule may have to be applied to exclude small volumes of otherwise ore-grade material remote from the main ore zone(s) if including it in the inventory destroys rather than adds value. Conversely, rules may need to include small unmined waste or low-grade areas within an area that is generally mined out if, for example, damaging stress concentrations could develop in the unmined pillars.

PLANNING AND OPERATIONAL CUT-OFFS

Most mines will operate with a number of cut-offs. These have different purposes and will be applied in distinct circumstances. The cut-offs to be used must be clearly defined, and their application specified in company documentation. It should be noted that there are few if any standard definitions used consistently throughout the industry, from company to company and country to country. The definitions that follow are therefore for consistency throughout this book. Companies could choose to use different terms and definitions; however, these descriptions will be typical of those used at mines.

A *planning cut-off* is used in developing mine plans and designs as well as in long-term studies, such as feasibility studies and strategy studies, to distinguish between ore and waste. It sets the overall mine strategy.[2]

An *operational cut-off* is used on a day-to-day basis by mine operators for making decisions as to what is ore or waste. Because of the variability in orebodies and our inability to predict perfectly what will be in the mine, not to mention product prices, it may be both necessary and legitimate to vary the cut-off from the long-term planning cut-off in the short term. The variations must be within the context and constraints of the overall plan, and that optimum long-term plan includes specifying the long-term cut-off; however, the reasons for the variation must be understood, and part of the aim of short-to medium-term mine planning will be to return to the planning cut-off. The operational cut-off must be derived within the context of the planning cut-off. If it is not, it may result in a definition of what is ore (and what is not) that may be a long way from optimal and very difficult to change.[3] This is discussed in more detail late in Chapter 3, which covers the dangers of break-even cut-offs.

Although related, planning and operational cut-offs are distinct. One or the other may have more prominence in different types of mines, but they must not be confused. For example, in an open stoping underground mine, the planning cut-off will tend to have prominence. Planning engineers will design stopes to the planning cut-off, and there may be little or no scope to change the cut-off used by the operators on a short-term basis: what is designed is what gets blasted and produced. Alternatively in a cut-and-fill mine, although the initial stope sills and infrastructure development may have been planned according to the planning cut-off – and in general the stoping operations will be controlled to comply with this – there is much more scope to vary the cut-off applied in the short-term; for example, to keep the mill supplied with lower grade but marginally profitable material below the planning cut-off when there is, for whatever reason, a shortfall of ore available above the planning cut-off.

The danger in the latter case is that, having identified that it is easier to 'fill the mill' at the lower operational cut-off, the longer-term cut-off can be changed to this lower cut-off. The planning cut-off grade has become the grade derived for the operational cut-off, so rather than moving back to the original planning cut-off after the short-term problem has been resolved, the mine remains on a lower cut-off strategy. If the higher planning cut-off delivered the optimal plan, short-term considerations will have been allowed to subvert

---

2. The planning cut-off could be thought of as analogous to a motorway or freeway: it is the main route to get from A to B. So long as we remain on the freeway, we will reach where we want to be quickly and efficiently. In the mining business, it represents the path to achieving the corporate goal – whatever that is specified to be – most effectively. We can of course go from A to B by way of side roads off the main freeway, but at the cost of additional time, increased fuel consumption, etc. So too we may finally deplete our orebody and close our mine, but if we have gone by way of suboptimal production rates, cut-offs and other options, we will not have achieved the corporate goals as well as we could have. We will have taken longer, spent more or earned less revenue than we might have. And just as a freeway will never be perfectly straight, having both its direction and gradient change to suit the topography and environment, so too will we expect that the planning cut-off will change over time to account for longer-term variations in, for example, geology and economic conditions.

3. Continuing the freeway metaphor, the operational cut-off may be thought of as the lane in which we are travelling. As we see traffic ahead slowing in the lane we are in, we change lanes into an adjacent faster-moving lane, so maintaining our efficient progress towards our destination. The freeway keeps us travelling on the right path, but we change lanes as problems and congestion manifest themselves in one course or another. The important point when changing lanes is that we do not lose track of the through route or find ourselves in a position in which there is no option but to exit the freeway.

the previously-identified long-term plan and become the long-term plan. This can only lead to suboptimal outcomes.

CUT-OFFS AND COMPANY GOALS – OR HOW THE JUNIOR ENGINEER DETERMINES THE CEO'S BONUS

Specifying the corporate and societal goals for an operation must be a primary part of the cut-off derivation process. Without specifying these goals, the goals implicit in the cut-off derivation process will become the de facto high-ranking corporate goals, whether they are recognised as such or not.

In many companies, cut-offs are derived at the mines by relatively junior mining engineers or geologists. In the absence of any other direction or by following past practices, they determine a break-even grade that is then applied as the cut-off for the next planning cycle. The *how* of cut-offs has become indistinguishable from the *what* – cut-off and break-even have become synonymous by default. Using a simple break-even grade as the cut-off merely ensures that the break-even is achieved. Break-evens will be discussed in more detail in the next chapter, but for now, we could say that simplistically the goal implicit in a break-even is that 'every tonne pays for itself'.

As already noted, the cut-off that delivers one goal will not necessarily deliver another. The author is not aware of any public company whose stated principal aim is to ensure that every tonne of ore pays for itself. More commonly we see goals that speak of 'maximising returns for shareholders' or similar. There is no reason why a cut-off that ensures that every tonne pays for itself should also maximise shareholder value; indeed, the experience of strategy optimisation practitioners suggests that it will not! So while senior executives are telling shareholders that they are seeking to maximise their returns, operating staff – albeit with the best intentions – are implementing mining strategies that are almost guaranteed not to deliver the promised returns to shareholders.

The corporate goals must therefore be integral to the cut-off and strategy evaluation process from senior managers through to the person on-site deciding what is ore and what is waste. In many companies the corporate goals are not transmitted down the line, which means decision-making that is aligned with the goals is difficult, if not impossible.

It should be noted that this assumes that the company is at liberty to select a cut-off that is aligned with and helps to deliver the corporate goals. In some jurisdictions, however, the way the cut-off is determined is mandated by law. In such places, the cut-off clearly needs to be whatever is specified, and any variations can only be what is legally permitted. Achieving the corporate goal will be constrained by the allowable limits within the law.

SUMMARY

It is important to establish key concepts and definitions from the start of the discussion on cut-offs and strategy optimisation.

Essentially, we must distinguish between what a cut-off is – a number that separates two courses of action, in particular the separation of ore and waste – and its purpose and method of determination. Break-even calculations are but one way of deriving the value for the cut-off. Break-even and cut-off are not the same. Much of the misunderstanding surrounding cut-offs stems from a failure to distinguish between the *what*, *why* and *how*.

We need to distinguish three classes of material:

1. *Rock* – the total ore and waste (or total material) before the distinction between ore and waste has been made.
2. *Ore* – the portion of the rock that we intend to treat and derive value from.
3. *Product* – the component/s of the ore that actually generate the revenue and hence the value. There are costs and production capacities associated with or driven by each of these.

The way in which costs behave is crucial to correctly determining the cut-off grade:

- *Variable costs* are those costs for which total dollars vary in direct proportion to the quantity of some physical parameter, such as tonnes or metres.
- *Fixed costs* are usually identified as costs not varying relative to physical activity; they are the same for any specified time at all levels of activity. Total dollars of fixed costs are variable with respect to elapsed time, rather than to physical quantity.
- A *lump sum cost* may be defined as a specified cost that does not change with either physical activity or time. Typically it is expended to provide some capability and will be a capital item, such as the construction or expansion of a part of the plant. It may also be the cost of development to expose and make available new rock.

The accounting distinction between capital and operating costs is irrelevant and potentially misleading for cut-off derivation. The critical issue is how costs behave:

- *Project capital* buys the capability to do something new or different and will typically be a lump sum cost.
- *Sustaining capital* maintains existing capabilities, and is typically irregular maintenance or replacement expenditure that is nevertheless driven by the wearing out of plant and equipment because of its use in operations. Although often expressed as an annual (fixed) cost in a stable, steady-state operation, sustaining capital will more usually be incurred because of some physical activity, such as tonnage that has been treated or handled. Sustaining capital should therefore be handled in the same way as variable operating costs related to the relevant activity.

Cut-offs may be applied to define the boundary between ore and waste without reference to the grade of ore thereby delineated. They may also be applied to the grade of the volume enclosed by a boundary that has been defined. There is no one correct method of application, and different mines may apply cut-offs differently for a number of valid reasons.

Planning cut-offs are used to set the long-term big-picture cut-off strategy. Operational cut-offs used on a short-term basis in operations may vary from the planning cut-off due to, for example, natural variations and fluctuations within the deposit, conditions in the plant, and economic conditions that vary from the long-term prediction. Planning and operational cut-offs must not be confused, however, and the operational cut-off should always be varying within the confines of the longer-term planning cut-off policy, which should be directed at achieving the corporate goals.

CHAPTER 3

Break-even Analysis – Accounting for Costs and Prices

A COMMON BUT LIMITED CUT-OFF MODEL

Many mining operations worldwide are operating with what may be called a *break-even cut-off*. What is meant by the term *break-even*? In its simplest manifestation as it applies to cut-off derivation, the break-even is the grade at which revenue obtained is equal to the cost of producing that revenue.

The implication of this for cut-off purposes is clear: any grade higher than this will generate more revenue than the costs to produce that revenue, while any grade lower will generate less revenue than costs. This makes a break-even grade an obvious one to use as a cut-off – any material with a higher grade is apparently worth classifying as ore while lower-grade material is not. This, however, is not necessarily true.

It is important at this early stage of the discussion to recognise that break-even and cut-off are not synonymous: cut-off is not break-even, and break-even is not cut-off. In many cases, however, the terms are used interchangeably (noted in Chapter 2), and much of the confusion surrounding cut-offs comes from this misapprehension. The previous chapter identified that the cut-off is simply a number. A break-even calculation is one way of deriving a number that could be used as a cut-off; however, it is not the only valid way, as later chapters will illustrate.

Break-even analysis is essentially what might be called a one-dimensional (1D) process. It is based on financial parameters only. Costs and prices, adjusted to account for metallurgical recovery, are compared. Although widely used in the industry, break-even is a very limited cut-off model. In particular, it does not explicitly account for the geology, the nature of the mineralisation and the capacities of the mining and processing plant.

Yet the simple break-even model can be useful so long as it is used within the constraints of the outcomes of a more comprehensive model.

The reality is that in practice, a simple break-even model is all that is used by many operations. This can lead to mine plans that are almost guaranteed to not deliver the company's stated goals. Indeed, there is no guarantee that an orebody delineated by a break-even cut-off derived according to typical industry practices will actually even deliver a profit.

THE SIMPLE BREAK-EVEN FORMULA

The simple break-even formula to determine the grade at which revenue obtained is equal to the cost of producing that revenue is:

$$\text{Break-even grade [gu/t]} = \frac{\text{cost [\$/t]}}{\text{product price [\$/gu]} \times \text{recovery}}$$

Conversion factors may be required to convert the currencies in which costs and/or prices are specified into the same currency and to convert the price into the unit in which the grade is expressed. For example, gold price is usually quoted in US dollars per ounce (US\$/oz), but in many countries, gold grade is expressed in grams per tonne (g/t). Since there are approximately 31.1 grams per Troy ounce, the gold price in US\$/oz must be divided by 31.1 to obtain the price per gram, so that the formula can be used to obtain a break-even grade in grams per tonne. So, for example, if the price is US\$1244/oz, this equates to US\$40/g. Including such conversions adds slightly to the calculation but does not change the inherent simplicity of the basic break-even formula.

Although the formula is simple, many companies spend a lot of time and energy arguing over what costs should be included and then extracting them from the corporate cost accounting systems. The principal aim of this chapter is to establish principles regarding *what* costs should go into break-even calculations, and perhaps more importantly, *why*. If the *why* is understood, defining the *what* should become relatively straightforward.

Ultimately, the costs are whatever the user chooses to specify; however, there are no industry-wide standard definitions. Each company has its own. In this book, *break-even* is used as a generic term. The author is, however, aware of at least two major mining companies that use the term to define specific costs to be included in the calculation. It is therefore essential, when dealing with break-even cut-offs, to identify clearly which costs are (and which are not) included in the cost term of the formula.

A simple ore–waste break-even calculation

Consider this simple break-even calculation example:

- product price   \$10/gu
- recovery   90 per cent, and therefore
- recoverable revenue   \$9.00/gu
- variable cost   \$22.50/t
- total cost   \$60/t

$$\text{Variable cost break-even} = \frac{\$22.50/t}{\$10/gu \times 90\%}$$

$$= 2.5 \text{ gu}/t$$

$$\text{and total cost break-even} = \frac{\$60.00/t}{\$10/gu \times 90\%}$$

$$= 6.7 \text{ gu}/t$$

The grade is expressed as grade units per tonne (gu/t); for example, per cent (or product tonnes per ore tonne) for base metals, or grams per tonne for precious metals. The metal price has been converted from the price per unit of metal as normally quoted to a price per grade unit to simplify the break-even calculation.

Figure 3.1 shows these break-even relationships graphically. Variable and total unit costs and unit revenues are plotted as straight-line functions of material grade. Costs per tonne are assumed to be independent of grade and are therefore the same at all grades, being \$22.50 and \$60.00/t, as in the data above. The revenue received increases at the rate of \$9/t/gu, as calculated above. As expected, the revenue line crosses the cost lines at 2.5 and 6.7 gu/t, the break-even grades calculated.

Figure 3.2 also shows these break-even relationships graphically but this time as net contributions, positive and negative. The costs already noted have simply been deducted from the revenue. At zero grade with no revenue generated, the contribution is negative and numerically equal to the cost per tonne. The contribution based on variable costs increases linearly at the rate of \$9/t/gu from –\$22.50/t at zero grade. Similarly, the contribution based on total costs increases at the rate of \$9/t/gu from –\$60/t at zero grade. As expected, the net contribution lines cross the zero contribution axis at 2.5 and 6.7 gu/t, again the break-even grades calculated previously.

FIGURE 3.1

Graphical representation of break-evens: revenue = cost.

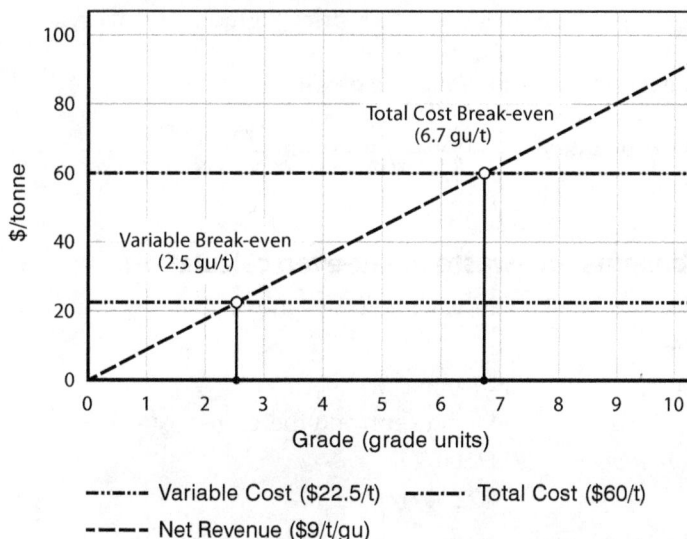

FIGURE 3.2

Graphical representation of break-evens: net contribution = 0.

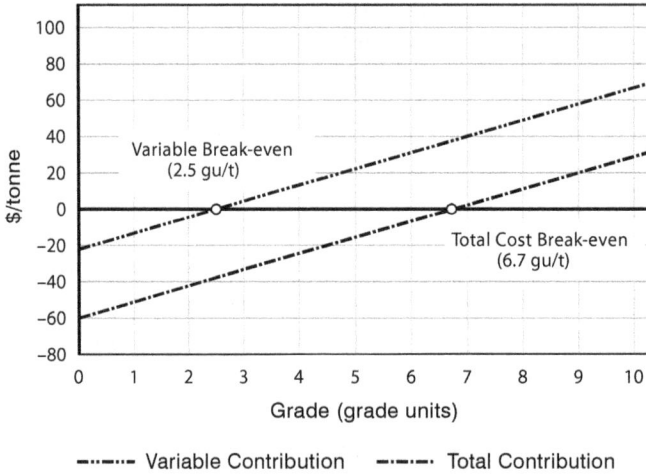

A MORE GENERAL DEFINITION OF BREAK-EVEN – EQUAL NET BENEFITS

The simple definition of break-even, as previously described, is the grade at which revenue obtained is equal to the costs of producing that revenue. A more general definition is the grade at which the benefit from one course of action equals the benefit from an alternative course of action.

Using this definition, the concept of the ore–waste break-even becomes the grade at which the benefit obtained from treating a block of rock as ore is the same as treating it as waste. This leads us to what is commonly known as a *cut-over* grade, where the benefit from treating ore by one process is the same as treating it by another. Typically this is applied to two distinct metallurgical processes, one with relatively high costs and recoveries, the other with relatively low costs and recoveries.

The *equal benefits* break-even occurs when:

$$\text{Price} \times \text{grade} \times \text{recovery}_1 - \text{cost}_1 = \text{price} \times \text{grade} \times \text{recovery}_2 - \text{cost}_2$$

Rearranging the terms of this formula, we obtain:

$$\text{Break-even grade [gu/t]} = \frac{\text{cost}_1 - \text{cost}_2}{\text{product price} \times (\text{recovery}_1 - \text{recovery}_2)}$$

An equal net benefits ore–waste break-even calculation

Let us reconsider the break-even calculation example, the key features of which are reproduced here:

- product price $10/gu
- recovery   90 per cent, and therefore
- recoverable revenue $9.00/gu
- variable cost $22.50/t

Assume now that the variable costs comprise the following components:

- treatment $10/t of ore
- mining $12.5/t of ore
- strip ratio 1.5 t of waste per tonne of ore
(therefore 2.5 t of rock mined per tonne of ore)
- thus, mining cost $5/t of rock

Therefore, the variable cost incurred by each tonne of rock if classified as:

- ore: $10 (treatment) + $5 (mining) = $15/t total
- waste: $5/t (mining only)

Break-even occurs when:

$$\text{Price} \times \text{grade} \times \text{recovery}_{ore} - \text{cost}_{ore} = \text{price} \times \text{grade} \times \text{recovery}_{waste} - \text{cost}_{waste}$$

$$\text{Break-even grade} = \frac{\text{cost}_{ore} - \text{cost}_{waste}}{\text{product price} \times (\text{recovery}_{ore} - \text{recovery}_{waste})}$$

$$= \frac{\$15 - \$5}{\$10 \times (0.9 - 0.0)}$$

$$= \$10/\$9$$

$$= 1.11 \text{ gu}/t$$

Figure 3.3 shows graphically the net benefits of classifying material as ore or waste. The net benefit of waste is negative – a cost of $5/t at all grades of material. The benefit of ore increases linearly at the rate of $9/t/gu from –$15.00/t at zero grade. At zero grade with no revenue generated, the net benefit is negative and equal to the cost per tonne of material that is both mined (at $5.00/t) and treated (at $10.00/t). The benefit-of-ore line crosses the benefit-of-waste line at a grade of 1.11, as calculated. Note that the net contribution from ore does not become positive until the grade exceeds 1.67, at which point the revenue covers the costs of both mining and treatment. However,

FIGURE 3.3

Graphical representation of ore–waste break-evens – equal benefits.

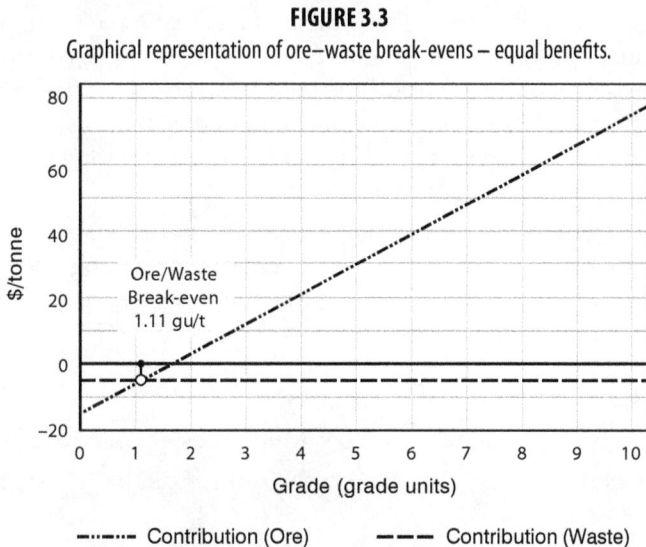

given the assumption that the mine is operating, the $5.00/t mining cost is a committed or sunk cost and the net revenue from treating a tonne of rock as ore need only cover the $10.00/t cost of treatment to make it worthwhile doing so. As noted in the preceding chapter, there is no guarantee that the value obtained when using the appropriate cut-off will be positive.

Our simple break-even calculation generated a value of 2.5 gu/t, compared with the equal benefits value of 1.11. What causes this difference? The distinctions in rationale and the arithmetic of the two calculations are self-evident, but what is perhaps more important for our understanding of cut-off derivation is the recognition that mines' cost systems often report all costs as a cost per tonne of ore, which may be satisfactory for many purposes but can be dangerously misleading when used for cut-off derivation without critical thought. As previously indicated, it is essential that costs be allocated to the physical activities that actually cause them to be incurred.

It can be seen from the derivation of the mining cost of $5.00/t of rock above that the $12.50/t total mining cost for ore is made up of $5.00 for the one tonne of rock that is classed as ore, plus $7.50 for the associated 1.5 t of rock classed as waste. The split of the costs of mining the total 2.5 t of rock between ore and waste is a function of the cut-off applied. If the cut-off were such that 0.5 t was ore and 2.0 t was waste, the cost per tonne of ore would be reported as $25.00/t. Conversely, if 2.0 t was ore and 0.5 t was waste, the cost per tonne of ore would be reported as $6.25/t. It is now evident that in the first simple calculation above, where the reported cost per tonne of ore was used in the break-even calculation, one of the main component costs used to determine the cut-off is itself a function of the cut-off. The value used is based on a prior assumption about what the cut-off is, generating the 1 t of ore/1.5 t of waste split.

This introduces a circularity into the calculation. If a set of iterative calculations were conducted, the process might eventually converge to the right answer; however, that would necessitate making use of the tonnage versus cut-off relationships for the block of rock to identify the ore–waste split at the cut-off calculated, from which one could recalculate the mining cost per tonne of ore, the break-even and so on. This iterative process is described in more detail later in this chapter.

There is no problem with this in principle, but in the development of cut-off theory we are at this stage considering break-even calculations that have been identified as 1D, considering only financial factors (prices and costs). The necessity to consider the geology, as expressed by the tonnage versus cut-off relationships, means we need to go beyond the 1D models' limitations and bring other parameters into the analysis. At first glance, this implies that the simple break-even cut-off model is inadequate for such a situation, and so it is if costs are expressed only per tonne of ore.

By using the more general definition and calculation of break-evens, however, and by correctly allocating costs to rock and ore, the circularity of having costs dependent on cut-off in the break-even calculation has been removed. A break-even cut-off grade can be derived by a simple 1D break-even calculation.

Cut-over calculation between two processes

Let us consider now two alternative processes. Mining costs remain at $5/t of rock and:

1. Process 1 has a 90 per cent recovery and costs $10/t of ore treated, as in the previous example
2. Process 2 has a 98 per cent recovery but incurs costs of $15/t of ore treated.

The break-even cut-over occurs when:

$$\text{Price} \times \text{grade} \times \text{recovery}_{\text{process 2}} - \text{cost}_{\text{process 2}}$$
$$= \text{price} \times \text{grade} \times \text{recovery}_{\text{process 1}} - \text{cost}_{\text{process 1}}$$

$$\text{Break-even grade} = \frac{\text{cost}_{\text{process 2}} - \text{cost}_{\text{process 1}}}{\text{product price} \times (\text{recovery}_{\text{process 2}} - \text{recovery}_{\text{process 1}})}$$

$$= \frac{\$(5+15) - \$(5+10)}{\$10 \times (0.98 - 0.90)}$$

$$= \$5.00/\$0.80$$

$$= 6.25 \text{ gu/t}$$

As determined previously, the ore–waste break-even for Process 1 is 1.11 gu/t. Applying the same calculation for Process 2 generates an ore–waste break-even of 1.53 gu/t but this is not used as a cut-off or cut-over. Combined with the derivation of the Process 1 - Process 2 cut-over of 6.25 gu/t, it can be seen that rock with a grade:

- less than 1.11 should be classified as waste
- greater than 1.11 and less than 6.25 should be classified as ore and treated by Process 1
- greater than 6.25 should be classified as ore and treated by Process 2.

Figure 3.4 shows graphically the net benefits of classifying material as waste and as ore treated by both Process 1 and Process 2. The first two lines are the same as in Figure 3.3. The net benefit of waste is –$5/t at all grades of material. The benefit of ore treated by Process 1 increases linearly at the rate of $9/t/gu from –$15.00/t at zero grade. Similarly, the benefit of ore treated by Process 2 increases linearly at the rate of $9.80/t/gu from –$20.00/t at zero grade, being the total cost of mining and treatment by Process 2. The Process 2 line crosses the Process 1 line at 6.25 gu/t.

FIGURE 3.4

Graphical representation of ore–waste break-evens and cut-over between two processes.

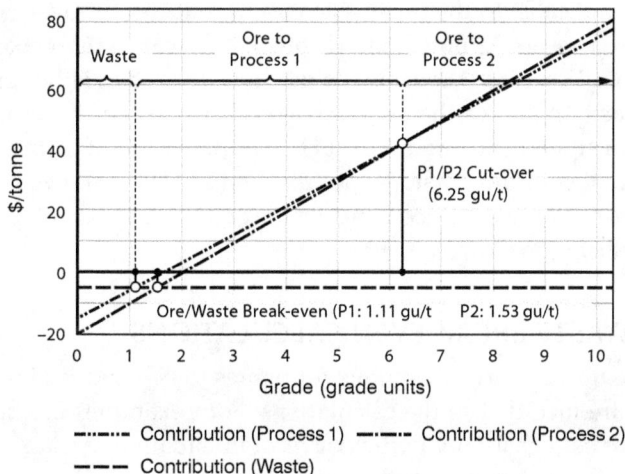

Note that the ore–waste break-even for Process 2 (1.53 gu/t) does not enter into the specifications of how rock is classified. In the absence of Process 1, rock with a grade above 1.53 would be classed as ore and treated by Process 2, but Process 1 provides better returns than Process 2 for all grades up to the identified cut-over of 6.25.

As mining progresses through the deposit, there is no guarantee that the distribution of grades will be such that both processes will operate at full capacity. If one process has spare capacity while the other is overloaded, the simple 1D break-even calculation does not suggest what to do with the excess material which, according to the cut-over value, should be processed by the overloaded process. It could be stockpiled until it can be treated by that process or treated immediately through the less profitable (for that grade of material) process with spare capacity. The simple 1D break-even cut-off model does not do everything we might reasonably require in order to maximise the cash generated.

Rock grades equal to cut-off

The preceding description of the split of rock between destinations does not address grades that are equal to the specified cut-off or cut-over. As with specifying costs to include in the calculation, there is no commonly agreed position on including or excluding material with grade equal to the cut-off.

In many cases this issue may be purely academic. For example, if grades are modelled to more significant figures than the cut-offs, the number of blocks in the geological model with an estimated grade the same as the cut-off is likely to be negligible and of no practical concern. The author would, however, suggest that it is pointless to classify material with the same grade as the ore–waste break-even grade as ore – it would then be processed for no net benefit.

If the cut-off and modelled grades have the same number of significant figures, both will probably have been calculated to a greater precision and then rounded. If the cut-off has been rounded up, it might be logical to say that rock with the same grade as the cut-off should be classified as ore. Conversely, if rounded down, you might specify the cut-off to be classified as waste. There will still be some potential for misclassification if the unrounded values for the rock grade and cut-off had been used. Again, this is likely to be negligible relative to other uncertainties implicit in the grade estimation and break-even calculation processes.

This rationale is applicable to the ore–waste cut-off, but the cut-over between the two processes is a different issue. As noted, the 1D break-even calculation takes no account of treatment plant capacities. Since the same net return is derived in either process, material whose grade is exactly equal to the cut-over would be sent to the process with spare capacity. If there is a significant amount of such material, it might be that some is sent to one process and the rest to the other. Again, the simple 1D break-even model does not account for all issues that need to be considered, and additional information must be brought into the decision-making process.

COSTS IN ORE–WASTE BREAK-EVEN CALCULATIONS

As stated, there are no standard definitions for names to be used for break-even grades and the costs that are included in the calculations. Four examples are described below, but these should not be seen as any particular recommendation for names or definitions; rather, they illustrate the range of costs that might be included in a break-even calculation:

- *mine operating break-even* – costs to be covered = total mining costs + total milling costs
- *site operating break-even* – costs to be covered = total mining costs + total milling costs + total site administration and services costs
- *full cost break-even* – costs to be covered = total mining costs + total milling costs + total site administration and services costs + head office charges + allowance for capital.

Many operations use at least two of these types of calculations for specifying cut-offs in a given situation. Companies tend to use a cut-off similar to the first of these as an operational cut-off for identifying marginal ore that can be treated opportunistically when there is spare capacity in the ore handling and treatment processes. Similarly, cut-offs similar to the third or fourth options are often used as the planning cut-off for identifying ore and waste in the long-term mine plan; however, the costs used in the break-even calculations are often in practice indicated with little understanding of what is being achieved.

As previously stated, many operations use the terms break-even and cut-off synonymously, so cut-off might be substituted for break-even in these definitions in company documentation. The author prefers in these circumstances to use the compound term *break-even cut-off* to indicate that the cut-off is to be specified by a break-even calculation. It should be clearly understood that in this book *break-even* is a generic term to describe the derivation methodology, with additional descriptors used to indicate the costs included in the calculation.

As can be seen from the formulas and example calculations, costs to be applied in break-even calculations ought to be the differential costs of classifying rock as ore rather than waste. A similar rationale applies when dealing with cut-overs between two processes. In the examples, costs that are common to both options (ore or waste, Process 1 or Process 2) effectively cancel out, leaving only the cost differential between the two options in the calculation.

What are the differential costs to be accounted for? If there is spare capacity in both the ore and product streams, the extra costs of classifying a piece of rock as ore are only the marginal variable costs of treating it as ore and dealing with the derived product. If either the ore or product stream is at capacity, choosing to treat a piece of rock as ore extends the operation's life; the extra costs of classifying it as ore are the variable costs of treating it as ore and handling the derived product, plus the fixed costs associated with the extension of the mine life.

The following subsections discuss the variable and fixed costs that arise in different parts of the operation and how they apply in break-even cut-off derivation.

Open pit – in the mine

When a decision is made to mine a pit to a particular size, the variable costs of mining (typically drill, blast, load and haul) are incurred for all rock removed from the pit, irrespective of its classification as ore or waste. All rock must be transported to the point where ore and waste streams diverge. The location of the rock in the pit – whether from deep or shallow areas – is therefore not relevant to the ore–waste decision.[1] Hence, open pit mining variable costs should typically not go into an ore–waste break-even calculation. This was seen in the more complex example, where the mining costs were included in the

---

1. Location, and hence haulage cost, is of course very relevant to determining the best mining schedule, but that is beyond the scope of the simple 1D process to specify a break-even grade.

pit mining variable costs should typically not go into an ore–waste break-even calculation. This was seen in the more complex example, where the mining costs were included in the costs of both ore and waste and cancel out in the resulting calculation. The exception, which follows directly from this rationale, is any differential cost of mining ore and waste. This could be the result of different:

- grade control procedures
- drill and blast patterns
- mining fleets
- haul distances.

The base mining variable cost should therefore be thought of as the cost of mining all the rock as waste, with any incremental cost of mining ore considered an extra cost that must be covered by the net revenue received from classing it as ore. The cost difference may be positive or negative. For example, drilling patterns for ore may be closer-spaced than for waste (generating a higher unit cost or positive incremental cost for ore), while haulage distances to waste dumps may be longer than for ore (resulting in a lower unit cost or negative incremental cost for ore).

Many readers may quite rightly raise an objection that, although the mining costs in general should not be included in the ore–waste break-even cut-off calculation, they nevertheless must be accounted for somehow or the operation might make a loss. In practice, this could lead to including these costs in the break-even cut-off grade, even though the preceding discussion indicates that they should not be. How is this resolved? That the problem arises at all indicates that the simple 1D break-even model of cut-off derivation is inadequate for generating a cut-off that can be used in long-term planning. The issue will be addressed and resolved in the more complex cut-off derivation models discussed in Chapters 4 and 6.

Underground – in the mine – excluded from break-evens

Costs to access new mining areas are typically independent of the amount of ore that is delineated for stoping within the new exposed rock. These are lump sum costs (described in Chapter 2) and are therefore not included in ore–waste break-evens. These excluded costs would typically be for:

- decline extensions downdip below existing stoping
- access and infrastructure for new orebodies
- footwall access development, to the extent that the strike length of the orebody does not vary with cut-off
- stope or production preparation that is not directly related to ore tonnage.

In the first two cases, there is no compulsion to access the mineralised material – the remote area could be left unmined, or the downdip extension could remain unaccessed and the mine closed instead. The decision to spend such project capital is justified by the additional value of having the area available to be stoped. Similarly, developing the footwall drive provides the choice of how much exposed strike length will be stoped, without compulsion to mine any or all of it. The initial development in a stope block, if it is simply providing the initial accesses and working faces such as sill drives in a cut-and-fill operation or raises in ore in a narrow tabular reef, is independent of what is subsequently stoped as ore from the rock made available.

Figure 3.5 illustrates how the strike length of the orebody and hence of access development may be independent of cut-off. Although the strike length of the material identified as ore may vary with cut-off, its spatial distribution is such that the accessed strike length is the same at all cut-offs considered.

FIGURE 3.5
Strike length of development unaffected by cut-off.

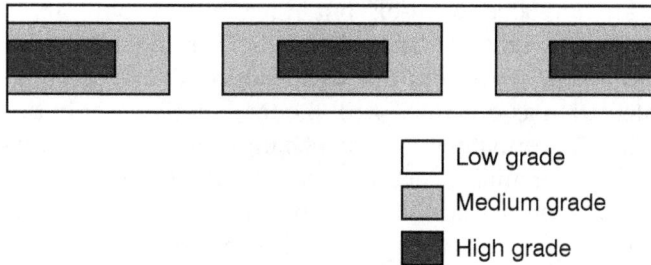

☐ Low grade
▨ Medium grade
■ High grade

The break-even grade for an area that may have been accessed by expending project capital does not change depending on whether the access has been developed or not; however, it is not uncommon in underground mines to see two different break-even cut-offs specified for an area, depending on whether it has been accessed or not. The difference between the two cut-offs purportedly accounts for the cost of accessing the area. This is fallacious reasoning. If the cut-off is going to be a particular value after the area is accessed, that was always going to be the appropriate cut-off to apply, whether the area was accessed or not at the time of the evaluation. The cost of the access development will have been sunk at the time the cut-off decision for the area needs to be finally made for mining purposes, so costs that will already be incurred at that time should be excluded from cut-offs being derived ahead of production for such areas.

As with open pit mining, many readers will quite rightly claim that the mining costs described here must be accounted for somehow or the operation might make a loss, and responding to this may lead to a decision to include these costs in the break-even cut-off grade, even though they should not be. Again, that the problem arises at all indicates that the simple 1D break-even model of cut-off derivation is inadequate for generating a reliable long-term planning cut-off. The issue will be addressed in Chapters 4 and 6.

Underground – in the mine – included in break-evens

Activities generating costs that do vary with ore tonnages, and which would therefore usually be included as ore-related costs in a break-even calculation, typically include:

- stope development in ore
- cross-cut stope access development in waste from strike access development
- strike or footwall access development in waste, to the extent that the strike length of the orebody varies with cut-off
- stope drilling and blasting
- ore haulage and hoisting
- backfill
- operating maintenance and sustaining capital costs for plant and equipment associated with the above.

Figure 3.6 shows a deposit for which the strike length varies for all cut-offs considered. Note that there may be a range of cut-offs over which the strike length does not change and a range where it does. Figure 3.7 shows an example similar to Figure 3.5, where the strike length of the mineralisation is unaffected by cut-off to the left of centre, but becomes variable at higher cut-offs to the right of centre. The variability may be continuous as illustrated in Figure 3.6, or have step changes as in Figure 3.7.

It is generally assumed that the listed mining costs are directly variable with ore tonnage and the unit costs are unrelated to cut-off, but in reality they may vary with cut-off. For example, if the width of the orebody varies with cut-off, the ore tonnes per metre of stope development will be a function of cut-off. This introduces the same sort of circularity into the break-even calculation as described with the total mining cost in the pit ascribed to ore rather than rock. Unfortunately, there is no simple mechanism in the underground situation to split this type of mining cost between ore and rock as there was in the open pit case. In the pit, mining costs are attributed to rock rather than ore for cut-off derivation, solving the problem; however, underground costs such as these are genuinely ore-related, though not in a linear way, rather than rock-related.

Depending on how variable the affected costs are with cut-off, this may or may not be a practical concern. If the variability is high, it may be necessary to perform iterative calculations for the appropriate break-even grade. This necessitates evaluating the effect of cut-off on the size and shape of the orebody, and takes the analysis beyond the simple 1D break-even model. Each iteration of the calculation is one-dimensional, but the process overall is not. Iterative calculations to converge on a consistent break-even cut-off are described below.

In an open pit as noted, all material within the pit limits is to be mined and only differential mining costs for ore would be included in break-evens. Where the rock has

FIGURE 3.6
Strike length of development variable with cut-off.

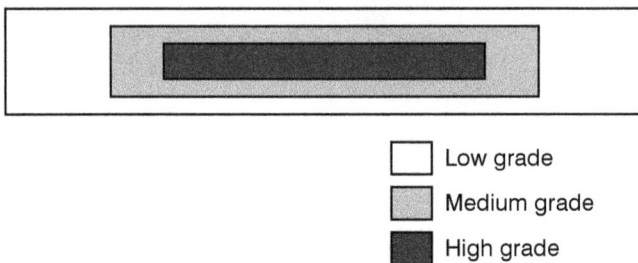

Low grade

Medium grade

High grade

FIGURE 3.7
Strike length of development partly unaffected by and partly variable with cut-off.

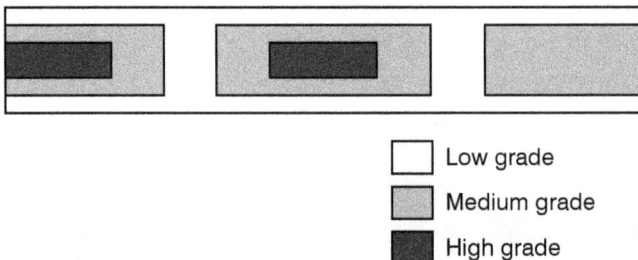

Low grade

Medium grade

High grade

come from in the pit is generally irrelevant. Apart from access development, underground material that is mined as ore would have remained unmined had it not been intentionally classified as ore. All mining costs downstream for those materials classed as ore are therefore typically ore-related; therefore underground location may be important. For example, if ore is hauled from the mine by truck, deeper zones will incur higher costs than those closer to the surface. The break-even grade for deep ore will be higher than for near-surface ore. But if all ore is dropped to a haulage level and hoisted in a shaft, there may be no cost difference depending on depth. There could, however, be additional costs associated with the lateral distance to be hauled – ore close to the shaft may have a lower break-even than more-distant material.

Many companies address this issue by conducting a break-even analysis on individual stopes, accounting for all the costs directly associated with each. Effectively, each stope could have a different break-even grade, but there are some dangers with this, which are covered in Chapter 7.

Processing – ore-related

Metallurgical activities generating costs varying with ore tonnages typically include:

- comminution (crushing and grinding) wear protection and media
- power (ore-related)
- maintenance of plant – for which wear is primarily driven by the amount of ore processed, such as
  - comminution plant
  - flotation cells/leach tanks
  - tailings thickeners, etc
- tailings-related activities such as thickeners, pumps and dam raisings (adjusted to account for any tailings removed from the stream for backfill)
- sustaining capital costs for ore-related plant and equipment.

Power is often a major component of ore-related treatment costs, and the bulk of it will typically be associated with grinding. It is often thought of as flowing from a specified energy requirement per tonne of ore to achieve desired grind sizes and liberation of the valuable mineral. There are two issues that have the potential to complicate this:

1. There may be ore types with different grinding energy requirements and hence milling rates. The simple break-even cut-off model may be inadequate to handle the impact of different blends of materials with various milling rates; however, it is at least feasible to calculate break-even cut-offs for different ore types, but how to then apply these is not so easy.

2. Mills will typically have a maximum-rated power draw, and may at times be operated at full power regardless of the throughput rate. In the extreme case, power could be thought of as a fixed rather than variable cost. It may be necessary to consider the mill's mode of operation and identify fixed and variable components of the power usage and hence costs. It should also be noted that, if power draw is constant but mill throughputs vary, in addition to the impact of different ore types noted, the grind size, liberation, recovery and various treatment costs will become functions of the feed rate. This implies that key inputs – costs and recoveries – are functions of other parameters not accounted for in the break-even calculation. Break-evens can therefore notionally be calculated for any particular set of values for these external parameters.

Yet using these break-evens to define ore and waste and subsequently specifying different treatment rate and power use strategies will invalidate the whole process, or alternatively require iterative calculations to converge on a break-even cut-off that is consistent with all the other operating parameters.

Processing – product-related

Activities that generate costs varying with product quantities usually include:

- reagents and reagent preparation
- power related to product processing and handling
- maintenance of plant for which wear is primarily driven by the amount of product handled, such as concentrate pumping, filtration and thickening
- sustaining capital costs for product-related plant and equipment.

Other product-related costs

Activities that generate costs that vary with product tonnages downstream of the treatment plant typically include:

- product transport
- roasting, smelting and refining of product
- royalties.

Royalties are usually classed as either:

- *Ad valorem* – these are usually expressed as a percentage of the net price after deduction of allowable costs, which are typically off-site realisation costs such as treatment charges (TCs) and refining charges (RCs) – collectively referred to as TC/RCs – and perhaps freight and transport from site to smelter. On-site operating costs associated with product (concentrate pumping, filtration and the like) are often not deductible for royalty purposes. Some jurisdictions charge *ad valorem* royalties on gross value of product.
- *Tonnage-based* – expressed as a simple dollar charge per tonne of product, or sometimes per tonne of ore.
- *Profit-based* – expressed as a percentage of net profit.

In some jurisdictions, royalty specifications may comprise a combination, or the rate specifications to be used may be more complex than a unit cost or single percentage.

Tailings costs were listed as ore-related costs. For precious metals operations, the amount of product extracted from ore is relatively small, and for all practical purposes the tailings tonnage is the same as the ore tonnage. In base metals operations producing a concentrate, the product tonnage may be significant and hence the tailings tonnage is much lower than ore tonnage. Break-even calculations are dealing only with costs for rock, ore and product. Tailings costs will be applied to all ore tonnes, which include concentrate tonnes. The simplest, most rigorous way to account for this in break-even calculations may be to include a cost credit in the product-related costs for the concentrate that is diverted from the ore stream and not incurring the tailings-related costs applied to all ore tonnes.

Mine cost accounting systems often express product-related costs in the mill (and frequently some or all of the downstream costs) as a unit cost per tonne of ore. Doing this is similar to expressing rock-related costs in an open pit mine as a cost per tonne of ore, as discussed earlier. This may be satisfactory for management control if head grades and hence the amount of product produced per tonne of ore remain constant. If cut-offs are under investigation, it needs to be recognised that the head grade is a function of cut-off, and hence the product cost per tonne of ore is itself a function of cut-off. In the same way that rock-related costs are related to rock, not ore, when dealing with open pit mining costs, so product-related costs reported as costs per tonne of ore must be extracted and reported separately as a cost per unit of product. This avoids the circularity of calculating a break-even cut-off using a cost that is a function of cut-off.

Product-related costs per unit of product are effectively a reduction in the net price received per unit of product. They should therefore be applied in the break-even calculation to reduce the price per unit of product in the denominator of the break-even calculation, not included as a cost per unit of ore in the numerator. The handling of product-related costs will depend on the way that they behave. Where costs are incurred because of the quantity of product (such as pumping, filtration and transport of product), they should be deducted as a cost per unit from the price per unit. Where product-related costs are driven by the value of the product (such as *ad valorem* royalties), they should be applied as a proportional or absolute reduction in price. The net value used as the price in the denominator of the break-even calculations, as shown earlier, will therefore be:

$$\text{Net price} = (\text{price} - \text{TC/RCs}) \times (1 - \text{royalty rate}) - \text{product-related costs}$$

Note that this formula applies when royalties are *ad valorem* based on the price received by the mine after allowing for realisation costs. Some jurisdictions may charge royalties on the gross value of the product. If royalties are actually levied as a cost per tonne of ore, they may be correctly included in the cost per tonne of ore in the break-even calculation. Simple profit-based royalties may be excluded from break-even calculations, as notionally at the break-even no profit is generated and hence no royalty is payable.[2] Where royalties are derived by more complex mechanisms, it may be appropriate but possibly not feasible to incorporate them in a simple break-even formula.[3]

Stockpiling and reclaim

Stockpiling and reclaim costs are costs that are associated only with ore. Depending on how a company chooses to look at it, this could be included in company reports as a mining cost (if it is an intermediate process before delivery to the mill for treatment, or because the mining fleet is being used to handle it) or as a treatment plant cost (if mining is deemed to finish at the point where the material is first dumped). Either way, there will be an additional cost per tonne of ore to be included in the break-even calculation.

Several future economic scenarios may be assumed to calculate break-even cut-offs to define the grade limits of various stockpiles for different categories of material, often referred to by such names as high, medium and low grade and mineralised waste.

---

2. Care should be taken to ensure that royalties behave in the way stated. In most cases this will not be an issue; however, the author is aware of one large mining operation whose royalty was expressed as a dollars-per-tonne-of-ore charge. The rate was varied by way of a complex formula which, when reduced to its simplest terms, showed the royalty to be in fact *ad valorem*.

3. Whenever it is difficult to fit an important cost or revenue component into a break-even calculation, it indicates that the simple break-even cut-off model is inadequate and a more complex model may be required.

Fixed costs in break-evens

As indicated, fixed costs need to be included in break-evens when classifying a piece of rock as ore will extend the life of the operation. This will occur if the ore or product stream in the production process is operating at capacity. The additional ore therefore extends the life of the operation and thereby incurs additional fixed costs for the extra time – these costs must be paid for by the revenue obtained from the additional ore, and hence included in the break-even calculations. A similar rationale applies when stockpiles continue to be treated after the mining operation is complete.

A break-even planning cut-off will usually include fixed costs, as the implied purpose is to identify the life-of-mine reserve: every tonne thus identified is essentially generating extra mine life.

Which fixed costs should be included in such cases? The rationale suggests only fixed costs that are incurred while the operation continues should be included. This implies that all site administration costs and fixed costs in the mine (if still operating) and treatment plant should be counted. In the case of a one-mine company, virtually all corporate administration costs (except ongoing exploration) are incurred because the mine is operating, and will cease when the mine ceases. But if the mine in question is part of a corporate group, closing the one mine may or may not impact on the overhead costs incurred in any one year or on the life of the group's combined operations. Each mine and group's circumstances must be considered to determine whether overheads will vary if more or less rock is classed as ore at the operation in question.

It should be noted that corporate overheads are frequently charged to the group's operations on the basis of ore tonnes, product quantities or product values. All of these are arbitrary processes; they do not imply that the costs are actually driven by or related to those parameters, for if they were they would be variable costs, not fixed. It is therefore irrational to assume for cut-off derivation that if a mine producing part of a group's total ore tonnage is closed, that proportion of fixed costs will be eliminated. The proportion by definition has to be less than that, and may even be zero. One needs to identify if, and to what extent, corporate fixed overhead costs should be included in any one operation's break-even calculations.

The same reasoning applies if, for example, an open pit mine closes but treatment of stockpiled ore continues. There is usually no simple common formula that relates site administration fixed costs to rock tonnes, ore tonnes and product quantities in order to identify the costs that will change if the life of the mine or the treatment facilities alters.

Capital costs in break-evens

Capital costs are significantly more expensive than operating costs if the effects of taxation and the time value of money are taken into account. Consider a reserve of 10 Mt being mined and treated at the rate of 1 Mt/a, an operating variable cost item of $1.00/t and a capital cost item of $10 M, typically representing $1.00/t. Table 3.1 shows the derivation of net cash flow and net present value (NPV)[4] for each of these, assuming a 30 per cent tax rate and a ten per cent discount rate. Some values in the column for year 0 and in the bottom row have been rounded.

---

4. As it is beyond the scope of this book to discuss discounted cash flow techniques, readers unfamiliar with concepts such as discount rate, net present value and equivalent annual value should refer to standard texts on project evaluation and discounted cash flow techniques.

TABLE 3.1

Effective costs of operating and capital expenditure.

Year	0		1	2	3	8	9	10
Ore tonnes			1000 kt	1000 kt	1000 kt	1000 kt	1000 kt	1000 kt
Operating costs	$1.00/t		$1 000 000	$1 000 000	$1 000 000	$1 000 000	$1 000 000	$1 000 000
Tax benefit	30%		$300 000	$300 000	$300 000	$300 000	$300 000	$300 000
Operating costs after tax			$700 000	$700 000	$700 000	$700 000	$700 000	$700 000
Net present value of net operating costs	10%	$4 300 000						
Capital cost		$10 000 000						
Depreciation			$1 000 000	$1 000 000	$1 000 000	$1 000 000	$1 000 000	$1 000 000
Tax benefit	30%		$300 000	$300 000	$300 000	$300 000	$300 000	$300 000
Capital cost after tax		$10 000 000	($300 000)	($300 000)	($300 000)	($300 000)	($300 000)	($300 000)
Net present value of net capital cost	10%	$8 200 000						
Equivalent annual value of net capital cost			$1 327 000	$1 327 000	$1 327 000	$1 327 000	$1 327 000	$1 327 000

The columns for years 4 to 7 have been collapsed for simplicity. The operating cost of $1.00/t generates a cash cost of $1 M per year, which in turn generates a tax credit of $300 000 per year (assumed to be realised in the year of expenditure), and hence a net cash cost of $700 000 per year. This can be seen to generate an after-tax NPV of some $4.3 M.

The capital expenditure is assumed to occur at time zero, with straight line depreciation for tax purposes over the ten-year life. This generates a depreciation charge of $1 M per year, similar to the cash operating cost, and hence generates the same $300 000 per year tax credit. The net cash flow for the capital item is therefore a $10 million outlay at time zero, followed by what is effectively an annual cash inflow of $300 000. The NPV of this cash flow pattern is $8.3 M or 1.9 times – that is, nearly double – the net cost of the operating cost. The last row of Table 3.1 shows the after-tax NPV distributed over the mine life as an equivalent annual value of some $1.33 M per year. The table indicates that the net after-tax cost of what is nominally a $1.00/t operating cost is $0.70/t; however, the net after-tax cost of a $1.00/t capital outlay is $1.33/t when the net spending (considering the time value of money) is expressed at the same time as the physical activities, similar to operating costs.

The ratio between capital and operating costs will depend on the tax rates, discount rates and inflation rates. Inflation is not shown in Table 3.1 but its effect is to erode the value of the depreciation charge over time, and hence increase further the net after-tax cost of capital expenditure relative to operating costs. Is it therefore sufficient to merely include capital in break-evens at the nominal cost (capital cost divided by the tonnage to which it relates) or should a suitable factor be calculated and applied to account for the extra cost of capital costs relative to operating costs when considering tax and the time value of money?

BREAK-EVENS AT VARIOUS POINTS ALONG THE PRODUCTION PROCESS

In an underground operation, it is not uncommon to find that mines will have a number of break-even cut-offs calculated for material at different points along the production process. This is logical because there are a number of sequential ore-related activities with underground mining, so we can reassess the ore–waste cut-off decision at each point. It is only when there is no potential to halt the movement of rock through the system that the irrevocable decision as to whether the rock is ore or waste is made.

At each point in the process, the costs associated with classifying the rock as ore must be compared with classifying it as waste. Typically the cost of waste will be zero at any point – we simply walk away at whatever stage that happens to be; however, there may still be some costs, such as making safe or rehabilitating the partly prepared material.

Working backwards, the ore-related costs that would go into these break-evens (depending on how advanced the production preparation is) would typically include:

- metallurgical treatment costs versus the cost of discarding as waste material that could be processed as ore
- backfill costs if fill is used
- haulage/hoisting costs and primary crushing costs in the mine, if relevant
- stope loading
- blasting
- stope drilling
- stope development
- waste access development, to the extent that these may be included in an ore–waste break-even calculation.

For an open pit operation, there will be one break-even cut-off-defining ore to be treated at any time, since an open pit operation lacks the sequence of ore-related production steps found in underground operations in which the ore–waste decision can be re-evaluated. There might be different break-even cut-offs for freshly mined and stockpiled material to be fed to the treatment plant, since there may be different costs for material treated from each source. If nothing else, direct feed is notionally already in trucks ready for tipping into the process stream, whereas stockpiled material must incur the additional costs of reclaim. There will also typically be stockpiles of potential ore between the mining and treatment operations, with different cut-offs applied. These may be constructed so that declining grade stocks provide potential feed to the plant if conditions change.

ITERATIVE BREAK-EVEN CONVERGENCE AND DIVERGENCE

It has been noted that some unit costs used in a break-even calculation are themselves functions of cut-off. This may manifest itself as a cost structure – a set of unit costs for activities comprising the planned underground production operations – that is different to the cost structure used to derive the break-even cut-off to generate that plan. This results in a break-even grade different from the initial break-even cut-off used. There is then an understandable desire to apply this new break-even grade as a revised cut-off and rework the reserve and mine plan. The same outcome may recur, and hence the calculation and mine plan revision may need to be repeated. The implicit aim of this iterative process will be to converge to the break-even that results in costs that generate the same break-even grade as the break-even cut-off used.

Figure 3.8 shows how this process may operate, and how there is no guarantee that it will actually converge as desired. The axes on the plot are the specified cut-off, either arbitrary or from a previous iteration, and the break-even grade resulting from using the

FIGURE 3.8

Iterative break-even determinations converging or diverging.

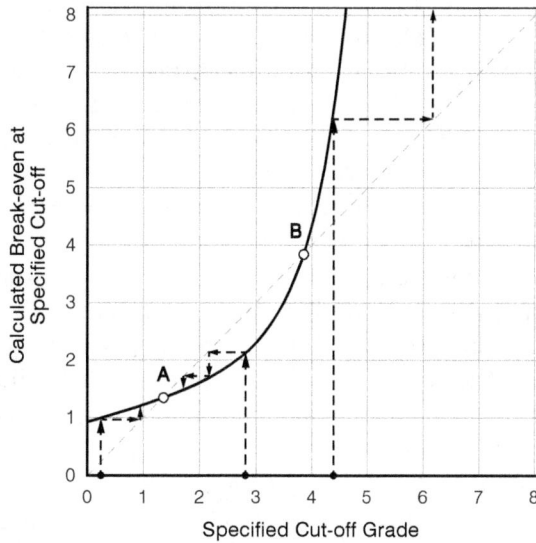

specified cut-off. The curved line plots the break-evens derived from the orebody costs defined by each of the cut-offs on the horizontal axis. The dashed straight diagonal line refers to where the values on the axes are equal; that is, where the break-even grade for an orebody is the same as the cut-off applied to generate that orebody.

The break-even versus cut-off curve must start and end above the reference line. At a cut-off of zero, which implies that all the mineralised rock is classed as ore, there will be a cost per tonne for the orebody defined and the resulting break-even must be greater than zero. At high cut-offs the orebody becomes small, but there will be costs that are not purely variable. As the cut-off used increases, the break-even will increase rapidly and at the theoretical limit where increasing the cut-off reduces the reserve tonnage to zero, the break-even will become infinite.

Depending on the nature of the mineralisation (which is strictly outside the bounds of the 1D break-even model of cut-off) the break-even versus cut-off plot may cross the reference line. If so, it must cross in two places as shown by the points marked 'A' and 'B' in Figure 3.8. It can be seen now that there are three situations:

1. At specified cut-offs less than at Point A, the break-even versus cut-off curve is above the reference line but converging to Point A with increasing cut-off. If an initial cut-off is specified in this range (the black dot at approximately 0.25 on the horizontal axis) the break-even resulting from the orebody would be approximately 1.0, indicated by the vertical dashed arrow. If it is now reassessed using that break-even as the cut-off (shown by the dashed horizontal arrow) and then vertically up again, a new break-even of approximately 1.2 results. If this process continues, the cut-offs and break-evens will increase and converge to Point A where the cut-off and break-even are the same, as is to be expected when any cost circularities used to derive the break-even have been resolved.

2. At cut-offs between those at Point A and Point B, the break-even versus cut-off curve is below the reference line. If an initial cut-off is specified in this range (the black dot at around 2.8 on the horizontal cut-off axis), the break-even resulting from the orebody

would be approximately 2.1, indicated by the vertical dashed arrow. Continuing the iterative process with the orebody defined by using the previous break-even as the cut-off, the successive cut-offs progressively reduce to converge to Point A.

3. At cut-offs greater than at Point B, the break-even versus cut-off curve is above the reference line and diverging from it with increasing cut-offs. If an initial cut-off is specified in this range, (the black dot at approximately 4.4 on the horizontal cut-off axis), the break-even resulting from the orebody would be roughly 6.2 (the vertical dashed arrow). Continuing the iterative process, the successive cut-offs progressively increase but diverge instead of converging, and eventually reach a value that reduces the tonnage of ore to zero.

In the special case where the break-even versus cut-off plot touches the reference line tangentially at one point, Points A and B become coincident, and the second situation described for cut-offs starting between Points A and B ceases to exist. If the break-even versus cut-off curve does not cross the reference line, the third situation applies at all initial cut-off values; the iterative break-even process will generate cut-offs that increase until the ore tonnage is reduced to zero.

Hence, there is no guarantee that the iterative break-even process is able to generate a consistent break-even cut-off. If it is feasible, whether the convergence is achieved or not will depend on the initial cut-off used in the process.

Both Point A and Point B generate consistent break-even cut-offs, but if convergence does occur, it will generate the lower of the two. The higher cut-off at Point B is equally valid, but will only be found if it is, by coincidence, the cut-off used initially.

PROBLEMS WITH BREAK-EVEN CUT-OFFS

Philosophically, the author's biggest objection to the use of break-even cut-offs is that it surrenders responsibility for specifying the corporate policy to the vagaries of the market place. As prices fluctuate, so too do break-even cut-offs. The company does not actively control what is ore and what is not, but passively accepts whatever the price fluctuations deliver. There are, however, important practical ramifications.

Achieving the corporate goals

As has been indicated, there are various cost components that are not included in an ore–waste break-even cut-off calculation; however, these costs must be covered in some way or the operation potentially makes a loss.

Applying a break-even cut-off simply identifies the ore tonnage with a grade above that cut-off. Yet it tells us nothing about the overall grade of that tonnage. There is no guarantee that the grade above the break-even that is required to cover the costs excluded from the break-even – and provide some desired profit margin – actually exists.

By way of analogy, let us consider the difference between a coral atoll and a volcanic island. Both grow from the sea floor but by different mechanisms. The coral atoll essentially stops growing when it reaches sea level, and sand can then accumulate to a relatively small height above sea level. The volcanic island, however, can continue to grow vertically and attain a great height after reaching the surface of the sea.

The sea level is analogous to the 'every tonne pays for itself' break-even cut-off, and simply defines the areal extent of the dry land. The coral atoll and the volcanic island might have the same plan area, which compares to the tonnage above cut-off. But it is the average elevation that determines how affected the island will be if the sea level rises – the average elevation is akin to the ore grade, which is what generates revenue. If the sea level rises, similar to adding the costs excluded in the break-even cut-off to achieve a minimum profit, the coral atoll may be completely flooded; although there are tonnes that 'pay for themselves', these tonnes do not cover the additional costs to ensure that a profit is made. The volcanic island, however, rises significantly above this cost requirement, and profitability may be assured under all realistic economic conditions.

It is important to note that achieving a level of profitability is totally different from ensuring that every tonne pays for itself. It is inappropriate to simply add these additional costs and required profit margin to the cost used to generate a break-even cut-off and then apply that higher cut-off to redefine what is ore and what is not. It is the average grade of the ore that pays for all costs and generates a profit; not every tonne mined covers those additional costs and profit margins.

This issue will be addressed in the more complex cut-off derivation models discussed in Chapters 4 and 6.

Marginal versus full cost break-evens

In the discussion of costs to include in break-evens, only marginal costs need to be included in the calculations if both the ore and product stream are operating below capacity; however, fixed costs must be included if either of these streams is at capacity. System capacities is outside the scope of the 1D break-even derivation, so the need to consider that at all indicates that the break-even model of cut-offs has limitations.

Let us accept that consideration of capacities does not 'break' the break-even cut-off model, but simply sets conditions for including or excluding certain costs. It could be argued that we can thereby generate satisfactory break-even cut-offs for the two situations and apply the appropriate one for the state of the operation at any point in time; however, the issue is not necessarily resolved.

If it happens to be that the tonnage available between the marginal and full-cost break-even cut-offs (referred to as the marginal tonnage) is insufficient for filling the capacity of either the ore or product circuits, the marginal break-even cut-off will be appropriate. Yet this will often not be the case – there will frequently be more marginal tonnage available than is needed to fill the capacities of the ore or product circuits. Should we therefore treat all the available marginal tonnage if there is some spare capacity in the ore and product circuits? The answer of course is no; if we did, the logical extension would then be to use a marginal break-even for all ore definition. We know – both from previous discussion and general industry-wide common sense – that using a marginal break-even as the overall cut-off is asking for disaster.

What is required, of course, is utilising the highest grade marginal material to fill the available spare capacity. The 1D break-even cut-off model is now definitely lacking: an analysis will require considering both the grade distribution and production system capacities. What this can lead to in practice, though, is a decision that if all the marginal grade material cannot be treated, none of it should be. Both extremes are suboptimal. Nevertheless, when the understanding of cut-off is that it is the break-even and the marginal cost break-even does not work, then the only other alternative is the full cost break-even, and there is no acceptable position between the two.

The author is aware of an operation where a full-cost break-even cut-off was calculated but, when applied, the nature of the resulting orebody was such that a production rate to fill the mill could not be maintained. The cut-off was then reduced to increase the size of the orebody so that the mill could be operated at its capacity. In the absence of a more comprehensive model of cut-off, company staff were uneasy: 'The cut-off should be X (the break-even), but we're actually using Y. It's wrong, but we have to do it!' The discussions of Lane's methodology in Chapter 5 and strategy optimisation in Chapter 6 will indicate that they were intuitively correct, but without the understanding of more complex cut-off models, were unable to recognise the optimum strategy as being correct.

It is also worth noting here that *marginal* actually means extra or differential costs. In practice, marginal costs are often assumed to be the same as variable costs, with the terms marginal and variable used interchangeably. This will be true when there is capacity in the ore and product streams, and additional material classed as ore need only cover its direct variable costs to add value; however, as noted, if either stream operates at capacity, additional material extends the life of the operation, thereby incurring more fixed costs. The same applies when stockpiles only are being treated after mining has finished. In these cases, marginal costs include both the variable and fixed costs.

Rising and falling prices

When prices are rising, the break-even cut-off reduces and proportionally more of the rock is classified as ore. In an underground mine, mineralisation that was previously waste in areas already accessed by development becomes ore, so the pressure to extend development into new areas is reduced. In addition, the ore tonnes per metre of development will increase with reducing cut-off, so the pressure on development is relaxed further. Similarly, in an open pit mine more of the exposed mineralisation will be ore and less will be waste. The stripping ratio reduces and a lower mining rate of total material is needed to maintain the same tonnage of ore though the mill. If prices remain high for a period of time, a new scheduling balance of access development, stope development and production of waste and ore mining will be established. In a desire to minimise spending, access development and overburden removal capital will not be spent until it is absolutely required.

The average grade of the ore produced must, however, decline. One possible outcome of this lowering of grade is that product targets are no longer met, even with the ore stream operating at capacity. Significant capital may be needed to increase ore treatment rates through the metallurgical treatment plant so that product targets can again be met.

All will be well until prices start to fall, and the larger or faster the fall, the worse the potential problem. Lower-grade material that was ore at low cut-offs will no longer be ore at the higher break-even cut-offs with lesser prices. The now sub-cut-off material will be removed from the production plan, resulting in a significant reduction in stocks of developed or exposed ore. Production will be rescheduled to maintain the planned rate from remaining sources above cut-off in the short-term, but there is potential for a major shortfall in available ore down the track. This results from the reduction in the rate of access development into new areas or overburden removal when prices were higher and cut-offs were lower, which will have restricted the amount of ore available above the new higher cut-offs in the medium term. At this stage, if previously developed but now sub-cut-off material is still available, it may have to be brought back into the production schedule to maintain planned ore production rates, but at lower grades. If it is no longer accessible, production targets may not be met.

It is hoped that such reintroduced low-grade material is at least above a marginal break-even cut-off. The author is aware of cases where the value of the low-grade material was below its direct variable costs of production, so additional production was actually losing money. If this were necessary to maintain the production rate above some minimum level necessary for the mill to operate, so be it, though in those circumstances periodic mill shutdowns and batch processing might have been a better solution. If, however, the additional material was simply meeting an arbitrary ore tonnage target, the potential cost of value-destroying management targets would have to be addressed.

Either way, a development or stripping rate crisis is now manifest. When prices were rising and cut-offs falling, the ore tonnes per metre of development or per tonne or cubic metre mined increased; however, with falling prices and rising cut-offs, the ratio will reduce. To maintain the planned ore production rate at the higher cut-off, the steady ongoing development or stripping rate must increase to a new level. On top of this, there must be a campaign of development or stripping to re-establish ore stocks so that production can be maintained. This will be happening at a time when prices are falling and cash availability is reducing. The pressure will be to reduce these costs at a time when the critical operational need is to increase them.

Simplistically, it is usually not possible to change the cut-off in an established operation without changing something else, such as the development or stripping rate as indicated. When prices are rising and break-even cut-offs are falling, the 'new' low-grade ore is gradually worked into the production schedule and the overall development or stripping rate is adjusted downwards without any perceptible problems. The necessary changes are made but they may be unrecognised, or, if acknowledged, applauded as a cost reduction. The increase in reserves is also viewed favourably; however, when prices fall and the break-even cut-off is increased, potentially major changes to the schedule need to be made almost immediately. The price falls that are the immediate cause of the problem may also result in a reluctance or inability to spend the extra needed – in the short-term, for re-establishing the developed or exposed ore stocks, and in the longer-term, for maintaining an ongoing rate higher than before the cut-off increase. Production shortfalls may therefore occur for several years until stocks are slowly rebuilt and the appropriate balance between development or stripping and production is re-established.

Many of the reported problems in the mining industry in recent times, when prices for many commodities have reduced significantly after a period of substantial rises, can be attributed to this type of scenario. This is driven directly by a lack of appreciation by industry decision-makers and market analysts that simple 1D break-even cut-offs, when used to develop long-term mining strategies, do not result in maximised cash generation. They can actually drive a mine into a perilous financial position when prices turn down.[5] This is discussed in more detail in Chapter 12.

---

5. For many years, while prices were rising and break-even grades were falling, industry analysts and mining company executives seemed happy to accept the increased reserves and lowered operating costs (though costs per unit of metal did not necessarily fall significantly due to large tonnages of lower-grade material being brought into reserves). In more recent times, reports of discussions at a number of mining financial conferences suggest that analysts are now chastising company executives for not delivering more cash when prices were high, charging them with 'deciding to reduce head grades'. With prices falling and what was previously considered 'ore' no longer being ore, analysts are complaining about production shortfalls. For those involved in strategy optimisation and long-term mine planning, all of this was predictable. It has happened in the past and will continue to happen again in the future as business cycles cause prices to rise and fall, unless things change. As long as the industry as a whole – mining executives, analysts who comment upon their decisions and technical staff who develop

SUMMARY

The implicit goal of a break-even cut-off is, in its most general terms, to ensure that every tonne classified as ore 'pays for itself' at the time it is treated. There are no industry-standard definitions of what this means or what costs go into a break-even calculation. Most companies using break-even cut-offs specify the costs that are to be included in perhaps several different break-even-style cut-offs.

An alternative definition of a break-even cut-off ensures that every tonne is dealt with by the course of action that generates most benefit. As a general principle, the costs used should be the differences associated with the two potential courses of action. If the decision is between classifying a piece of rock as ore or waste, it is only the differential costs of ore that need to be considered.

In the case of an open pit, all planned material is mined from the pit regardless of whether it is ore or waste. The differential costs will therefore be the variable costs of dealing with ore and product through the treatment circuits of the metallurgical plant and downstream to the point of sale. Mining costs will typically not come into this calculation, except where there are different mining costs associated with ore and waste – perhaps as a result of different drill and blast patterns or haulage distances.

Underground, virtually all activities are associated with ore. The main exception is major waste access development, which is exposing new mineralisation within which ore to be produced can be delineated. This major access development typically does not depend on the amount of mineralisation exposed. It is a function of the lateral or vertical distance to be covered to access a remote zone of mineralisation or to expose the next sublevel block. Because the cost of this access does not depend on the amount of ore delineated, it does not change according to how much of the mineralisation is classified as ore rather than waste, and is therefore not a differential cost for a break-even calculation. Some access stope preparation development, though associated with and maybe even mined in ore, is also not directly related to the tonnage of ore to be extracted from a block and should similarly be excluded from a break-even calculation.

If there is spare capacity in both the ore and product circuits in the plant, material may be included in the production plan if its grade is above the break-even cut-off derived using downstream ore and product-related variable costs; however, not all such material can necessarily be added to the ore stream. Only the highest grade material that fills the capacity shortfall should be mined and treated, but the simple break-even cut-off mechanism does not facilitate this – it is typically an all-or-nothing scenario analysis.

There is no guarantee that the average grade of an orebody defined by a break-even cut-off will deliver a profit. Generating a profit is completely different from merely ensuring that every tonne pays for itself. The level of profitability comes from the average grade of material above cut-off. A simple break-even cut-off takes no account of the grade distribution of the mineralisation, nor the various stages of the mining and treatment

---

the plans that seek to implement corporate strategies – continues to believe that cut-off and break-even are synonymous, the situation will stay the same.

The author suggests that company executives did not 'decide to lower head grades', but passively accepted that cut-offs and hence head grades had to fall because the price was rising and cut-off is assumed to be a break-even grade. The key issue is that break-even cut-offs are not focused on maximising cash returns. Ounces and tonnes in reserves, the measures that managers and analysts appear to have focused on in the past, are negatively correlated with cash generation. This critical issue with break-even cut-offs is discussed further in Chapter 14.

process; however, many of the problems identified with break-even cut-offs have indicated that these must be considered. Unless they are specifically brought into the cut-off model, attempts to account for them will be arbitrary and suboptimal.

Using a break-even grade as a cut-off effectively surrenders responsibility for specifying what is ore and what is waste to uncontrollable changes in the market. In principle, a mining company should consciously decide what is ore and what is waste in order to achieve its corporate goals. Break-even cut-offs are incompatible with this.

CHAPTER 4

Mortimer's Definition – Accounting for Geology

MORTIMER'S DEFINITION DEFINED

In a paper describing grade control practices in South African gold mines in the late 1940s, Mortimer (1950) established an important though little recognised advance in cut-off theory. He identified that what was classified as ore had to meet two criteria:

1. the average grade of rock must provide a certain minimum profit per tonne treated
2. the lowest grade of rock must pay for itself.

In this book, these two goals comprise Mortimer's Definition. Break-even cut-offs, discussed in the previous chapter, have an implicit goal that 'every tonne pays for itself', but as noted this does not guarantee achieving a level of profitability. Mortimer, however, in describing what is to be classed as ore, has stated a goal that is more closely aligned with the company's true goals: that some specified profit target is to be achieved. Yet he recognises that ensuring that every tonne pays for itself is also important, stating it as the second part of his definition. What ore is – the cut-off decision – has been defined in terms of profitability. Mortimer clearly states the goals upfront and these therefore drive the method and calculations by which the cut-off will be derived. Mortimer's Definition may be thought of as a two-dimensional (2D) cut-off model, because as well as making use of costs and prices, like break-evens, it also accounts for the geology, in particular, the grade distribution.

CALCULATIONS FOR MORTIMER'S DEFINITION

The grades associated with each part of Mortimer's Definition may be determined by break-even calculations, but since each part comprises a distinct goal, different 'costs' will be associated with each.

To ensure that the average grade of rock provides a minimum profit per tonne, the 'costs' to be covered in the break-even calculation need to include:

- total mining costs
- total milling costs
- total site administration and services costs
- depreciation and amortisation
- head office charges
- interest on debt
- the required profit margin.

Note that while all or most of the rock-related mining costs would be excluded from an *equal benefits* ore–waste break-even cut-off calculation, they need to be included in the calculation of a profit, and so they are legitimately included in this break-even calculation. The purpose of any break-even calculation must be clearly specified to ensure that the appropriate costs are included. The calculation for the first part of Mortimer's Definition is not defining a cut-off to distinguish between ore and waste; it is identifying the average ore grade that delivers the required profit. This will be used to derive the cut-off that distinguishes between ore and waste, but is not itself that cut-off.

To ensure that the lowest grade of rock pays for itself – the second part of Mortimer's Definition – the costs to be covered in the break-even calculation should include the direct variable costs associated with classifying a piece of rock as ore. As previously discussed, this typically includes:

- the differential costs associated with mining ore, relative to the costs of waste
- variable costs associated with ore and product, including sustaining capital
- fixed costs, depending on what meaning the company applies to 'pays for itself'.

This second calculation in Mortimer's Definition is an ore–waste break-even cut-off, and it effectively sets the lower bound of possible cut-offs. It cannot be any lower or the material classified as ore will generate a loss. It is perhaps an equal benefits differential costs break-even calculation as described in the previous chapter.

Both parts of the Definition lead to break-even calculations, and the first part – with more cost components – will have higher costs and therefore a higher break-even grade. Many readers may be questioning why we have both parts of the definition and not just the first. They may also be wondering how geology comes into it to make it a 2D cut-off model.

To answer the first question, we note that the second break-even grade of Mortimer's Definition (that every tonne pays for itself) sets the minimum criterion for a block of rock on its own to be included in the inventory of material that could become ore. It could be seen as a boundary cut-off, which is used to define the limits of material above that grade without taking any cognisance of the grade distribution of the enclosed material with higher grades. It suffers from all the problems identified for 1D break-even cut-offs; however, the break-even grade delivering the first part of Mortimer's Definition can be thought of as a volume cut-off. It is the minimum average grade of all the material classified as ore for a planning time frame, not just a single block or of only one stope, and is the minimum head grade required to deliver the specified minimum profit.

Upon reflection we can deduce qualitatively that, if the grade distribution has a lot of high-grade material above the every-tonne-pays-for-itself break-even grade, the average grade of all that material could easily be above the minimum profit break-even grade. Conversely, if there is not a lot of material higher than the minimum profit grade, the average grade of all the material above the every-tonne-pays-for-itself break-even grade may be below the minimum profit grade. For the average grade of what is finally classed as ore above the minimum profit grade, we must drop some of the lower-grade material out of the ore inventory. Although it pays for itself on a standalone basis, it drags the average grade below what is required.

This leads to the second question regarding geology. As mentioned in Chapter 2, in situations where both boundary and volume cut-offs are used, the two cut-off values are typically the same; however, when applying Mortimer's Definition, the volume cut-off (the minimum required head grade) will usually be higher than the boundary cut-off (ensuring that every tonne pays for itself). If we were to generate an initial ore inventory at that boundary cut-off and then apply the higher volume cut-off to the stopes or mining blocks in that inventory, some would have an average grade that was too low and would be excluded. If we were to examine those, we would find that some contained higher-grade portions above the volume cut-off that ought to be in the final inventory. If a higher boundary cut-off were applied to eliminate the lower-grade material, the higher-grade portions would then comprise a revised stope or mining block with a grade above the volume cut-off.

If we similarly examined stopes or mining blocks in the initial inventory that were above the volume cut-off, we would also find material with a grade between the initial boundary cut-off and the higher boundary cut-off applied to the lower grade block to get its average grade above the required volume cut-off. If we took no further action, this lower-grade material would be included in the final inventory merely because it happened to be adjacent to enough higher-grade material. Also, material of the same grade has been excluded from the final inventory merely because it was not adjacent to higher-grade material.

But this is illogical. We do not, in principle, mine low-grade material in some areas simply because there is sufficient high-grade material nearby to raise the overall average grade. If that were to be done, then taking it to its logical extreme we would dilute all high-grade material down to some desired head grade, even using waste if necessary. Rather, there should be a balance between head grade and boundary cut-off so that the lowest-grade material mined from all areas is the same, regardless of where zones of higher-grade material might be. For consistency, therefore, the same higher boundary cut-off should be applied to the whole resource, not just to low head-grade blocks.

The geology must be examined to identify what boundary cut-off will deliver the required head grade. This can be done by examining either a table or a plot of tonnes and grades versus cut-off. The following section will illustrate this graphically.

MORTIMER'S DEFINITION ILLUSTRATED

Figure 4.1 shows a typical plot of tonnage and grade versus cut-off curves. The horizontal axis is the boundary cut-off, expressed generically in grade units per tonne. The vertical axis on the left shows the ore tonnage above cut-off, represented by the line reducing from a total of 8 Mt at a cut-off of zero (the total tonnage of rock) to almost zero at a cut-off of 10 gu/t. The grade of the material (also in generic grade units per tonne) is reported on the right vertical axis, and increases from approximately 2.7 at zero cut-off in the total

FIGURE 4.1

Tonnage–grade curves – every tonne pays for itself.

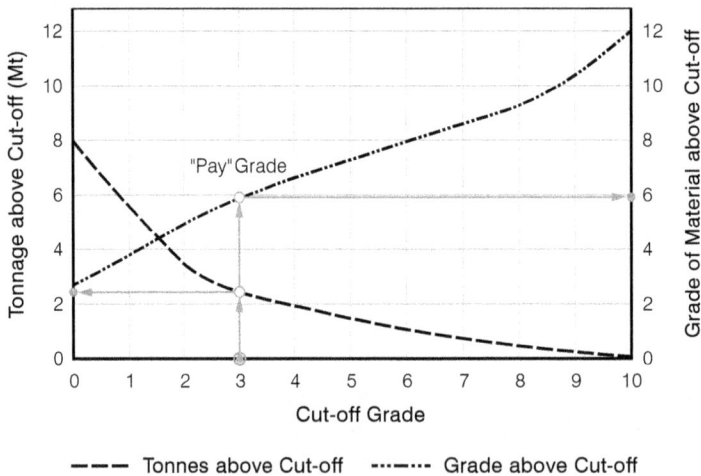

rock tonnage to 12 in the small tonnage above a cut-off of 10. This is indicated by the line rising from left to right.

Let us start with the second part of Mortimer's Definition – that every tonne mined as ore must pay for itself – and suppose that we have completed the second break-even calculation suggested and identified that the cut-off required is 3.0 gu/t. Moving up to the tonnage and grade lines, and then left and right to the tonnage and grade axes, we see in Figure 4.1 that this cut-off will deliver some 2.5 Mt of ore at an average grade of 6.0 gu/t. So far we have not done anything more than a simple ore–waste break-even cut-off calculation and identified the resulting ore tonnes and grades.

Let us assume that we have done the first break-even calculation suggested to determine the grade that delivers the required profit margin, and that grade is 8.0 gu/t. It is not necessary that every tonne of ore delivers this profit, just that the average grade does. Figure 4.2 shows how this break-even is applied using the grade versus cut-off curve to identify the cut-off grade that delivers this head grade. Starting at 8.0 on the grade (right vertical) axis, we move left until we intersect the grade versus cut-off curve. From here we travel down to the cut-off axis where we note that, in this example, the required boundary cut-off is 6.1 gu/t.

The two parts of Mortimer's Definition have now provided these possible cut-off values: 6.1 and 3.0. To ensure that we meet the conditions of both parts of Mortimer's Definition, we must select the larger of these two: 6.1. This ensures that we simultaneously deliver the required profit margin while every tonne is at least paying for itself.

The process has been illustrated graphically here, but it can be done equally well by inspecting the tabulated tonnage and grade versus cut-off data, with interpolation between data records if appropriate.

In the scenario described, both the 'minimum profit' grade and the cut-off that delivers it are higher than the every-tonne-pays-for-itself break-even cut-off grade, but this need not be the case. Suppose that the scenario is for a relatively new mine with high depreciation and amortisation charges, and being the only mine operated by the group must bear all head office charges. Suppose also that some years later, the mine in question is nearing the end of its life. The capital has been fully depreciated and the company now has a number

FIGURE 4.2

Tonnage–grade curves – average grade provides a specified profit.

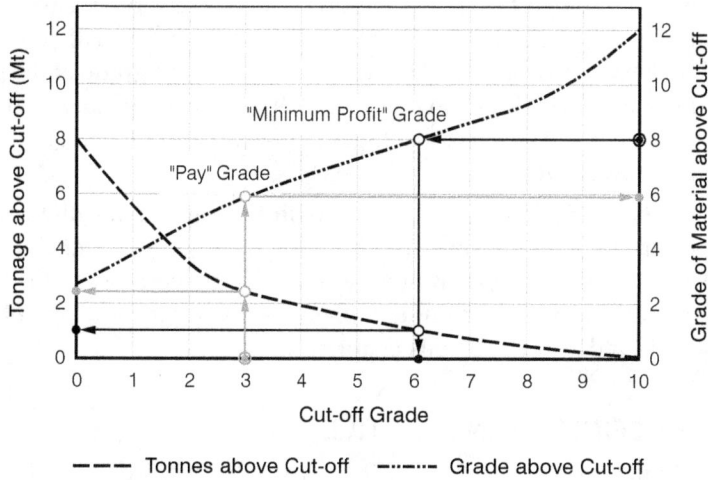

—— Tonnes above Cut-off   ——··— Grade above Cut-off

of operations to share head office costs, and does not require this operation to contribute a large profit margin to justify its continuation. Applying the same minimum profit break-even calculations as before, and assuming prices and site costs have not changed, the reduction in three major cost components in the break-even calculation results in a required average mined ore grade of 5.0 gu/t. Figure 4.3 shows the new situation. It can be seen now that the cut-off that delivers a head grade of 5 is of the order of 2.0, which is less than the every-tonne-pays-for-itself cut-off of 3.0. Again, the larger of the two cut-offs – 3.0 – should be selected.

It is interesting to note that the cut-off that delivers the required profit via the grade versus cut-off curve will vary depending on how much high-grade material is in the volume of rock under consideration. Suppose that the small tonnage above a cut-off of 10 grade units has a grade of 15 rather than 12 as illustrated in the grade versus cut-off curves

FIGURE 4.3

Tonnage–grade curves – average grade provides a specified profit (reduced costs).

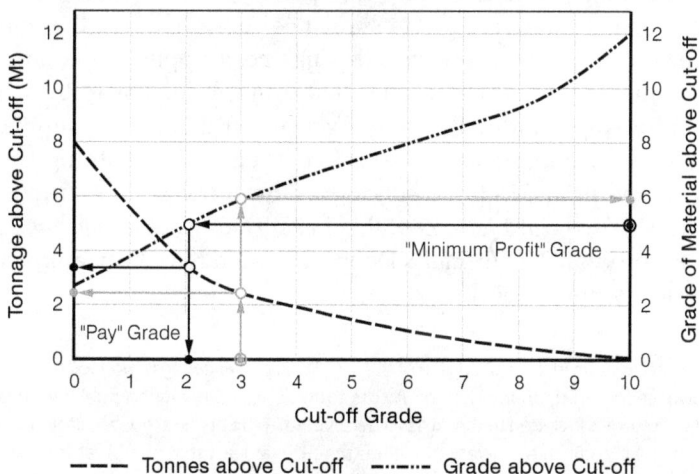

—— Tonnes above Cut-off   ——··— Grade above Cut-off

in the figures in this chapter. Although the distribution of tonnages and grades below a cut-off of 10 may be the same as before, the higher grade in the high-grade portion will result in the average grade at all lower cut-offs being higher than shown in the figures. The grade versus cut-off curve will be higher, and the cut-off that will deliver the required average grade of 8.0 will therefore be lower. Conversely, if the grade of the material above a cut-off of 10 is lower than 12, all average grades will be lower than shown. The grade versus cut-off curve will be lower, and a higher cut-off will be needed to deliver the same minimum profit head grade.

Depending on the distribution of high-grades in the deposit and the relativities of the two Mortimer cut-offs, it is possible that the mineral resource might never deliver the desired level of profitability, even though a mineable orebody is defined by the every-tonne-pays-for-itself cut-off, as was noted when discussing the analogy of the coral atoll and the volcanic island in the previous chapter.

MORTIMER'S DEFINITION IN PRACTICE

In practice, volume cut-offs (if used) will typically be applied to individual stopes or relatively small contiguous regions defined by a boundary cut-off during the initial generation of a mining reserve; however, the minimum profit grade required by Mortimer's Definition needs to be delivered by the blend of material from multiple areas to be processed within a particular period of time. The minimum profit grade and hence volume cut-off applicable to a Mortimer analysis may vary over time, even if the every-tonne-pays-for-itself break-even cut-off does not change. This implies that the tonnage–grade relationships developed for a Mortimer-style analysis should be for an inventory of mining units identified by boundary cut-offs only and not prefiltered by applying volume cut-offs to smaller mining units. Different boundary and volume cut-offs can be applied to the combinations of mining units scheduled for concurrent mining and processing to obtain the best cut-off policy over time. There are no absolute rules: applying volume cut-offs to the component mining blocks might be appropriate in some circumstances; however, the author suggests that these should be lower than the lowest minimum profit-grade derived for Mortimer analyses over the life of the operation, so that mining blocks that might add value are not excluded from the reserve inventory.

The author has encountered situations where staff at a mine have identified that the break-even cut-off grade fails to deliver a profit-generating head grade, and have completed calculations to derive a grade similar to those here for the minimum profit grade of Mortimer's Definition; however, this has been applied as a boundary cut-off rather than a volume cut-off, which implies that every tonne has to be at or above this grade, not just the average grade.[1] By applying a cut-off that is now significantly higher than it might need to be, the orebody is effectively reduced to nothing and the process is discarded as inappropriate. In the absence of better processes and understanding, the cut-off is arbitrarily raised until an acceptable head grade is obtained by trial and error. The more rigorous process of Mortimer's Definition could have been applied successfully if it had been correctly understood.

---

1. The sample cut-off calculation in the first edition of the *SME Mining Engineering Handbook* referred to in the preface includes an allowance for return on investment. A cost equivalent to the initial capital investment multiplied by the required rate of return is included in the costs to be covered in the break-even calculation, but this is implicitly applied as a boundary cut-off (that is, every tonne has to return at least that much) rather than a volume cut-off (the average grade has to generate that return), as suggested by Mortimer.

Having said that, it is entirely appropriate for a company to decide that every tonne mined must generate a minimum profit if that is what it wishes to do. The cut-off then derived would be a pure 1D break-even grade based on the simple definition of break-evens described in the previous chapter. If that did not result in an impractical orebody, the author suggests that there would be a lot of pressure to bring marginal material that still 'made money' into the schedule, which would potentially push the effective cut-off down towards what would have been the Mortimer-style cut-off anyway. The point of discussing Mortimer's Definition as defined here is to indicate that, once we move beyond the break-even concept of every tonne covering specified costs, the introduction of targets focused on profit leads to an extension of the cut-off derivation process to consider the nature of the geology – the grade distribution – in addition to costs and prices.

It should be noted, however, that there will be an unresolved circularity. Mortimer's Definition makes use of unit costs. For the minimum profit break-even calculation, the total unit cost will include the cost per tonne of costs that are not directly related to ore tonnes, which will therefore be a function of the cut-off. The iterative processes to converge on an appropriate cut-off described in Chapter 3 may therefore be required.

While the specifications of the grades required for the two parts of Mortimer's Definition are related to costs, prices and profits, it is worth noting that practical constraints in some treatment plants may require a minimum and/or maximum head grade, which could also be accounted for using the processes described. The required head grade would then be the larger of the practical minimum and the minimum profit grade.

SUMMARY

Mortimer's Definition uses break-even calculations to identify grades that ensure that:

- the average grade of ore provides a specified minimum profit per tonne treated
- the lowest grade of ore pays for itself.

Both of these grades are determined by break-even calculations, but with different costs included. The required profit margin is treated as a cost for this purpose.

The corporate goal is explicitly stated to achieve a minimum level of profitability while ensuring that no material treated generates a loss. Compare this with the implicit every-tonne-pays-for-itself goal of a simple break-even cut-off.

The second component of Mortimer's Definition is focused on cost and product price, and generates a simple break-even cut-off as described in the previous chapter.

The first component of Mortimer's Definition accounts for the nature of the mineralisation. The break-even grade required to achieve the defined level of profitability is not used directly as a cut-off but is the required head grade ore treated. The grade versus cut-off relationship within the material to be evaluated during a specified time is used to identify the cut-off that will deliver this required head grade.

The cut-off to be used will be the larger of the two derived for the components of Mortimer's Definition. Although the required 'minimum profit' head grade will be greater than the every-tonne-pays-for-itself break-even, the cut-off delivering that head grade will not necessarily be higher than the every-tonne-pays-for-itself break-even.

Perhaps one of the key understandings provided by Mortimer's Definition is the distinction between costs that must be borne by every tonne and those that must be paid for by the average tonne and the impact it will have on the cut-off derivation method. In Chapter 3, this distinction was noted by identifying that a simple ore–waste break-even should include only those differential costs that are incurred because of the decision to classify a block of rock as ore instead of waste. Costs such as capital development and waste stripping were specifically excluded because they do not vary with the ore tonnage. Rather, these are lump sum amounts expended to make more rock available, then allowing delineation of ore within that rock mass. Those costs must still be covered somehow if the operation is to be profitable. Mortimer's Definition indicates that such costs must be covered by the average grade of what is classed as ore, but not necessarily by every tonne that contributes some value.

As with break-even cut-offs, neither of the Mortimer Definition cut-offs generated accounts for the capacity of various stages of the mining and processing plant. Yet the geology of the deposit via the grade distribution relationships is now accounted for to ensure that some specified minimum level of profitability is achieved by the operations, making this what we might call a 2D process. This is a significant improvement on simple 1D break-even cut-offs, but still not in any way guaranteeing the maximisation of any value measure.

CHAPTER 5

Lane's Methodology – Accounting for System Capacities

DEMYSTIFYING LANE

Ken Lane published the foundation for shareholder value-based cut-off optimisation, first as a technical paper (Lane, 1964) and later as a text book (Lane, 1988, 1997). In developing our understanding of cut-off models with increasing numbers of parameters or dimensions, Lane's methodology could be described as three-dimensional (3D). As well as the financial and geological dimensions of break-evens and Mortimer's Definition, Lane accounts for a third – the production system's capacities to handle the three materials identified earlier: rock, ore and product.

Lane's methodology views the cut-off purely from the mine owner's point-of-view. It identifies that a resource's value is ultimately the value of the cash that was generated from it, and taking account of the time value of money, explicitly sets out to maximise the net present value (NPV) of the resource.[1] For break-even cut-offs the goal is implicitly ensuring that every tonne of ore 'pays for itself'. Using Mortimer's Definition, the goals are explicit: every tonne has to pay for itself, and the ore's average grade has to deliver a minimum profit. Similarly, for Lane's methodology there is an explicit goal, which is to maximise NPV. It should be noted, however, that there is no guarantee that the maximum NPV will be positive.

---

1. Despite mines being essentially in business to make money for their owners, focusing on NPV – probably the best single-value surrogate for quantifying cash generation – is often criticised for leading to 'high-grading' of the deposit. Whether maximising NPV is an appropriate goal is not the main issue of this chapter – corporate goals will be discussed in Chapter 10. This chapter is simply concerned with presenting a further step in the evolution of cut-off theory: including system capacities in developing an optimised cut-off policy.

Lane, being a mathematician, developed his theory using rigorous mathematical processes. From discussions with mining industry professionals, it is this author's experience that the majority who attempt to understand Lane's book give up when they reach the mathematical derivations of his formulas, missing the non-formulaic part of the theory completely. The formulas may, however, be derived logically without resorting to complex calculations. This chapter is this author's attempt to 'demystify Lane'.

ESTABLISHING UNDERLYING CONCEPTS

Lane and other workers in the field in the 1970s and 1980s introduced several important concepts to cut-off theory development, which are described in this chapter. These must be understood if a Lane-style cut-off optimisation is to be carried out correctly.

Simple Lane versus complex Lane

At the most fundamental level, Lane identifies that the value of an operation is a function of the remaining reserve, time and strategies adopted. Maximising value will result in a plan that specifies the optimum strategies to be adopted.

Optimum strategies will include items such as sizes of final pits and phasing of mining, maximum mining and treatment rates, cut-offs, and sequencing and scheduling of various deposits or parts of deposits. Lane's writings are directed largely at optimising cut-offs. For most processes described by Lane, strategic options other than cut-off are in effect assumed to be fixed, and the cut-off is optimised within that framework of other locked-in decisions. This is the way that most commentators see Lane, and for the purpose of this book, this approach will be referred to as *simple Lane*.

Within this framework of predetermined mining and treatment rates and mining sequences and the like, we will see that once all the required inputs are specified, a series of break-even calculations and simple algorithms can be used to derive the optimum cut-off/s. This can be done for a single time period or over the life of the operation in much the same way as simple break-even grades and Mortimer-style cut-offs; though certainly the implementation of the process is more complex than for the cut-off models described in the preceding chapters. Simple Lane may therefore be thought of as being available to a competent technical expert using, for example, standard spreadsheet software, and not requiring access to specialist optimisation software.

Lane's basic statement of the problem includes all strategic options available to the company, so his theory is not limited to cut-off optimisation. Whereas Lane's writings present readily available algorithms for cut-off optimisation if other strategic decisions are taken as given, the same cannot be said for optimising those other strategic factors. The author is aware of one software package on the market that includes proprietary algorithms that reportedly optimise other strategies. We will refer to the general application of Lane's theory to strategic decisions as *complex Lane*.

For building our understanding of cut-off theory, this chapter will deal almost exclusively with simple Lane. The 'Complicating Factors' section towards the end of this chapter will, however, take an evaluator beyond the simple towards complex Lane. Chapter 6 will go on to describe how optimisation of strategic decisions can be approached in principle without recourse to proprietary algorithms.

Lane's critical concepts

In his 1964 paper (and later his book published in 1988), Lane identified the concepts of:

- *Optimum cut-off* – set to achieve some optimised goal (in this case, maximising NPV).
- *Cut-off policy* – a planned sequence of cut-offs over time. Using an optimum cut-off policy will maximise the remaining NPV in every year of the mine's life. Constant cut-offs, typically used at operating mines, are often incompatible with this.
- *Balancing cut-off* – ensures that two of the three production system components (for rock, ore and product) are operating at capacity. This is a function of the geology and plant capacities, and is not related to costs and prices, but will often be the optimum cut-off to apply.
- *Opportunity cost* – this has always been understood qualitatively across the industry. For example, one can mine down to the marginal break-even to 'fill the mill' if there is a shortfall in ore supply, but this is only valid so long as low-grade material is not displacing higher-grade feed. The opportunity cost quantifies the extent to which such displacement can economically occur, and Lane's methodology automatically accounts for it in the cut-off optimisation process.

As Lane's theory and cut-off optimisation methodology are easiest to comprehend in the context of an open pit mine, we will first look at it in that situation. With underground mining, there are significant differences in thinking and these will be explained once the rationale has been developed for open pit mining.

Major underlying assumptions

Lane's methodology in its simplest form assumes a predefined mining sequence that generates a single stream of material (ie rock) from the mine. This also presupposes that the final pit limits have been identified. Ore will be separated from the rock, with product subsequently extracted from the ore. The methodology therefore does not attempt to optimise the final size of an open pit, mining sequences or schedules; these are givens.

Maximum handling rate capacities of the material flows – rock, ore and product – are also specified, though these may include capacity changes over time.

Definitions

In developing his theory, Lane uses common industry terms with specific meanings, some of which differ from their normal usage. These must be understood to avoid confusion. Lane identifies three production stages as mining, treatment and marketing.

Mining constitutes all the processes dealing with rock (undifferentiated ore + waste) and Lane (1988) refers to this as 'mineralised material'. This is potentially misleading, as some readers might take this to mean only the parts of the rock mass that contain mineralisation. The reality is that material that is totally barren waste must also be mined in an open pit and must be accounted for in the theory. Rock is all mined material, not just material containing mineralisation.

Treatment covers all the processes dealing with ore.

Marketing covers all the processes dealing with product, and can include smelting and refining. Because of the way that product is accounted for in the theory, marketing comprises all the processes that are dealing with product, not just selling.[2]

Most mines deal with all three material components as they distinguish ore from waste, and separate some form of product from the ore. The exceptions will be mines where the total material moved is classified as ore, or where the ore as mined is sold directly to a buyer without any additional processing by the mining company.

To apply Lane's methodology correctly, it is important to identify transitions between rock, ore and product. Let us now consider how these transitions are identified in open pit and underground operations.

An open pit example

Consider a low-grade disseminated base metal sulfide orebody being mined as an open pit with no stockpiling of low-grade ore. The implication with the valuable component being disseminated is that the same mining methods and mining fleet are used for all the material mined in the pit: there is no distinction in what we colloquially call *mining* between ore and waste. The no-stockpiling condition (which is not uncommon for mines in mountainous regions) implies that the ore–waste decision, once made, is irrevocable. This helps us to understand the development of the theory initially, but these conditions will be relaxed later in the discussion.

Although in practice grade control procedures will have identified which regions are ore and which are waste and these will be drilled and blasted separately, for the development of cut-off theory this early knowledge is irrelevant. In one sense, the operators do not need to know whether they are dealing with ore or waste; their mining practices are not impacted by this distinction. For cut-off derivation purposes, operations in the pit, which we would usually call *mining*, generate rock only. All in-pit operations are therefore mining as defined by Lane. Separating ore from rock does not occur until a truck has to branch to the treatment plant or the waste dump on surface after it has exited the pit.[3] Mining, whether the normal meaning of the word or Lane's definition, is the same thing.

Crushing, grinding and all other ore-related processes are *treatment*. Lane's specific usage corresponds with the normal use of the term.

Product becomes separated from ore at the flotation banks where concentrate overflows into the launders and the ore stream becomes physically separated into concentrate and tailings streams. For now we will assume that the tonnages of tailings and ore feed are the same, so the tailings stream may be thought of as a continuation of the ore stream after the product stream has been separated from it.

---

2. In Lane (1964) mineralised material is simply referred to as *material*, treatment is referred to as *concentrating* and marketing is referred to as *refining*. Lane identifies two product stream subcomponents: concentrate moving from concentrator to refinery, and product from refinery to market.

3. In some operations, this is the case in practice. For example, in some uranium mines classifying a truckload as ore or waste won't occur until the truck has passed through a detector to measure the level of radioactivity and the truck's destination is signalled to the driver. In most mines, however, ore and waste are identified prior to drilling and blasting, so that they are not mixed when blasted.

Concentrate pumping, thickening, filtration, drying and transport are all processes dealing solely with product, and are therefore classified by Lane's terminology as *marketing*. While smelting, refining and selling are identified by Lane (and correctly by many users of the theory) as marketing, Lane's omission in discussing the inclusion of the concentrate (or other product) handling activities within the concentrator or the mill has caused confusion to many of his readers.

The concentrator or mill – because of its primary purpose of separating product from ore – must have some activities that are ore-related and classed as treatment, and other activities that are product-related and hence classified in a Lane-style cut-off derivation as marketing. The mill is not treatment and treatment is not the mill. It is not uncommon in practice to find a milling constraint expressed as a tonnage of ore per unit of time that is constant up to a specified head grade, above which the milling constraint reduces with increasing grade. This is a practical recognition that, above a certain grade, the product-related parts of the circuit will be overloaded if treatment proceeds at the maximum rate at which ore could be handled at lower grades; hence, the treatment rate of ore must be reduced so as not to overload the product circuits. Similarly, all costs in a base metals concentrator are, in many mine cost-reporting systems, recorded as a cost per tonne of ore. This ignores that some costs are driven by ore tonnages and others by product quantities. The average cost per tonne of ore can therefore vary with grade, even though the underlying cost driver relationships have not changed.

So while for a simple open pit operation the rock–ore interface is easy to identify and the cut-off-specific meanings of the terms mining and treatment correspond with the normal meanings, the ore–product split in the concentrator, while fully appreciated operationally and technically, is not always well recognised as the treatment–marketing interface when using those terms in their cut-off-specific context. Often this interface is perceived to lie in the downstream part of the process flow, perhaps starting with concentrate transport after the concentrate has left the physical boundaries of the mill. The product constraint specified will then only be based on the limitations of the downstream concentrate-handling processes. But the real constraint on concentrate production or handling will frequently be located within the mill, and may be lower than the downstream concentrate handling capacity, so the product or marketing constraint will be overstated.[4] Additionally, because the starting point of the product stream is not correctly identified, costs in the concentrator will be allocated incorrectly between ore and product. For a Lane-style analysis, we need to distinguish between ore and product-related processes, and hence costs and capacities associated with each stream.

Note that for an open pit, rock and ore are mined (in the colloquial sense) simultaneously. Ore is merely reclassified rock, and is separated from the rock stream by directing trucks carrying rock to different ore and waste destinations. In exactly the same way, product is reclassified ore that is separated from the ore stream for further processing, if appropriate, and sale, while the remaining ore is disposed of as tailings.

---

4. From time to time one will see a technical paper that declares: 'We are doing a Lane-style cut-off optimisation. We can sell everything we produce; therefore we do not have a marketing constraint'. This will usually be incorrect, and indicates that the team conducting the analysis has missed the point. If a mine is not one of the relative few that does not separate a product from the ore stream, there logically has to be a limitation on the amount of product that can be dealt with somewhere in the production process. For a base metals operation, it will typically be one of the concentrate pumping, thickening, filtration, drying and transport activities noted in the text, or, at a gold mine, one of the activities associated with the activated carbon into which gold is adsorbed.

More complex open pit scenarios

A relatively straightforward extension of the open pit example described to consider different costs for material classed as ore during mining activities in the pit. For example, ore may use a closer-spaced drilling pattern to break the ore finer, use different mining equipment and have different haulage distances relative to waste, as described in Chapter 3. In these cases, the differential cost between ore and waste will need to be classified as an ore-related treatment cost. All rock mined will incur a basic mining cost, assuming it is classed as waste, and what is classed as ore then incurs the differential cost as an ore-related cost. The difference may be positive or negative: tighter drilling patterns and smaller, more selective mining fleets may incur additional costs for ore over those for waste, while shorter haul distances for ore could generate a negative cost differential for ore.

When considering capacities, problems may arise. At the simplest level for an operation with different-sized ore and waste mining fleets, the capacity for the mining process (that is, mining rock and removing it from the pit) will be the sum of the capacities of the two fleets; the capacity of the ore fleet in the pit will generate a treatment (ore-related) constraint, and if this is less than the ore-handling constraint in the treatment plant, the treatment-limiting constraint will be in the mine, not in the mill. This will not usually prevent the use of Lane's methodology.

The problem is exacerbated when the fleet capacity is a function of the cut-off used. This can occur if there are significantly different haulage distances for ore and waste. With different distances, the haulage rates (expressed in, for example, tonnes per hour) will be different for ore and waste. If changing the cut-off used generates more of the material with the lower haulage rate, the overall tonnage capacity of the fleet will be lower than previously. Lane's methodology in its simplest form cannot handle capacities that are functions of the cut-off they are being used to determine; a more complex analysis is required, but this may still fit within the conceptual limits of a Lane-style cut-off policy optimisation and does not necessarily invalidate the approach. Often such cut-off-dependent changes can be handled within the iterative processes that will be described for a complete life-of-mine cut-off optimisation, so long as the relationships required have been correctly formulated in the optimisation model being used. This is discussed shortly.

Other complex mining requirements may extend this. For example, if the smaller, more selective fleet is also required to be used for mining a certain amount of waste to assist with the efficient separation of ore and waste in the pit, a portion of the smaller fleet's capacity must be assigned to waste and the ore capacity will be that much lower; however, as in the previous paragraph, this quantity may be a function of cut-off. One may also come across situations where the acid-generating potential of waste to be dumped must be countered by the mining of additional acid-neutralising rock from a quarry outside the pit limits. The requirement to do this will depend on the waste tonnage to be dumped and its acid-generating potential, and this requirement will reduce the total rock mining capacity of the fleet in the producing pit. This will be a function of the cut-off applied, and again a more iterative analysis is required, but the methodology will remain robust.

An underground mine example

Consider now an underground mine with trucks hauling and tipping waste from development into stopes and hoisting ore in a shaft.

Most activities in an underground mine are dealing directly with ore. All these ore-related activities will therefore, by Lane's terminology, be *treatment*. For the example described, production drilling, blasting and shaft hoisting operations are only conducted for ore so these activities are therefore treatment.

It's already been noted that in an underground operation, rock is typically the total mineralised material. It is separated into ore and waste by planners creating designs for stopes: the rock within the design limits is extracted as ore while the remainder remains *in situ* as waste. In an underground mine, it is major access development that exposes rock so that it can be separated into ore that is stoped and waste that remains *in situ*. Major waste access development is therefore *mining* by Lane's definition, and is frequently the only mining activity in an underground mine for cut-off derivation purposes. It will usually be the extension of a shaft or decline to access deeper ore, or lateral development to access a new mineralised zone some distance from current resources. The key distinguishing feature of waste access development classed as mining is that its quantity does not depend on the amount of rock exposed or ore defined. Following similar logic to that in Chapter 3 – identifying what costs should and should not be included in a break-even cut-off – footwall access development that is independent of cut-off would also be classed as mining.

Lane (1988, 1997) suggests that the mining component of development includes 'driving headings, raises, etc and forming stopes in preparation for the extraction of ore.' The author considers this definition to be ambiguous or perhaps specific to certain mining methods, and suggests that mining development is any for which quantity does not depend on the amount of ore exposed by the development in question (typically major access development in waste); any that depends to some extent on the quantity of ore is treatment development. '(H)eadings and raises' may be correctly classed as mining if they form part of the overall access and infrastructure openings outside the orebody, but will often be treatment if they are in the orebody to form stopes.[5]

Some waste development may, however, be ore-associated and therefore classified as treatment. Footwall access development that varies with cut-off (for example, if the strike length of the orebody varies with cut-off) and stope access cross-cuts in waste, the number and hence total length of which will vary with cut-off, would be classed as treatment-related activities.

It should be noted that the relationships between treatment development metres and tonnes of ore may not be purely variable. For example, if an orebody expands in plan both along strike and perpendicular to strike as the cut-off is lowered, the length of strike development for both footwall access and strike drill drives may be directly proportional to the length of the orebody; however, because of the expanding width, the metres per tonne of ore may have two terms: a constant reflecting the proportionality between development and orebody length, and a term that is perhaps inversely proportional

---

5. Lane's classification of particular types of development as mining as quoted here may be valid for certain mining methods, such as South African narrow tabular reef mining for gold and platinum group metals. Each mine and mining method must be considered on its own merits.

to tonnes to account for the greater tonnages per metre of development as ore widths increase with decreasing cut-off.

If these relationships cannot be accommodated by formulas, the simplistic response is that the theory fails to adequately address all the real-world issues and cut-off models with more dimensions may be required; however, in many cases, these more complex interrelationships can be handled by the iterative processes involved in applying Lane's methodology in practice, which are described shortly, and the implementation's complexity does not invalidate the underlying principles.

Underground operations introduce other complexities, but careful thought will often resolve these. For the example given, both ore and waste are trucked to tips. To correctly classify this activity, it is useful to note that ore and rock-associated activities are both temporally and spatially separated in an underground mine. In a pit, waste and ore are mined together simultaneously, but underground, today's access development exposes tomorrow's ore, and today's ore is a result of yesterday's development. The rock-associated access development carried out during a particular period is the result of a policy decision that production will occur in the region being exposed in the future, and is not related to today's ore: it is therefore not relevant in determining what is today's ore.

In order to identify the optimum cut-off for the available material that may potentially be classed as ore, it is necessary to classify the mining waste development costs as mining costs. Additionally, because of the policy to mine this waste to expose future potential ore, the waste loading and hauling requirements must be deducted from the total loading and hauling capacity, with the remaining balance becoming the effective ore-handling or treatment constraint for determining what is ore today.

It may be useful to consider the treatment development costs as those that are in or adjacent to the orebody and are directly related to making the selected mining method work properly. Mining development costs are those that are needed to expose the mineralisation and would still be necessary, even if nothing was stoped from the area made available and production moved on to the next block of mineralisation exposed.

It could be argued that all mine development is required to make the method work and should therefore be classed as treatment development. The semantics become important. While it is necessary to have connections between operating levels so that activities related to the mining method can occur as required, the development between levels is merely access. The mining method does not depend on the nature of this development (for example, whether the access is by inclined drives or by raises), nor in many instances the quantity of this development, nor its routing through the rock mass. This is perhaps the distinguishing feature of mining development. For treatment development, on the other hand, the nature of the development – sizes, shapes and their placements relative to the orebody boundary – is critical to the mining method description.

This discussion has shown how the rock–ore or mining–treatment interface might be dealt with in an underground mine, but these will not be universally applicable. Each mine's circumstances must be examined individually to separate mining and treatment operations and hence distinguish their costs and define capacities.

APPLYING LANE'S METHODOLOGY IN PRACTICE

Having established some important concepts, let us now move on to discussing how Lane's methodology can be applied in practice. As indicated, this discussion assumes *simple Lane*, with all strategic options apart from cut-off predetermined. In particular, the mining sequences result in a predefined sequence of rock coming from the pit. Capacity limits for mining rates and treatment rates have also been decided. The optimisation process will seek to determine cut-offs that maximise NPV within these constraints.

It was seen in previous chapters that the 1D break-even process, given a specification of costs to use, will generate one number that could be used as the cut-off. In Mortimer's Definition, a 2D process, two numbers are generated and the larger of the two is selected as the cut-off. Lane's methodology, a 3D process incorporating financial and geological data and mining and processing capacities, generates six potential cut-off values, one of which will be optimal at any point in time. A simple two-stage process is applied to reduce the six, firstly to three and then those three to the one to be used. The values of the six will most likely change over time as both geological and economic conditions change. The cut-off to use, and its value, will also potentially change over time.

To develop Lane's methodology, we'll use algebraic symbols to represent the variables in the analysis. The notation in the following subsections is the same as that used by Lane (1988, 1997), so a reader familiar with Lane's work can make the necessary correlations.

Physical parameters

Since cut-off derivation deals with three types of material (rock, ore and product), we require variables for the associated stages in the overall production process. Lane denotes these and other physical quantities as follows:

- M = maximum mining rate (expressed as units of rock per time period)
- H = maximum treatment rate (expressed as units of ore per time period)
- K = maximum marketing rate (expressed as units of product per time period)
- y = metallurgical recovery or yield (ie, product recovered/product contained in ore)
- g = each of the six potential cut-off grades initially identified by the process
- G = the three intermediate cut-offs and one final cut-off grade obtained during the process of reducing the six, firstly to three and finally to the one cut-off to be used.

It should be remembered that, for an underground operation, the ore stream bottleneck and hence the process defining the maximum ore tonnage that can be treated (the treatment capacity) may be in the mine. In some circumstances this may be the case in an open pit.

Cost and revenue parameters

We'll also require variables to represent the costs associated with each stage in the production process – mining, treatment and marketing. Lane denotes these and other financial quantities as follows:

- m = mining variable cost (per unit of rock)
- h = treatment variable cost (per unit of ore)
- k = marketing variable cost (per unit of product)
- f = fixed costs (per time period)

- p = product sale price (per unit of product)
- F = opportunity cost (per time period – discussed shortly in more detail).

The lower case letters 'm', 'h' and 'k' are also used as subscripts with cut-off variables 'g' and 'G' to identify the process stages under consideration.

Chapters 2 and 3 detail the types of costs that would be included and excluded in each of these components. As noted there, the accounting distinction between operating and capital costs is irrelevant for cut-off determination. Sustaining capital should therefore be included in each cost component according to how the costs behave and which physical parameters are driving them.

Capital development costs

It could be argued that ongoing capital development in an underground mine is sustaining capital and should be included in the cut-off derivation as a tonnage or time-based cost. It cannot be included as cost per tonne of ore: if it were, the ore tonnage exposed by capital access development would depend on the cut-off, which would make the process circular – a cut-off would have to be assumed to derive the cut-off. Lane (1998, 1997) suggests that capital development should be included in the mining cost, in which case it would need to be applied to the tonnes of rock exposed or as a fixed cost, which assumes that the duration of mining has been determined. This may also depend on assuming a cut-off to identify the time over which the capital must be expended to make the rock available for production. The author suggests that this is generally not the right way to deal with capital development expenditure.

It is certainly appropriate to consider this cost in the break-even calculation used to determine the head grade in order to deliver the minimum required profit with Mortimer's Definition. For a normal full-cost break-even cut-off, and in using Lane's methodology, capital development spending enables producing from the area made available, and is a lump sum cost not dependent on the amount of mineralisation (or rock) exposed and ultimately classified as ore and stoped. Expenditure is justified because the difference between the optimum mine plan's value with the area made available and the optimum plan without it is greater than the cost of the access development. The optimum plan, though it may be generated before the capital is spent, considers what would be done after the development is complete, at which stage the development cost would be a sunk cost that is irrelevant to decisions about what constitutes the optimum mine plan. The planner's task is to make the most of all assets available, such as fleet and fixed plant, as well as capital development that accesses mineralisation. As with all project capital, the access capital development costs are not part of a break-even cut-off derivation process.

What is 'opportunity cost'?

This is not a textbook on discounted cash flow (DCF) techniques, but to understand opportunity cost, a brief introduction is necessary for readers unfamiliar with DCF concepts.

Let us assume that we invest $100 today at an interest rate of ten per cent. In one year's time, we will have earned $10 interest, and our investment will have a balance of $110. All things being equal, $100 now therefore has the same value for us as $110 in a year's time. If we have $100 in a year's time rather than now, it is the same as having $100/1.1, or $91 now, rounded to the nearest dollar. This is known as the present value – the value to us now of a cash amount at some time in the future. So, if we could have $100 now, but for whatever reason its receipt is deferred for one year, we have effectively lost $9 of present value. This loss of value due to the delay in receiving it is the essence of the opportunity cost.

We could think of our operation as generating a sequence of parcels of rock. Some proportion of each of these will be classified as ore to generate product and hence some (we hope) positive net cash flow. We can therefore also think of the results of the operation as a sequence of net cash flows associated with each sequential parcel of rock. Our aim is to identify the strategies to maximise the value of that sequence of cash flows.

We may identify optimum strategic plans for many years into the future; however, the reality is that we only have to make decisions for the next parcel of rock to be dealt with in the sequence. Future strategies that we have identified are generally just provisional. After we have mined, treated and generated revenue and profit from the product, we will notionally be able to make decisions based on the knowledge we then have regarding the resource remaining.

With our current assumptions about future conditions, the resource remaining after dealing with the next parcel of rock will have a maximum value. From both a DCF and a cut-off optimising point-of-view, the operation can be represented simply by the next parcel of rock that we are dealing with, plus the receipt of a lump-sum cash benefit that represents the value of the resource remaining after we have dealt with the next parcel of rock. Any decisions that increase the length of time it takes to deal with the next parcel of rock will defer receiving the rest of the resource's value. As previously indicated, this results in a reduction in the value of the rest of the resource to us in present value terms.

This loss of value is the opportunity cost. The implication is that additional tonnage of rock from the mining parcel being evaluated that is classified as ore will add material to both the ore and product streams. If one of those streams is already operating at full capacity, this will extend the duration of dealing with that parcel of rock, and therefore defers receiving the rest of the resource's value. To make that treatment economic, the additional value obtained must cover not only the marginal cash costs of treating the ore, handling the product and extending the period, but also the opportunity cost - the time-value-of-money cost of deferring receipt of the value of the rest of the operation.

Lane (1988, 1997) presents an extensive discussion of opportunity cost. An approximation of the opportunity cost (F) is given by the formula:

$$F = rV - dV/dt$$

where:

r $\quad$ = discount rate[6] (expressed as a decimal fraction)

V $\quad$ = NPV

dV/dt = the rate of change of NPV with time

The first component, rV, is approximately the loss of value due to the deferral in receiving the NPV of the rest of the operation. This assumes that the NPV does not change

---

6. It is beyond the scope of this book to discuss the derivation of discount rates. For this the reader is referred to any standard text on project evaluation or discounted cash flow techniques.

with deferring, implying that economic conditions stay the same in the future. The lower the discount rate, the better the approximation.[7]

The second term is the rate of change of NPV with time as a result of changing economic conditions, expressed mathematically as the differential of NPV with respect to time, dV/dt. If economic conditions deteriorate – for example, if the price is predicted to fall – moving the mining and treatment plan back into a period of worse conditions will result in a reduction in the NPV of that plan. In these circumstances the change in value is negative: the value becomes lower with deferral. The inclusion of dV/dt in the opportunity cost formula with a negative sign converts this change in value into a positive, or extra, cost.

It is, however, possible that deferring the project for a year moves the production plan into a period in the future with higher prices. In that case, the NPV will increase as a result, producing a negative component that will reduce the total opportunity cost. If dV/dt numerically is greater than rV, the overall opportunity cost may become negative, representing an opportunity gain rather than cost.

It is also possible, even with profitable projects, for the remaining NPV to be negative at some point. For example, this can occur towards the end of the mine life when the major remaining cash flows are payments for mine closure, redundancy and rehabilitation. Deferral of these closure payments will also result in a negative opportunity cost.

Finally, it should be noted that because the discount rate (r) is typically expressed as an annual rate, both rV and dV/dt have units of dollars per year. F is therefore equivalent to an extra annual fixed cost, expressed in dollars per year.

BASIC PRINCIPLES OF FINDING THE OPTIMUM CUT-OFF

Having established a vocabulary, a notation and an understanding of how costs are to be allocated to cost parameters, we can now establish the optimum cut-off. Three of the six potential cut-offs to be derived are known as *limiting cut-offs*. They are derived by break-even formulas for each of the three production stages considered independently to be the limiting factor on production. The other three are known as *balancing cut-offs*: using these will result in two of the three production process stages operating at their capacity limits concurrently.

The analysis addresses an increment of rock to be mined, rather than a specified time frame;[8] however, it might be convenient to use the existing schedule to identify these rock increments, each of which will be referred to as a *mining step*. It should be remembered that applying simple Lane assumes a prespecified mining sequence. The mining steps are therefore the increments or parcels of rock that are generated sequentially by the mining process. Although the steps might initially represent sequential years, quarters or months, the cut-off optimisation process may potentially change each step's duration.

---

7. The loss of value is more accurately given by $V.[1 - 1 / (1 + r)t]$. If $t = 1$ year, this reduces to $rV / (1 + r)$, so rV will overestimate this component of the opportunity cost somewhat. Since both r and V are approximations based on forecasts and the time of deferral will almost certainly not be 1, rV can be demonstrated to be a reasonable approximation for the time value of money loss over a range of deferred times, at least up to one year.

8. In practice it is feasible to construct a model to conduct a Lane-style evaluation using time periods rather than rock increments; however, for developing this methodology we will consider sequential increments of rock, as illustrated in Lane (1988, 1997).

What is important is that by definition steps are sequential and do not overlap; mining steps must be mined in their entirety so that the sequence is unchanged.

Mining steps are therefore not the same as the pushback or cut-back stages normally used in open pit mine planning. Rather, a mining step might comprise rock from two or more pushbacks or cut-backs; for example, ore from the bottom of the pit from the first cut-back, a mix of ore and waste from the middle levels of the pit from the second cut-back and waste only from the upper benches of a third cut-back. For cut-off derivation using simple Lane with mining sequences predetermined, it is irrelevant what pushback stage or where in the pit the rock emanates from. Rather, the task of the cut-off optimisation process is to identify the best cut-off to apply to all material that reaches the rim of the pit simultaneously. (In full strategy optimisation or complex Lane, optimising the sequence is also part of optimisation – but that is for a later chapter.)

The *same time* is the time frame for the planning purposes involved, and is by definition the time taken to deal with a mining step. Therefore, mining step tonnages might be relatively small for short-term planning, but become progressively larger in the future for long-term planning where plans are less precise. The primary attribute of each mining step is its total tonnage. The maximum mining rate will then set the minimum time to deal with the step; however, the actual time could be greater, but will only be identified at the completion of the cut-off optimisation process at which point the cut-off to be applied, the actual mining, treatment and marketing rates, and hence the duration of the mining step, will all be specified. The implication is that what is mined within a mining step is assumed to be homogeneous. If not, redefining the steps might be appropriate to achieve a satisfactory approximation of homogeneity within each step.

The aim of the cut-off optimisation process is to maximise the value of each mining step while accounting for its impact on the rest of the resource's value, which then maximises the combined value of all steps. As noted, maximisation of NPV does not necessarily imply that the NPV is positive: the maximum NPV may be the least negative value.

LIMITING CUT-OFFS

Three of the six potential cut-offs to be derived are known as limiting cut-offs. These are derived by break-even formulas considering each of the process stages to be the limiting factor on production. When each stage is considered to be the rate-limiting bottleneck in the overall process, the other two production stages are assumed to not be constraints: their capacities are ignored and may be infinite. This is, of course, unreal but this lack of realism is resolved when the six potential are reduced to the one optimum cut-off.

Lane (1988, 1997) derives these formulas by a rigorous mathematical process. It is this feature in the early chapters of his book that causes many mining industry technical staff to stop reading and consequently fail to grasp many of the important concepts he has developed. To demystify Lane, therefore, the following subsections derive these formulas by simple logical discussion rather than by mathematical manipulations.

The formulas generated below are presented using the same notation as Lane uses, so that the reader can make appropriate correlations between this book and Lane's if desired.

Mining-limited cut-off (g_m)

The mining-limited cut-off is derived by assuming that the maximum mining rate (for removal of rock from the pit or extending underground access development to previously unexposed parts of the resource) is the overall constraint on the production process. If the mining rate is the constraint, mining will proceed at the maximum rate rather than any slower rate, which would reduce the value.

The rock associated with the mining step must be completely mined (so that subsequent steps can also be mined). Mining variable costs are therefore incurred regardless of what is classed as ore. It is a principle of economic analysis of alternative options that factors such as common costs will not influence the relativities of each option and can therefore be ignored in the comparison. The total mining variable costs, being the same for all cut-offs, are irrelevant to the decision and are therefore not included in the calculations to support it. These are therefore excluded from the mining-limited break-even calculation. We could, if we wished, perform a full-equal-benefits-type of break-even calculation to explicitly include mining costs and see them cancel out, as shown in Chapter 3; however, having established the principle that only differential costs need to be included, we will simply exclude them.

The tonnage of rock in the mining step is known and the maximum mining rate is specified. Together these will determine the time taken to deal with the step. Given that mining is the overall rate-limiting activity, the duration is therefore fixed and does not depend on cut-off. All fixed costs are incurred for that duration regardless of what is classed as ore, and are therefore also irrelevant to the cut-off decision and excluded from the mining-limited break-even calculation.

The revenue from what is classified as ore must therefore only cover the variable treatment and marketing costs associated with that ore and the product derived from it in order to break even. Any grade higher than the break-even will make a positive contribution to cash flow and hence NPV.

The variable cost per tonne of ore is the treatment-variable cost (h). As described in Chapter 3, the net revenue per unit of product in the ore is obtained by deducting the variable cost per unit of product (k) from the price per unit of product (p). The difference is then multiplied by the metallurgical recovery (y) to derive the recoverable net revenue for a unit of product contained in the ore.

The mining-limited cut-off is therefore given by:[9]

$$g_m = h/(p-k)y$$

This is similar to a traditional marginal break-even cut-off, and indicates the lowest grade of material to be treated when the mine is unable to fill the mill. Note, however, that product-related costs are deducted from the price in the denominator, not included in the cost per tonne of ore in the numerator. Its rationale is fully explained in Chapter 3. Note also that the numerator of the formula has units of 'dollars per tonne of ore' and the denominator 'dollars per unit of product'. The resulting units for the cut-off grade are therefore 'units of product per tonne of ore', a grade unit, as is to be expected.

---

9. It should be noted that the formula as presented here is not strictly mathematically correct. As it is written, the recovery 'y' is a factor in the numerator of the calculation. To place it correctly in the denominator where it belongs – to generate the recoverable net revenue per unit of product in the ore – there should be a second set of brackets around the product of terms in the denominator; that is: $g_m = h/[(p-k)y]$.

    Similar inconsistencies exist for other limiting cut-off formulas. To retain Lane's style of presentation, the author has not added these additional brackets. Readers should note that, in these formulas, all terms to the left of the '/' are in the numerator, and all terms to the right are in the denominator.

Treatment-limited cut-off (g_h)

As for the mining-limited case, the rock in the step under consideration must be completely mined. Mining variable costs are again incurred regardless of what is classed as ore, and hence are irrelevant to the cut-off decision and do not come into the break-even calculation. However, with treatment being the overall rate-limiting process, the amount of the rock classed as ore and the maximum treatment rate will determine the minimum time taken to deal with the mining step. In the extreme case where the cut-off were so high that none of the rock were classed as ore, the duration of the step would be zero, implying an infinite mining rate. As noted, the unreality of this will be resolved when all the six potential cut-offs are considered to settle on the single optimum cut-off.

Given that the duration of the mining step depends on cut-off, fixed costs are incurred for the time that results from what is classed as ore. Fixed costs, therefore, are now marginal costs and relevant to the cut-off decision. The value of the ore should cover:

- variable treatment and marketing costs (as for the mining-limited case)
- fixed costs incurred during the time taken to treat the ore
- the time-value-of-money opportunity cost of deferring the NPV of the rest of the operation by the time taken to deal with the mining step under consideration.

The net revenue calculation in the denominator is the same as for the mining-limited formula. Ore-related costs in the numerator include not only the treatment-variable costs (h), but also the fixed costs since they now depend on cut-off. These consist of both the normal cash fixed costs (f) and the time-value-of-money opportunity cost (F), which as described is also a cost per year (or whatever the relevant unit of time is). Given a treatment rate limit of 'H', every extra tonne of rock classed as ore extends the mine life by '1/H' years (or time units). The additional fixed costs associated with that extension of life are therefore represented by '(f+F)/H'.

The treatment-limited cut-off is therefore given by:[10]

$$g_h = \{h+(f+F)/H\}/(p-k)y$$

All costs in the numerator are expressed in dollars per tonne of ore. Since the ore stream is the production bottleneck, fixed costs expressed in dollars per unit of time are converted to dollars per tonne of ore by dividing by the treatment limit, H, which is expressed in ore tonnes per unit of time. Expressing fixed costs as a cost per tonne of ore by dividing the fixed costs per year by the production rate is common at most mines.

Marketing-limited cut-off (g_k)

A similar rationale applies when the marketing operations, dealing with product, constrain the overall production process.

As with the mining and treatment-limited cases, the rock in the step under consideration must be completely mined. Mining variable costs are again incurred regardless of what is ore, so are irrelevant to the cut-off decision and break-even calculation.

With marketing being the overall rate-limiting process, the amount of recoverable product in the rock classed as ore and the maximum marketing rate (to deal with the product recovered from the ore) will determine the minimum time taken to deal with the

---

10. Or more correctly mathematically: $g_h = \{h+(f+F)/H\}/[(p-k)y]$.

mining step. Again, in the extreme case if the cut-off were so high that none of the rock were classed as ore, no product would be generated and the mining step duration would be zero, implying an infinite mining rate.

Given that the duration of the step again depends on cut-off, fixed costs are incurred for the time that results from marketing the product recovered from the ore, and therefore are relevant to the cut-off decision. The value of the classified ore must therefore cover:

- variable treatment and marketing costs (as with mining and treatment-limited cases)
- fixed costs incurred during the time taken to market the product recovered
- the time-value-of-money opportunity cost of deferring the NPV of the rest of the operation by the time taken to deal with the mining step under consideration.

For the treatment-limited cut-off, an extra tonne of ore treated extends the life by $1/H$ years, and the term $(f+F)/H$ is added to the numerator of the mining-limited formula to represent the extra fixed cost incurred per tonne of ore treated. However, when the maximum marketing rate – the rate of dealing with product – is the constraint limiting the total production process, every extra unit of product recovered from rock classed as ore extends the mine life by $1/K$ years (or time units). That extra unit of product must therefore bear not only the associated direct variable costs, but the additional fixed costs associated with that extension of life. The fixed cost incurred is given by '$(f+F)/K$', which has units of dollars per unit of product. For the marketing-limited cut-off, $(f+F)/K$ is a product-related cost and is deducted from the price (p) in the denominator of the break-even formula in addition to the direct product-related variable cost (k).

The marketing-limited cut-off is therefore given by:[11]

$$g_k = h/\{p-k-(f+F)/K\}y$$

Relationships between limiting cut-offs

The remaining NPV and hence the opportunity cost (F) will usually be positive, therefore the total fixed cost (f+F) will be positive.

Relative to the mining-limited cut-off, the treatment-limited cut-off has an additional cost term in the numerator. The treatment-limited cut-off is thus usually greater than the mining-limited cut-off. Similarly, relative to the mining-limited cut-off, the marketing-limited cut-off has an additional cost term reducing the denominator. The marketing-limited cut-off is also usually greater than the mining-limited cut-off.

If towards the end of the life of the operation the remaining NPV and hence the opportunity cost (F) becomes negative, it is possible that the total fixed cost (f+F) will become negative. In this case, the relativities between the mining-limited and the other two limiting cut-offs will be reversed. The practical implication is that if there are large closure costs, it may be rational to reduce the cut-off to less than the marginal break-even defined by the mining-limited cut-off, thereby extending the life and deferring the incurrence of the closure costs.

The relativities of the treatment and marketing-limited cut-offs cannot be specified simply by comparisons of the break-even formulas.

---

11. Or more correctly mathematically: $g_k = h/[\{p-k-(f+F)/K\}y]$.

Examples of limiting cut-offs

The following inputs are used to derive mining-, treatment- and marketing-limited cut-offs (g_m, g_h and g_k respectively) and the limiting cut-offs are derived by simple substitution of these values in the formulas.

Physical parameter assumptions

- Maximum mining rate   M = 5.0 Mt/a of rock
- Maximum treatment rate   H = 2.0 Mt/a of ore
- Maximum marketing rate   K = 150 000 t/a of product
- Metallurgical recovery   y = 90 per cent

Cost and price parameter assumptions

- Mining variable cost   m = \$5.00/t of rock
- Treatment variable cost   h = \$7.50/t of ore
- Marketing variable cost   k = \$40.00/t of product
- Fixed cost   f = \$16.0 M/a
- Product price   p = \$800/t of product

Opportunity cost assumptions and derivation

- Discount rate   r = 10 per cent/a
- Net present value   V = \$120 M
- dV/dt = 0 (ie economic conditions are forecast not to change in the future)
- Opportunity cost   F = rV – dV/dt = \$12 M/a

Mining-limited cut-off derivation

g_m = h/(p–k)y

        = 7.5/[(800 – 40) × 0.9]

        = 0.0110

g_m = 1.10 per cent

Treatment-limited cut-off derivation

g_h = {h+(f+F)/H}/(p–k)y

        = {7.5 + (16 M+12 M)/2 M}/[(800 – 40) × 0.9]

        = 0.0314

g_h = 3.14 per cent

Marketing-limited cut-off derivation

g_k = h/{ p–k–(f+F)/K}y

        = 7.5/[{800 – 40 – (16 M + 12 M)/150 k} × 0.9]

        = 0.0145

g_k = 1.45 per cent

These cut-off values will be used to derive the optimum cut-off.

The impact of opportunity cost and location of the production bottleneck

Opportunity cost is a component of both the treatment- and marketing-limited cut-offs. The treatment-limited cut-off g_h, for example, is derived by this formula:

$$\{h+(f+F)/H\}/(p-k)y$$

When substituting the values of the parameters, this becomes:

$$\{7.5 + (16\,M + 12\,M)/2\,M\}/[(800 - 40) \times 0.9]$$

The total cost in the numerator is $\{\$7.5/t + (\$16\,M + \$12\,M)/2\,Mt\}$, which evaluates to $21.50/t. The opportunity cost component is $12 M/2 Mt, or $6.00/t, which equates to approximately 28 per cent of the total cost or approximately 39 per cent of the $15.50/t of cash costs. This is a significant component of the cut-off. In fact, for a high-value long-life project, an opportunity cost in excess of 100 per cent of cash costs is not unrealistic in the early years of the operation's life.

There is often an emotive argument advanced that all material with a grade above a break-even cut-off should be mined, since it makes money. If ore above full cost break-even has been prepared for mining but there is adjacent material with a grade below the full cost break-even but above the marginal break-even, the argument is that the value of this marginal material will be lost forever by not being mined at the same time as the above-break-even material. So what is the value of that marginal material that we might sterilise if we don't mine it now?

If this lower-grade material is displacing higher grade ore, the unpopular answer is: nothing at all – it is rather a significant value destroyer. The fixed and opportunity cost components of Lane's treatment- and market-limited break-even cut-offs now show us the additional costs that need to be covered by each tonne of material added to the ore stream if it displaces higher-grade material and extends the mine life. However, if it is not displacing higher grade ore, the common, popular answer is that it is worth treating if its value exceeds marginal cost or, in Lane's terminology, if its grade is greater than the mining-limited cut-off.

Lane's mining-limited and treatment-limited conditions, and the formulas derived to generate cut-offs applicable in each case, are formal and rigorous representations of what are colloquially referred to as *mill-limited* and *mine-limited*. These are situations where additional ore is and is not displacing higher-grade material. But, what are colloquially, organisationally and geographically known as the mine and the mill are not necessarily mining and treatment in the strict sense required for cut-off derivation. In an open pit operation they often will be, but in an underground operation much of what occurs in the mine is treatment. To speak of an operation as mine-limited in the way this term is commonly used may be misleading when it comes to assessing what break-even cut-offs can be applied to material potentially added to the ore stream. Although the constraint may geographically be underground in the mine, it is essential to identify whether the output-limiting activity is in:

- the rock (mining) stream – the rate of access development or waste stripping is constraining the exposure of mineralisation that may be then classed as ore or waste
- the ore (treatment) stream – limited by activities such as drilling, loading or hoisting, or by scheduling constraints that limit available ore sources at any particular time.

Mine-limited does not necessarily mean mining-limited, and the mining-limited cut-off formula is therefore not always the appropriate one to use just because the constraint is in the mine. We would do better to avoid the common terminology of *mine-limited* and *mill-limited* and instead use *rock-limited* and *ore-limited*.

Finally, if there is an ore shortfall and a marginal cut-off may be necessary to fill the mill, there is one more unpopular question that must be asked: why do we not have higher grade ore available now? If it is simply part of the normal geological variability and uncertainty associated with mining, and therefore a short-term fluctuation from average conditions that will right itself naturally in the near future, then there is no major concern. But perhaps something has gone wrong with the planning or decision-making in the past. For example, deferring access development or exploration to identify further resources to be brought into the mine plan may mean that in the longer term there will regularly not be sufficient ore above the optimum cut-off available. In this case, we need to learn from such events, generate plans to bring ourselves back into a position where we can implement the optimum mine plan – which includes the optimum cut-off policy – and put procedures in place to ensure that short-term expediency does not lead to such problems again in the future.

Once we identify that there is spare capacity, this does not necessarily mean we can classify as ore all the material down to the marginal break-even, or in Lane terms, the mining-limited cut-off. Lane's methodology, however, now gives us a process to identify what the cut-off should be in this case. It will often be what Lane identifies as a balancing cut-off, which is the subject of the next subsection.

BALANCING CUT-OFFS

Balancing cut-off concepts

We've identified that Lane's methodology generates six potential cut-offs, three of which are limiting cut-offs derived as break-evens by assuming that each production stage is separately constraining the overall production rate. The other three are known as balancing cut-offs, which occur when two production stages operate at capacity.

Consider the mining and treatment stages of the overall process. For each stage as the throughput-limiting stage of the overall process, curves of value versus cut-off could be generated as illustrated schematically in Figure 5.1.

The mining and treatment capacities limit value generation to the two curves, considering each stage on its own. The values on the curves are the maximum values at each cut-off, generated by the process stages operating at their capacities. Process stages can operate below their capacities, in which case the full value potential will not be achieved, indicated by the two points in Figure 5.1 illustrating one stage operating at capacity and the other at less than capacity.

When two stages are being considered together, the two curves constrain value generation potential to the area that is under both curves. The maximum value potential is defined by the lower of the two curves at any cut-off, as illustrated in Figure 5.2.

As indicated in the figure, the point where the two curves cross is physically where both stages of the process are operating at their limiting capacities, at which point they are said to be in balance. The cut-off at this point is the balancing cut-off, which in the case

FIGURE 5.1

Value versus cut-off curves for mining and treatment process stages.

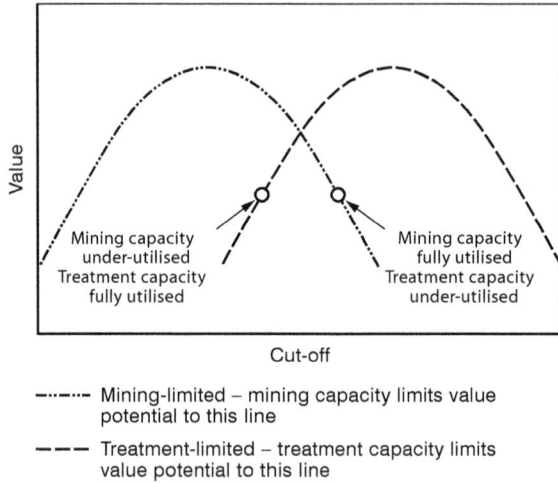

--·--·-- Mining-limited – mining capacity limits value
potential to this line

— — — Treatment-limited – treatment capacity limits
value potential to this line

FIGURE 5.2

Value versus cut-off curves for mining and treatment process stages, showing the balancing cut-off.

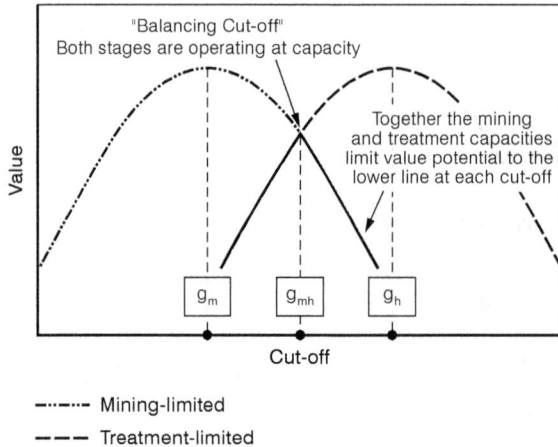

--·--·-- Mining-limited

— — — Treatment-limited

illustrated is the mining–treatment balancing cut-off denoted by g_{mh}. For limiting cut-offs a single subscript is used to indicate the process stage that is assumed to be the overall constraint; the double subscript identifies the two process stages that are in balance for a balancing cut-off.

Figure 5.2 also shows that the mining and treatment-limited cut-offs generate the maximum values of the mining and treatment-limited curves.[12]

_____

12. Lane (1988, 1997) does not specifically generate these curves but does derive their formulas with rigorous mathematics. He then applies the principles of differential calculus to obtain the first derivatives of the value curve functions, which are the formulas that specify the gradients of the curves. These are set to zero to identify the maximum, and the resulting equations solved to derive the formulas for the three limiting cut-offs, which, when used, will maximise the NPV in each case. These same formulas are derived by logical argument rather than rigorous mathematics in the preceding subsections of this chapter.

An important feature of the balancing cut-off is that it may deliver maximum value. As will be seen, the balancing cut-off depends solely on the capacities of the process stages and the distribution of value, or tonnage and grade versus cut-off relationships that exist in the material to be mined in the time frame considered. Neither of these controlling factors is in any way related to price or cost. The balancing cut-off is derived from physical parameters only, completely independently of financial factors.

General principles for deriving balancing cut-offs

With three process stages, there are three pairs of capacities that may be balanced:

1. mining and treatment (rock and ore), for which the balancing cut-off is g_{mh}
2. treatment and marketing (ore and product), for the balancing cut-off of g_{hk}
3. mining and marketing (rock and product), with a balancing cut-off of g_{mk}.

To find the balancing cut-offs, tonnage and grade versus cut-off tables or curves are inspected. Over a range of cut-offs, the tables or curves required will show:

- tonnages of ore above each cut-off
- average grade of the material above each cut-off
- average recoverable grade of the material above each cut-off
- ratio of tonnage of material above each cut-off to total rock
- recoverable product in the material above each cut-off
- ratio of recoverable product to total rock.

Some of these would be normal components of a conventional tonnage–grade data set, whereas others may be thought unusual. The reasons for needing these will become evident in the following explanations of the derivation of balancing cut-offs. These tonnage and grade values should be for extracted ore, after ore loss and dilution, and accounting for mining selectivity, not the *in situ* resource, as it is the tonnes and grades that are actually mined and processed that use and fill the capacities of the various production process stages. These data sets should be generated for the each increment of rock, or mining step, not for the global resource. Changing tonnage–grade relationships will result in changes of the balancing cut-offs even if production capacities do not change.

Balancing cut-offs for each pair of stages are obtained by finding the cut-off where the ratios of each pair of material components – rock, ore and product – are in the same proportions as the process capacities for each pair of components – mining (M), treatment (H) and marketing (K).

Mining–treatment balancing cut-off (g_{mh})

The mining–treatment balancing cut-off will be the cut-off that results in both the mining and ore processes operating at capacity. This occurs when the ratio of ore (that is, rock tonnes above cut-off) to total rock equals the ratio of treatment capacity (H) to mining capacity (M). Using the same assumptions for deriving limiting cut-offs:

$$H = 2 \text{ Mt/a and } M = 5 \text{ Mt/a}$$

$$H/M = 2 \text{ M} / 5 \text{ M} = 40 \text{ per cent}$$

The mining–treatment balancing cut-off (g_{mh}) will therefore be the cut-off that demarcates the best 40 per cent of the total rock; that is, with the best grade. Using the same tonnage–

grade curves as were used for developing Mortimer's Definition in Chapter 4, it can be seen in Figure 5.3 that, for a mining step of 8 Mt of rock, the best 40 per cent will be 3.2 Mt. From the curves, this will be generated using a cut-off of 2.3, and the corresponding grade of the ore above this cut-off will be approximately 5.3.

At this cut-off, generating 3.2 M ore tonnes from 8 Mt of rock and applying the stated mining and treatment capacities (5 Mt/a and 2 Mt/a respectively), both the ore and rock tonnages will take 1.6 years to be dealt with. The same durations are required for both process stages and they are indeed in balance.

FIGURE 5.3

Tonnage and grade versus cut-off curves illustrating the mining-treatment balancing cut-off.

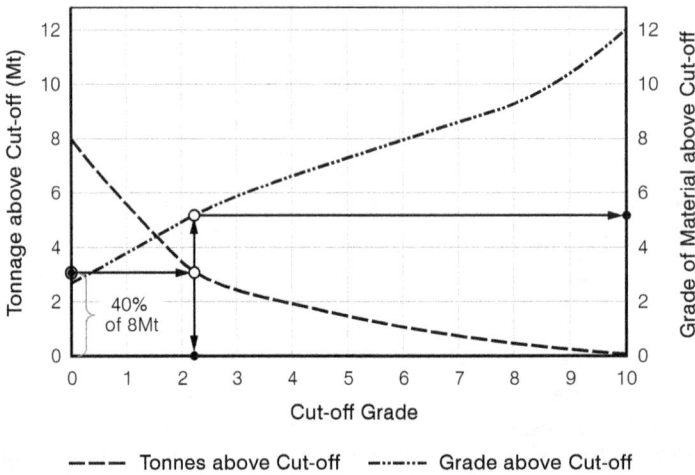

Treatment–marketing balancing cut-off (g_{hk})

The treatment–marketing balancing cut-off results in both the ore and product processes operating at capacity. This occurs when the ratio of recovered product to ore – that is the recoverable grade – equals the ratio of marketing capacity (K) to treatment capacity (H). Using the same assumptions for deriving limiting cut-offs:

$$K = 150\ 000\ t/a\ and\ H = 2\ Mt/a$$

$$K/H = 0.15\ M\ /\ 2\ M = 7.5\ per\ cent$$

The treatment–marketing balancing cut-off (g_{hk}) will therefore deliver a recovered grade of 7.5 per cent. The recovered grade must be used rather than the head grade or actual average grade of the ore because it is the recovered product that will be handled by the product circuit and restricted by its capacity, not the product contained in the ore. Figure 5.4 shows the same tonnage–grade curves as were used for developing Mortimer's Definition in Chapter 4, with the assumed metallurgical recovery of 90 per cent applied to the actual grade to obtain the recoverable grade curve.

It can be seen that the required recoverable grade will be obtained using a cut-off of approximately 6.7. The head grade will be 8.3 per cent and the corresponding ore tonnage will be only approximately 800 000 t.

FIGURE 5.4

Tonnage and grade versus cut-off curves illustrating the treatment-marketing balancing cut-off.

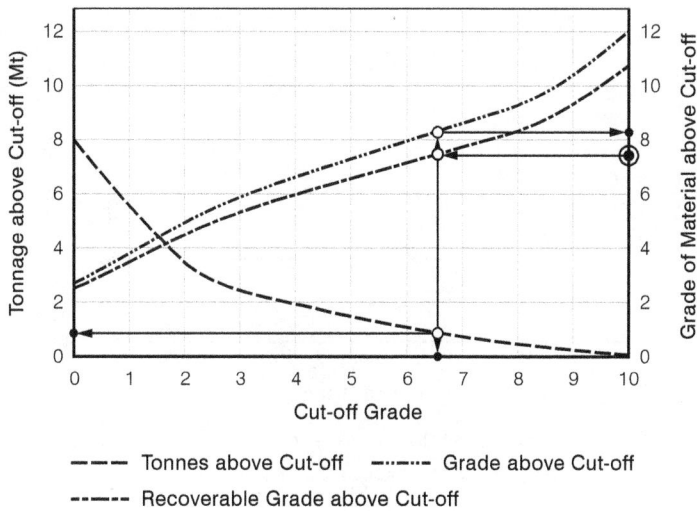

Key:
——— Tonnes above Cut-off   —·—·— Grade above Cut-off
——·—— Recoverable Grade above Cut-off

Mining–marketing balancing cut-off (g_mk)

The mining–marketing balancing cut-off is where the ratio of recovered product to total rock equals the ratio of marketing capacity (K) to mining capacity (M).

The ratio of product to the total rock in which it is contained is a grade, but it is not the grade with which we are most familiar, the grade of product in ore. As the cut-off applied to a given tonnage of rock is increased, the amount of ore must decrease. So, too, must the amount of product contained in the ore, but at a slower rate so that the ratio of product in ore – what we typically call the ore grade – rises with increasing cut-off. Although it is unusual, it is nevertheless legitimate to derive a grade value for product contained in rock rather than ore – what we might term the rock grade. Although the amount of product in the rock component above cut-off will decrease with greater cut-off, the tonnage of rock remains constant, so the rock grade will reduce with increasing cut-off. As with the treatment–marketing balancing cut-off, it is the recovered product that will take up capacity in the product stream of the process. To derive the mining–marketing balancing cut-off we must therefore, as for the treatment–marketing balancing cut-off, derive a recovered rock grade using recovered product.

Using the same assumptions for deriving limiting cut-offs:

$$K = 150\ 000 \text{ t/a and } M = 5 \text{ Mt/a}$$

$$K/M = 0.15 \text{ M} / 5 \text{ M} = 3 \text{ per cent}$$

Figure 5.5 shows the same tonnage–grade curves used for Mortimer's Definition in Chapter 4, with the metallurgical recovery of 90 per cent applied to the contained product at each cut-off to create the recovered grade in rock curve. As can be seen, this decreases with increasing cut-off, as expected from the rationale previously described.

The required recoverable rock grade is three per cent, and it can be seen that the recoverable rock grade is less than this at all cut-offs, including zero. The implication is

FIGURE 5.5

Tonnage and grade versus cut-off curves illustrating the mining-marketing balancing cut-off.

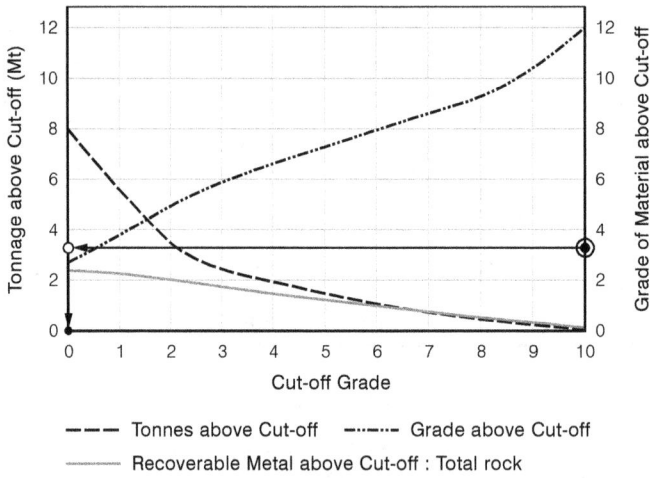

that, even if all the rock in this mining step were classed as ore, mined at the maximum mining rate and treated, there would not be sufficient product to fully utilise the capacity of the product stream. One could hypothetically extrapolate the recovered rock grade curve to the left, increasing with reduced cut-off, and identify a negative cut-off grade that would deliver this balance. This is mathematically reasonable in the abstract, but it has no physical meaning. We therefore consider the balancing cut-off to be zero.

To illustrate the generation of a positive mining–marketing cut-off, consider the situation if K were reduced from 150 000 t/a to, say, 110 000 t/a. The recoverable rock grade to generate the mining–marketing balance would be 2.2, and as can be seen in Figure 5.6, g_{mk} would be of the order of 1.3.

FIGURE 5.6

Tonnage and grade versus cut-off curves illustrating the mining-marketing balancing cut-off with lower product stream capacity.

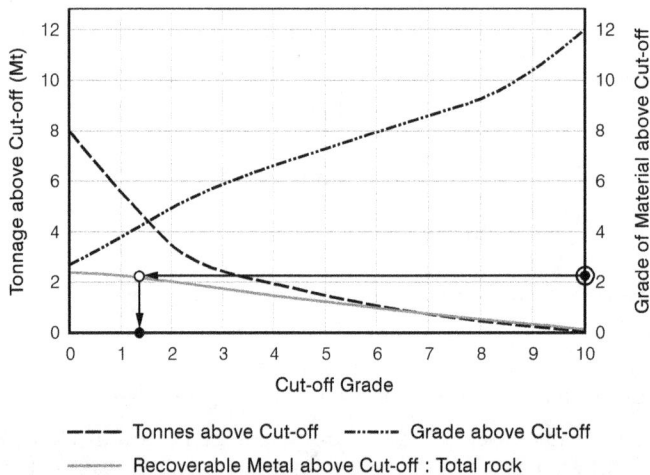

COMPLETING THE ANALYSIS

It was noted that, using Mortimer's Definition, two numbers are generated and the larger of the two is selected as the cut-off. Lane's methodology, a 3D process, generates six potential cut-off values – three limiting and three balancing – one of which will be optimal at any point in time. Lane presents a simple two-stage process where the six are reduced to three, then to the one optimum value to be used.

The process of finding the optimum cut-off

The first stage of reducing the six potential cut-offs to one considers each of the three pairs of production stages. The optimum cut-off for a pair of stages is the middle-ranked value of the two limiting cut-offs and the balancing cut-off associated with the two stages being considered. (The reason why this is so will be illustrated graphically.) With three pairs of stages to consider, this process results in three intermediate optimum cut-offs, one for each pair of stages. The final optimum cut-off is then the middle-ranked value of the three intermediate optimum cut-offs. Upper case 'G' is used to denote these intermediate and final optimum cut-offs. So for:

- mining and treatment: G_{mh} = middle-ranked value of g_m, g_h and g_{mh}
- treatment and marketing: G_{hk} = middle-ranked value of g_h, g_k and g_{hk}
- mining and marketing: G_{mk} = middle-ranked value of g_m, g_k and g_{mk}
- all three production stages: G_{mhk} = middle-ranked value of G_{mh}, G_{mk} and G_{hk}.

Completing the analysis – numerically

For the examples developed in the preceding subsections:

- limiting cut-offs calculated were: g_m = 1.10, g_h = 3.14, g_k = 1.45
- balancing cut-offs derived were: g_{mh} = 2.3, g_{hk} = 6.7, g_{mk} = 0.0.

  The optimum cut-offs for pairs of stages are as follows:

- G_{mh} = middle value of g_m, g_h and g_{mh} = middle value of 1.10, 3.14 and 2.3 = 2.3
- G_{hk} = middle value of g_h, g_k and g_{hk} = middle value of 3.14, 1.45 and 6.7 = 3.14
- G_{mk} = middle value of g_m, g_k and g_{mk} = middle value of 1.10, 1.45 and 0.0 = 1.10.

  The optimum cut-off for all stages is then given by:

- G_{mhk} = middle value of G_{mh}, G_{mk} and G_{hk} = middle value of 2.3, 3.14 and 1.10 = 2.3.

  For this set of circumstances with six potential cut-offs that range from 0.0 to 6.7, the optimum is 2.3, which happens to be the mining–treatment balancing cut-off g_{mh}.

Completing the analysis – graphically

It is easy to see why using the larger of the two potential cut-offs rule works for applying Mortimer's Definition. But why should the somewhat more complex but still simple mechanical process described reduce the six potential cut-offs to one using Lane's methodology? Using the same information, we will now see how the process works geometrically because of the generalised shapes of the value versus cut-off curves for each of the three production stages.

Figure 5.7 shows the graphical derivation of the intermediate optimum cut-off for mining and treatment process stages.

As shown in Figure 5.2, the mining and treatment-limited cut-offs define the maximum values of the value versus cut-off curves for the two stages. The balancing cut-off identifies the cut-off value where the two curves cross, since at this point both stages are operating at full capacity. For these two stages, the balancing cut-off lies between the two limiting cut-offs. The general shapes and relativities of the curves must therefore be as shown in Figure 5.7. The maximum value obtainable at any cut-off is the lower value of the two curves at that point; the value cannot be any higher as that would imply that a stage is operating at greater than its capacity. With this configuration, the maximum value has to be at the balancing cut-off, as shown in Figure 5.2. The geometrical relationships in these circumstances require that the middle value of the three relevant cut-offs, the balancing cut-off – g_{mh} – must be the cut-off that delivers maximum value, and hence is the intermediate optimum cut-off G_{mh}.

Figure 5.8 shows the graphical derivation of the intermediate optimum cut-off for treatment and marketing process stages.

Again, the limiting cut-offs define the maximum values of the value versus cut-off curves for the two process stages, and the balancing cut-off identifies the cut-off value where the two curves cross, since at this point both stages are operating at full capacity. For treatment and marketing stages, the balancing cut-off derived is greater than both the limiting cut-offs. The general shapes and relativities of the curves must therefore be as shown in Figure 5.8. Again, the maximum value obtainable at any cut-off is the lower value of the two curves at that point. Although mathematically it might be possible for the curves to cross in two places – that is, at two different cut-offs – the physical reality of the balancing cut-off implies that in the special case of these value versus cut-off curves they can only cross once. The curve symmetries imply that the maximum value of the curve with the lower maximum value will occur at a cut-off between the balancing cut-off

FIGURE 5.7

Graphical derivation of intermediate optimum cut-off for mining and treatment.

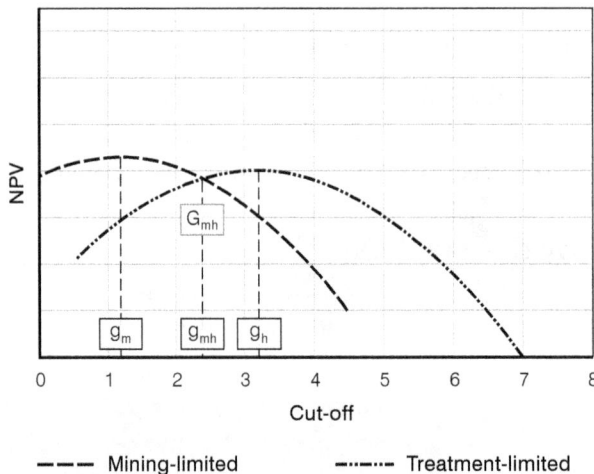

and the limiting cut-off of the higher value curve.[13] The geometrical relationships in these circumstances require that the middle value of the three relevant cut-offs must deliver maximum value, but with the configuration shown, the maximum value has to be at the higher of the two limiting cut-offs – that is, the middle value of the three cut-offs, g_h – and this is therefore the intermediate optimum cut-off G_{hk}.

Figure 5.9 shows the graphical derivation of the intermediate optimum cut-off for mining and marketing process stages. The two curves are the same as for these two

FIGURE 5.8
Graphical derivation of intermediate optimum cut-off for treatment and marketing.

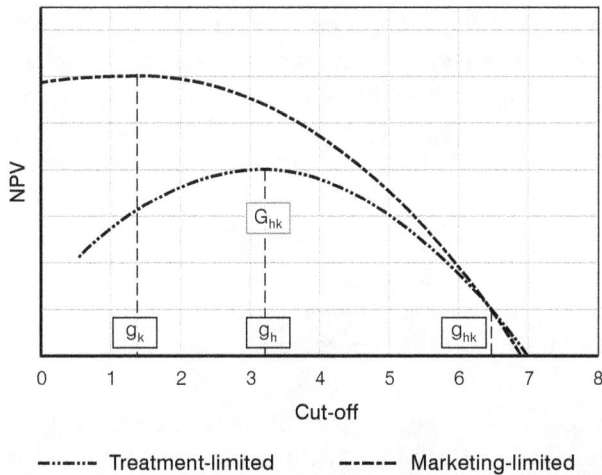

FIGURE 5.9
Graphical derivation of intermediate optimum cut-off for mining and marketing.

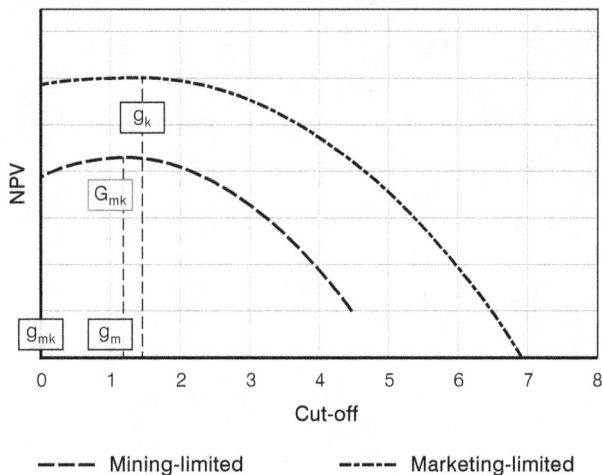

13. Lane (1988, 1997) explains that it can be shown from the mathematical forms of these curves that they must be of this geometrical form. The argument in this text is not mathematically rigorous, but qualitatively provides an explanation of why the geometries of the curves are as they are. Users of the methodology can, however, be assured of the mathematical rigour underlying the processes used.

production stages in the two previous figures illustrating the mining–treatment and treatment–marketing balancing cut-offs.

It has been noted, when deriving the balancing cut-off for mining and marketing, that it is not feasible for the two stages to be in balance, given the assumptions used for the example developed. Therefore, the two value versus cut-off curves do not intersect, at least within the physically practical range of non-negative cut-offs. It was seen that, even if the mining process were operating at capacity at a cut-off of zero, there would be insufficient product in the ore feed to fill the product stream capacity. This implies that mining is the constraint, which in turn implies that the value is higher if marketing is the constraint than if mining is the constraint at all cut-offs. This can be seen in Figure 5.9. It was also indicated, that, although the balancing cut-off is taken to be zero, by projecting the trends of the tonnage and grade versus cut-off data, the balancing cut-off could notionally be considered negative. Applying similar logic to geometrical relationships of the treatment and marketing, with a zero or negative balancing cut-off and two positive limiting cut-offs, the lower of the limiting cut-offs will be the cut-off delivering the smaller of the maximum values, which as described will be the mining-limited cut-off. This is the optimum cut-off for the combined two stages, and is also the middle value of the three. The mining-limited cut-off g_m is the intermediate optimum cut-off G_{mk}.

Figure 5.10 shows the final graphical derivation of the optimum cut-off for the processes of mining, treatment and marketing. The value versus cut-off curves are as shown in the three preceding figures, with the intermediate optimum cut-offs G_{mh}, G_{hk} and G_{mk} indicated.

Since each of the three curves plotted shows the maximum value that can be generated when each process stage is individually the throughput-constraining process, the maximum value of the overall operation must be the lowest of the three values at every cut-off. For the case illustrated, this is the treatment curve for cut-offs lower than the mining–treatment balancing cut-off gmh, and the mining curve for higher cut-offs. In this case, g_{mh} is G_{mh}, and this is also the cut-off that delivers the maximum value, and therefore is the overall optimum cut-off G_{mhk}.

FIGURE 5.10
Graphical derivation of optimum cut-off for mining, treatment and marketing.

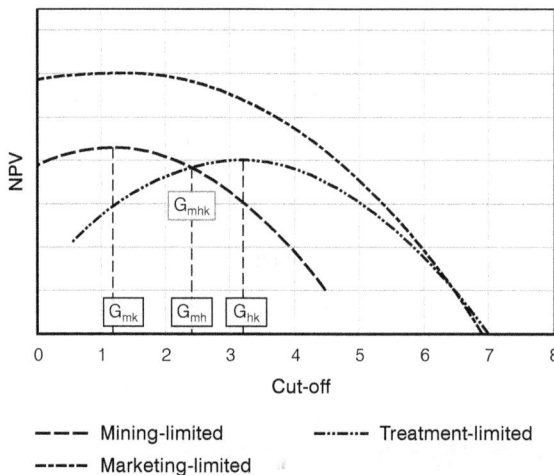

It can be seen that this is the middle cut-off of the three intermediate optimum cut-offs. Considering the geometries involved will indicate why the middle value will be the overall optimum cut-off delivering maximum value. Similarly, considering the shapes of the curves and their geometrical relationships will indicate why this will be true for all possible combinations of the six potential cut-offs generating the three intermediate cut-offs and one overall optimum cut-off.

It is not essential to derive curves as illustrated to determine optimum cut-offs. Although Lane (1988, 1997) derives the general mathematical form of these curves, generating the actual curves does not form part of the process required to find the optimum cut-offs. This is simply done using cost and price data, as well as tonnage and grade versus cut-off data that derive actual cut-off values, as illustrated in the previous sections.

PRACTICAL PROCESSES TO DERIVE OPTIMUM CUT-OFFS

At the time when he was developing his methodology, Lane had to work with large corporate mainframe computers. Now the simple process can be applied by anyone using modern spreadsheet software, especially for simple single-metal resources. There are two separate but related processes to determine the short-term cut-off for a single mining step, increment of rock or short-term planning period and the long-term life-of-mine cut-off policy (that is, the planned sequence of optimum cut-offs) for all mining steps and hence for the life of the operation.

Short-term single-period optimum cut-offs

Short-term optimum cut-off derivation for a single mining step, or, for instance, the next month's short-term plan, is straightforward. Current cost and price information and the tonnage and grade versus cut-off relationships for the rock increment to be mined are used exactly as in the previous example.

While the process and calculations are simple, obtaining the required data may not be. It may be necessary to rework reports to reallocate costs to their correct classifications of fixed and variable, and to the mining of rock, ore treatment and marketing of product. Capital cost records may need to be reviewed to allocate capital to project and sustaining categories, the latter also assigned to material types in the same way as operating costs.

Mining, treatment and marketing capacities should be known. The first two do not usually present problems for an open pit operation; however, for an underground operation, identifying the maximum mining rate – the rate at which mineralised material (rock) is exposed to allow subsequent stoping in what is ultimately ore – may not be easy to define. It should also be noted that this is not necessarily the rate at which access development is *actually* scheduled to advance, but the rate at which it could be scheduled, given the existing mining fleet, labour force and system capacities. The same may, however, apply to the stripping rate in an open pit.

The maximum marketing rate may be poorly understood in either type of operation. As noted earlier, it is defined by the process that limits the throughput of product wherever it occurs in the overall treatment and selling process. Depending on the type of product, the constraint may apply to another material containing the product such as a concentrate for a base metal; in this case the concentrate tonnage constraint must be converted to a metal product constraint using the concentrate grade to effect the conversion.

Long-term plans may need to be examined to obtain the NPV and current discount rate used to determine the opportunity cost (F) required for the treatment and market-limited cut-offs. For a short-term analysis, it is often satisfactory to assume that economic conditions are not changing in the future; that is, that dV/dt is zero, so the first term of the opportunity cost formula – the simple product of discount rate and NPV – will be a sufficient estimate of the opportunity cost.

Tonnage and grade versus cut-off relationships must also be obtained for the next increment of rock to be treated. As indicated earlier, Lane's methodology as originally formulated deals with increments of rock, not time periods (though the underlying principles can be applied to time periods as well). If the intention is to develop the optimum cut-off for the next short-term planning increment – say, a week or a month – the analysis might start with the assumed rock to be mined during that period; however, the application of Lane's methodology may result in a different duration when the optimum cut-off is applied using the existing mining, treatment and marketing constraints.

If a short-term Lane-style optimum cut-off is to be calculated regularly, technical staff should not have to reallocate costs in order to do so. Regardless of how the costs might be reported for other purposes, they should then also be classified and reported in the categories needed to correctly calculate Lane-style limiting cut-offs. Similarly, geological operating procedures must be put in place to generate the required tonnage and grade versus cut-off data for the next increment of rock, whatever size that might be, on a regular basis.

Long-term multiperiod optimum cut-offs

For determining the long-term optimum cut-off policy, similar processes must be conducted for each increment of rock or mining step. As indicated, Lane's simple methodology assumes that mining sequences have already been determined, and the increments of rock are therefore mined successively with no overlaps. The pushback or cut-back stages used in normal open pit mine planning are irrelevant – the rock in each mining step may be made up of rock mined in two or more pushback stages in the same time frame.

Although the Lane-style analysis deals with rock increments, time periods in an existing schedule may be used to define the rock in each mining step, but as with the short-term process, the final durations for the mining steps may be different from those used initially to generate the steps.

Calculations for each mining step will require the same data as for the single-increment short-term process. Costs, prices and capacities for the production stages may vary with each mining step, but not within a step. As before, costs must be provided according to Lane's definitions of what is included and excluded with each cost category.

The NPV for each step will not be provided from an externally derived mine plan as a piece of input data. Rather, the derivation of the optimum cut-off policy, which also determines the optimum duration of each step, is in fact determining the optimum long-term plan. The NPV values must therefore come from this same process. As a general rule a set of iterative calculations will be required, working forwards mining step by step in each iteration. An initial feasible plan must be generated by some means.

Forecast economic conditions – prices and costs, etc – are applied to generate NPV (V). Those conditions are then moved forward by a year to derive an estimate of the value for V if the operation were deferred by a year. These two values of V are used to estimate dV/dt and hence the opportunity cost (F). The optimum cut-off is then calculated for the first iteration, potentially changing the duration of any or all mining steps, hence changing the remaining NPV for each step. The process must then be repeated, using the resulting NPV values to derive values of F to be used in the treatment and marketing-limited cut-offs in the next iteration. This process is repeated until it has converged satisfactorily to a stable set of cut-offs and mining step durations over the life of the operation.

If prices, costs and processing capacities change over time, the situation becomes more complex. We determine the average prices, costs and capacities to apply to each mining step; however, as successive iterations of the cut-off optimising process change the durations of mining steps, the average prices, costs and capacities for each step will also change. The process described is iterative to account for the fact that changing durations will change the remaining NPVs and hence opportunity costs for each mining step. It is therefore possible that other altered parameters, such as various physical relationships that are cut-off dependent, may be accounted for by this iterative process.

In the special case where prices and costs do not shift over time, changes in capacities are aligned with mining step boundaries rather than time[14] and there is no stockpiling of lower grade material, it is theoretically possible to complete the Lane-style cut-off optimisation in one calculation, working backwards from the last mining step. Since these conditions do not change over time, it does not matter when each mining step occurs: the economic conditions will be the same regardless. It is therefore reasonable to derive the optimum cut-off for the last mining step, generate its NPV, use that to determine the opportunity cost to use for the second-last step, optimise the second-last step and so on, back to the first step. In the more usual practical case where there are changes in prices and costs over time and stockpiling of low-grade material for treatment later, the duration of each step affects the timing and hence the economic conditions for the last step. The end point is unknown and the single-step working-backwards process is not applicable; the iterative working-forwards process is necessary.

If future economic conditions are predicted to be stable, the iterative process will usually converge reasonably quickly to a stable plan. If there are significant changes predicted, it is possible that the process will not converge, but rather oscillate between a couple of cases that flip iterations from the one to the other. The study team may then have to manually intervene to specify an appropriate final plan.

THE EFFECT OF PRICE AND COST CHANGES ON CUT-OFFS

For many associated with the mining industry, cut-off and break-even are synonymous terms. Break-even generally varies in direct proportion to cost changes, and in inverse proportion to price changes. If prices double or costs halve, the break-even will halve. If the break-even is assumed to be the cut-off, the cut-off has halved; however, if cut-offs are derived to be Lane-style optimum cut-offs, the aim of which is to maximise NPV, what impact might price and cost changes have?

---

14. If mining steps have been delineated so that each is homogeneous, aligning capacity changes with mining step boundaries is a more rational approach than arbitrarily specifying a date that may fall within a mining step, since production and economic conditions are assumed to be unchanging throughout a step, but may change from step to step.

We have seen that a Lane-style analysis generates six potential cut-offs for each mining step evaluated, one of which will be the optimum. Three of these are limiting cut-offs that account for prices and costs, but although these are break-even calculations, the responses to price and cost changes may not be what is expected.

As already derived, the mining, treatment and marketing cut-off formulas are:

- $g_m = h/(p - k)y$
- $g_h = \{h + (f+F)/H\}/(p - k)y$
- $g_k = h/\{p - k - (f+F)/K\}y$

If there is a price rise, p increases. Since p is in the denominator of all formulas, the three cut-offs will reduce. Because the product-related costs are deducted from prices the change in cut-off will be more rapid than an inverse-proportional change.[15] This is the primary effect of a price change and the only effect for the mining-limited cut-off.

For treatment and marketing-limited cut-offs, there is a secondary effect. A price rise over the life of the operation will also increase its value, so the NPV and hence the opportunity cost (F) will increase. For the treatment-limited cut-off, F is in the numerator, and an increase in F will increase the cut-off. For the marketing-limited cut-off, F is in the denominator as a deduction from price. The increase in F will therefore reduce the denominator and, as for the treatment-limited cut-off, increase the cut-off. This secondary effect will offset the primary effect, at least to some extent, but it may more than compensate for the primary effect, so that a price increase may result in an increase in the treatment or marketing-limited cut-offs.

The net result of the primary and secondary effects of a price rise is therefore that the mining-limited cut-off will reduce, but the treatment- and marketing-limited cut-offs may reduce or increase. As a minimum, the movement in the latter two cut-offs will be less than is normally expected for a classical break-even.

Cost reductions will have the same effect. This is, of course, to be expected: both a price rise and a cost reduction will increase the margin, and qualitatively, that should lead to the same general outcomes, regardless of how that margin increase was obtained.

The exact mechanism, of course, will be somewhat different. If there is a cost reduction, any or all of the cash variable and fixed cost components h, k and f will reduce. Reduced components in the formula numerators will reduce the corresponding cut-offs, while reduced costs in the denominators, where they are deducted from the price, will increase the denominators and reduce the cut-offs. Regardless of which cost components reduce, the primary effect of the cost reduction will be a reduction in the three limiting cut-offs.

As for a price increase, a cost reduction over the life of the operation will also increase the operation's value, so the NPV and opportunity cost (F) will rise. As for a price increase, there is the secondary effect of a cost reduction that will increase the treatment- and marketing-limited cut-offs. Similarly, the net result of the primary and secondary effects of a cost reduction will cause mining-limited cut-off to decrease, but the treatment- and marketing-limited cut-offs may increase or decrease.

The other three Lane cut-offs are balancing cut-offs, which are totally independent of prices and costs. If one of these is the optimum, increases or decreases in the limiting cut-

---

15. If cut-off were plotted as a function of price, a simple break-even cut-off would approach infinity asymptotically as the price approached zero. With product-related costs deducted from price in the denominator of the limiting cut-off formulas, the cut-off will approach infinity as the price approaches the product-related costs, not zero.

offs may not result in any change to the cut-off, unless the relativities of all six potential cut-offs change so that one of the other six cut-offs becomes the optimum cut-off.

It is therefore impossible to state the effect of price or cost changes on optimum cut-off without completing a full analysis.[16] Price rises and cost reductions may result in optimum cut-off reductions or increases, or no change at all, and the effects may vary.

The author suggests that it is likely that, in a long-life operation with a relatively consistent grade distribution, debottlenecking of the mining and treatment processes over time will result in a balance being established between at least two of the production processes, and that the balancing cut-off thereby created will then remain as the optimum cut-off regardless of price and cost changes over time. To put this another way, if constraints do not change, prices and costs can change significantly but the optimum (balancing) cut-off will not change. This is quite counter-intuitive for those who have always believed that cut-off and break-even are synonymous – that if the price goes up, the cut-off has to go down, and vice versa.

If not rigorous proof, consider as a logical explanation the curves in Figure 5.2. If we postulate that we spend additional capital to increase the treatment capacity, the value versus cut-off curve for treatment will rise. The intersection point will move up the mining curve, and the balancing cut-off will therefore reduce; however, because the mining value versus cut-off curve is flattening with reducing cut-off, the additional value for each increment of treatment capacity will lessen. At the point where the balancing cut-off is the same as the mining-limited cut-off – that is, where the treatment curve has moved to the point where it passes through the maximum value of the mining curve, and the mining-limited cut-off becomes the optimum cut-off for the two stages shown – there is no additional value gained from investing in additional treatment capacity.

In reality, the treatment plant expansion would not be pushed to that limit – the relationship between additional value and the cost of expansion would show that the expansion would become uneconomic before the treatment capacity is increased to the point where the balancing cut-off is the mining-limited cut-off. The optimum cut-off would therefore continue to be the balancing cut-off, greater than the mining-limited cut-off. The further implication is that, if the optimum cut-off during longer-term steady-state operations is a limiting cut-off, the capacities of the various stages are mismatched and at least one stage has been oversized.[17] Unless another production stage can be debottlenecked to add further value, excess capital will have been spent on the oversized plant stage and the value destroyed.

---

16. From time to time one will see an article in the technical press suggesting that, when prices are high, cut-offs should be increased rather than reduced to increase output and cash in on the higher price. Simple examples may be presented, sometimes almost apologetically, to show the effects of two cut-off policies, demonstrating that an increased or elevated cut-off may generate better outcomes than a break-even cut-off.

    The author suggests these writers have intuited that the best value to use as the cut-off will not necessarily be the break-even (which of course will reduce when prices increase), but they do not have sufficient understanding of the issues to be quantitative rather than qualitative. Lane's formulas confirm the logic of this possibility, and the complete process provides a quantitative methodology to determine the best strategy.

17. This will often be a practical outcome of an operation transitioning from open pit to underground – the treatment plant on surface will have been constructed to handle a relatively large tonnage at a low grade. With higher costs per tonne of ore underground, and given the practical limitations on the rate at which the underground operation can be opened up, it may be impossible to mine at a rate sufficient to fill the mill. It may never be practical or economic to increase the mining rate (the rate of access development that exposes the rock) and the optimum cut-off will therefore be and remain the mining-limited cut-off.

If the optimum cut-off is a balancing cut-off, it is illogical to change the cut-off – for example, because changes in costs and prices have changed the break-even grade – unless one of the capacities in balance is also altered. This will require a conscious decision either to operate below the capacity of one of the processes involved or spend capital to increase a capacity. If reducing a capacity or, more accurately, operating below capacity to accommodate a change in cut-off from a balancing optimum to a break-even, value will by definition be destroyed. We will have moved, for example, to one of the underutilised capacity situations illustrated in Figure 5.1. Spending capital to increase capacity must be justified by an improvement in value resulting from using the additional capacity, or again, value will be destroyed.

COMPLICATING FACTORS

There are a number of real-world situations that may make the simple Lane-style cut-off policy derivation more difficult. Some, with varying degrees of ingenuity, can be handled by the iterative processes described, but others may require a change to complex Lane or full strategy optimisation, which will be described in the next chapter.

Costs, prices and capacities varying over time

As noted, if prices and costs change over time, the situation becomes more complex. Calculations identify the averages for the various mining steps as their durations change, but in principle these can be handled by the iterative process described. Processing capacities shifting over time may likewise be handled iteratively if the rates and times are specified, a feature of simple Lane; however, this author advocates that, unless capacity changes are linked to some external factor, such as predicted changes in prices or costs, it is likely that they will be prompted by changes in the nature of the material dealt with. This would suggest that the timing of such capacity changes should not be specified as dates, but rather by reference to the boundaries between mining steps.

We can generate optimum cut-off policies for different sets of capacities – for example, should we expand the mill from one capacity to another? Mining or treating different quantities because of changes in capacities will change the value of the operation. Such evaluations are common, but they are typically evaluated using the same cut-offs for each option. But different capacities will potentially change the optimum cut-off policy and hence impact on the value of the operation. The value gained from capital spent to change capacities can only be identified by comparing the values of the optimum mine plans, including their different cut-off policies, for each set of capacities. Simply changing tonnes mined or treated and failing to account for these changes in cut-off policies may give rise to a poor investment decision if the value of an optimal plan for one set of capacities is compared to a suboptimal plan for a different set of capacities. Capacities must be appraised in separate Lane-style cut-off optimisation evaluations, and the set that generates the maximum value, using its optimised cut-off policy, identified.

Simple Lane does not optimise system capacities – that is a task for complex Lane or strategy optimisation. We may evaluate a number of options – for example, for mining and treatment rate capacities – but unless we employ suitable optimisation methods, there is no guarantee that by simply evaluating a few alternatives, the option with optimum capacities has been included in the cases evaluated. So while the optimum cut-off policies for the capacities evaluated may have been identified, the overall optimum strategy for both capacities and cut-offs may not have been.

Capacities varying with cut-off

The capacities of the three stages of the production process are integral to a Lane-style cut-off optimisation. They are part of the derivation of all three balancing cut-offs and two of the three limiting cut-offs. It is possible for a capacity to be a function of the cut-off. This will introduce complexities that might or might not be able to be dealt with using a simple Lane analysis.

For example, in an open pit mine, the mining capacity (M) may be the truck haulage capacity. If the haulage distances vary for ore and waste, truck productivity and the rock-handling capacity of the fleet will depend on the proportional split of rock into ore and waste, which is defined by and dependent on the cut-off.[18] Haulage distances to stockpiles and allocation of loading and trucking fleets to stockpile reclaim introduces further complexities. The mining capacity, M, only comes into two of the three balancing cut-offs: the rock–ore and rock–product balances. It is possible to derive the mining capacity for each cut-off in the tonnage and grade versus cut-off data tables for each mining step and include it in those tables. Instead of comparing the rock–ore and rock–product ratios at each cut-off with the capacities' constant ratio, the comparisons would be with capacity ratios varying with cut-off within the data tables. Graphically, rather than identifying the cut-off at which the material type ratio (expressed as a grade) crosses a grade value (representing the ratio of capacities) on the grade axis, the intersection would be with another line plotting the capacity ratio as a function of cut-off.

In a similar way, the treatment capacity (H) may be the milling capacity, which is the product of the maximum plant operating hours and the milling rate of the ore. The milling rate will be a function of the ore hardness, which may be a function of the head grade that in turn depends on the cut-off. The two balancing cut-offs, including the treatment capacity, could be accounted for similarly to the mining capacity; however, the treatment capacity is also included in the treatment-limited break-even cut-off formula. If the relationship between cut-off and capacity can be expressed as a formula, this could be substituted, generating a more complex formula that potentially requires solving numerically. If the capacity cannot be expressed as a function of the cut-off, it would be necessary make use of the tabulations of tonnages and grades versus cut-off. One would derive the treatment capacity for the head grade of the material above each cut-off in the table, and then calculate a hypothetical limiting cut-off using each capacity. The treatment-limiting cut-off would be the cut-off where the hypothetical cut-off is the same as the initial cut-off in the table used to calculate it.

Polymetallic orebodies

Lane (1988, 1997) presents formulas for polymetallic or multiproduct orebodies that have (p–k)y terms for each metal component, introducing multiple grade/cut-off grade dimensions into the analysis. Lane also indicates that search techniques may be needed for the best solution, rather than the analytical and graphical processes described for the single-product case, because of the increased number of dimensions. This is classified as complex Lane or strategy optimisation.

---

18. This effect may often be small, to the point of being insignificant for the evaluation, but the author has encountered a situation where the acid-generating potential of the pit waste created a need to mine additional acid-neutralising waste from elsewhere. The cut-off selected therefore had a significant impact on the rock-handling capacity of the fleet in the main pit where a simple Lane-style analysis might have been applied.

Dollar value or metal equivalent grades use[19] is often suggested as a way around this problem. The product effectively becomes dollars or equivalent metal tonnes, and this may be satisfactory for the mining and treatment stages of the production process where costs and capacities are associated with rock and ore tonnages. For the marketing stage dealing with product, it will often be difficult to express the marketing constraint (K) when grade is a metal equivalent. The capacity must be a single value derived by combining the capacities for each product stream in a way that represents the capacity of the single equivalent metal process stream. There is also the problem of determining a product-related variable cost (k) when the product is expressed as a metal equivalent comprising two or more constituents.

Using an equivalent dollar value may remove some of these problems, depending on how the dollar equivalent has been derived. If generated as a net smelter return (NSR) type of value, the product represented by the grade is effectively net dollars after recoveries and product-related costs, so the (p–k)y calculations are already included in the dollar value. There is still, however, the problem of identifying a capacity (K) for the product expressed as net dollars per period of time for the mining–marketing and treatment–marketing balancing cut-off derivations as well as the marketing-limited cut-off calculation.

In some cases (for example, where precious metals are contained within a base metal concentrate), it may be feasible to revert to the base metal component when dealing with the three cut-offs that include product considerations. We would then have some components dealing with dollar value grades and others dealing with metal grades. Depending on the situation, it might be feasible to specify cut-offs appropriate to both dollar values and metal grades, with material classified as ore satisfying both criterion. There is no theoretical justification for this – it may be a good solution in some cases, and not apply in others. It may also require the use of software that can identify the optimum blend of material that satisfies both criteria, but this is well beyond the limits of the simple Lane cut-off optimisation process.

Multiple mining areas, mines and mills

A number of mines feeding one mill can derive consolidated tonnage and grade versus cut-off relationships to find balancing cut-offs; however, Lane's methodology, within the bounds of simple Lane, is based on an underlying assumption of a single rock stream passing through one set of rock mining, ore treating and product marketing facilities. Multiple mines and/or mills will almost certainly have capacity constraints applied to different components of the various rock, ore and product streams. Furthermore, different cost structures in parallel parts of the operations may invalidate the derivation of Lane-style limiting cut-offs for the combined material flow.

It was noted in Chapter 3 that the cost per unit of ore may vary with location in underground mines, leading to different cut-offs throughout the mine. Lane's methodology does not handle this; however, it may be feasible to account for these cost differentials in dollar value grade descriptors.[20] Lane's methodology in its simple form applies to a prespecified stream of rock. In other words, the mines schedule has been fixed before the cut-off optimisation begins. With multiple mines and/or mills in the analysis, the mix of materials from different sources to various destinations might change for the

---

19. These are discussed in more detail in Part 2, Chapter 9.

20. These are discussed in more detail in Part 2, Chapter 9.

best outcome. Lane's methodology does not handle this easily. Searching through a range of mixes may be the only way to arrive at a solution, and complex calculations of multiple options or specialist optimisation software may be needed.

Open pit versus underground

Lane's methodology is relatively easy to apply in an open pit mine with a single product. For a specified pit size, the mining sequence can be specified,[21] and the optimum cut-off will then be dependent on capacities, geology and economics. There is an assumption that the prespecified mining sequence is independent of cut-off, and this is often the case in reality. In addition, today's ore is a subset that is separated out of today's rock, as the latter is mined.

In an underground operation, however, the cut-off defines the size and location of the various ore lenses that will make up the mine; hence, the mining sequence may be highly dependent on cut-off. Today's rock (that is, access development) exposes ore that may be mined at any time in the future, but not today, while today's ore was made available by yesterday's development. Furthermore, today's stoping may sterilise sub-cut-off material that might otherwise have become payable at a time in the future.

There are some styles of underground mining that are readily amenable to applying Lane's methodology. In general, this will occur where there is a single production front advancing in one direction, such as downdip or along strike, with access development maintaining a lead ahead of production and no return to extract additional ore from parts of the resource that it has already moved through. This effectively establishes a single fixed mining sequence. Although rock mining and ore treatment are temporally and spatially separated and ore is not merely separated out of mined rock, as for an open pit, Lane's assumptions apply. For example, a sublevel caving (SLC) operation has a production front that is moving downdip, but cannot obviously advance faster than the development rate of new sublevels. Nor generally will it be desirable for the production front to be advancing slower than the development, as this will lead to increased costs – either from a time value of money consideration with development being done before it is required, or from the need to rehabilitate development that has been standing open unnecessarily, often in a highly stressed environment. There is therefore a clear balance between the rates at which mining is exposing new rock and treatment is producing ore from the SLC.

Anecdotal evidence suggests that it is not uncommon to find in such cases that a break-even cut-off is calculated, a suitable ore production rate is nominated for the resulting orebody and the development rate is then specified to achieve the same vertical advance rate, maintaining the required balance between rock mining and ore production. This makes the calculated break-even the de facto mining–treatment balancing cut-off also. It may be that a better solution is to develop the decline access at maximum rate, thereby raising the balancing cut-off and working through the resource faster, but potentially generating higher grades earlier and increasing the value.

Where there are multiple independent mining areas, or in cases where the mining sequence becomes dependent on the size of the orebody by the cut-off grade applied,

---

21. The specified sequence may not, however, be optimal, but simple Lane does not optimise the mining sequence. Some of the commercially available pit optimisation software applications do have modules – often optional at extra cost – that derive an optimal or near-optimal mining sequence.

Lane's simple methodology may cease to be applicable and more complex optimisation methods required.

Similarly, costs that may vary with cut-off, such as development costs per tonne of ore in an underground mine, may make Lane's methodology unsuitable, but in many cases this may be adequately handled in the iterative calculations required to derive a life-of-mine optimum cut-off policy.

Mining method constraints

Depending on the nature of the mineralisation, cut-offs may need to be applied to mining blocks rather than to mine-wide tonnage increments. For example, the geological conditions in some open stoping operations may mean that, once the stoping limits have been established, they cannot be changed without causing operational problems. The production schedule may then change to a number of mining blocks being mined in parallel, but starting and ending at different times. This type of requirement is similar to the stages used in open pit mine plans, but whereas the open pit plan is rearranged into sequential steps, this may be unfeasible in an underground operation. Parallel ore streams are not handled by the simple form of Lane's methodology.

Stockpiles

Stockpiling of mineralisation with a grade below the current cut-off is often viable with an open pit,[22] with stockpiled material becoming additional rock that can be reclassified as ore along with the rock that is freshly mined at a later date. There are costs associated with building, maintaining and reclaiming a stockpile.

Difficulties in applying Lane's methodology may arise when mining capacities need to be considered. Drilling and blasting capacity, and perhaps that of the in-pit loading fleet, will impose a limit on the amount of new rock mined; however, if the same trucks are used to haul both newly mined rock from the pit to waste dumps, stockpiles and treatment plant and reclaimed stockpile material to the treatment plant, the mining capacity may be a function of the cut-off. This may depend on the tonnage and grade versus cut-off relationships in both newly mined and stockpiled material and haulage rates associated with each. The problem is not insurmountable: there is no reason why the tonnage and grade versus cut-off curves or tables that would be used to find the two balancing cut-offs cannot include mining capacities that also vary with cut-off rather than remain constant, as described earlier.

Reclaimed stockpile material is ore, so the cost of reclaiming it is an ore-related treatment cost, even if the reclaim is being done by the rock-mining fleet and recorded as mining costs in the company's cost reports.

Lane (1988, 1997) presents a formula for calculating the stockpile break-even grade. It is similar to Lane's treatment-limited cut-off formula, but includes the cost of reclaim as part of the cost per tonne of ore. The recovery term (y) is for the stockpiled material, which might suffer some degradation during the years it is stockpiled. This grade is the lower boundary of material that should be stockpiled in any year, and represents the lowest

---

22. Stockpiling of underground material is less common – we would not usually incur the costs of mining material simply to put it on a stockpile; however, where the mine can produce more potential ore than the mill can handle and there is a need to mine through lower-grade material to reach higher grades, production and stockpiling of that lower-grade material may be appropriate if separation of material of different grades is feasible.

grade that could be economically treated at the end of the operation's life. The upper bound will, of course, be the cut-off applied to run-of-mine material. The construction of the stockpile is therefore easily defined.

However, the optimum stockpile withdrawal strategy cannot be determined in the same way as newly mined material using Lane's methodology without stockpiling, as described earlier in this chapter. Part of the reason is that, as indicated previously, simple Lane assumes a single stream of mined material delivered in sequential steps, from which ore is separated. The details of the material in each mining step are known before the cut-off optimisation process begins. Stockpiling not only adds a second potential source of ore, but the tonnage is not fixed – it can fluctuate over the life of the mining operation – so the tonnage and its grade distribution are functions of the cut-off policy applied to the newly mined material. Lane describes briefly how an appropriate algorithm will draw material for treatment from both newly mined material and stockpiles, such that the net contribution from the lowest grade material from each is the same. The capacities for mining, treatment and marketing – that is, for handling rock, ore and product – must also be balanced when dealing with these components of both the newly mined material and stockpiled material. These more complex algorithms fall within the realm of complex Lane or strategy optimisation, though some of the issues identified may be handled by the iterative processes required for simple Lane.

SUMMARY

Lane's methodology takes account of costs and prices, the grade distribution in the rock mass and the capacities of the various stages of the production process. In principle, Lane's methodology can be applied to optimise any strategic decisions, including cut-off. To develop an understanding of the parameters that influence cut-off derivation, we refer to simple Lane, which seeks to optimise cut-offs given a specified size of mine, mining sequence and capacities for dealing with rock, ore and product. The explicit goal of a Lane-style cut-off optimisation is maximising NPV; however, there is no guarantee that the maximum NPV will be a positive value.

Six potential cut-offs are derived for each sequential increment of rock (or mining step). These increments are the constants in the analysis. The optimum cut-off policy will determine the duration of each mining step and hence the overall life of the operation. Changing the assumptions about costs, prices or capacities may change the duration and overall life.

Three of the six cut-offs are limiting cut-offs, derived by assuming that the processes dealing with rock (identified by Lane as mining), ore (identified as treatment) and product (identified as marketing) are each independently the constraint on throughput. Limiting cut-offs are derived by break-even formulas. Since each increment of rock must be mined regardless of the amount of rock that is classified as ore, the mining variable costs will be the same for all possible cut-offs. Costs common to all options will not impact on the cut-off decision and do not need to be included in the analysis. Mining variable costs such as drilling, blasting, loading and hauling do not enter into any of the limiting cut-off calculations.

The mining-limited cut-off calculation includes only ore- and product-related variable costs. As well as these costs, the treatment- and marketing-limited cut-offs also include fixed costs and opportunity cost. The latter is the time value of money cost of deferring receipt of the NPV of the rest of the operation by treating additional material from the mining step under consideration as ore, as well as accounting for the impact of any changes in economic conditions in the period within which the displaced material would be produced. When the ore or product stream is assumed to be the capacity-limiting process, any additional material classed as ore from the rock increment will extend the mine life and hence incur additional fixed costs and opportunity cost, which must be paid for by the grade of the additional material processed.

The other three cut-offs are balancing cut-offs. These are determined for each of the three pairs of two of the production stages, and ensure that the two stages in each case are operating at their capacity limits. These cut-offs are physical functions of the grade distribution in the rock in each mining step and the capacities of the production stages considered. They are in no way defined by costs and prices.

Having generated all six potential cut-offs, there is a basic two-stage process to reduce the six, firstly to three and finally to the one optimum cut-off to be used.

For determining a short-term single-period optimum cut-off, the process derives the necessary cost data from mine cost accounting systems, though it may require effort to ensure that costs are attributed to the correct quantities. The opportunity cost is usually extracted from an existing long-term plan. For determining a long-term optimum cut-off policy – the sequence of optimum cut-offs over time – an iterative process uses NPVs remaining at the end of each mining step in one iteration to determine the opportunity costs for the next iteration. This process normally converges to a stable cut-off policy, though there is no guarantee.

It is likely that in a long-life operation with a consistent grade distribution over time, debottlenecking of the mining and treatment processes will result in a balance between at least two of the production processes, and the balancing cut-off created will then remain as the optimum cut-off regardless of price and cost changes over time. If constraints do not change, prices and costs can alter significantly, but the optimum (balancing) cut-off will not change. If the optimum cut-off is a balancing cut-off, it is not logical to change it (for example, because changes in costs and prices have changed the break-even grade) unless one of the capacities in balance is also changed. This must be justified by an evaluation of the change's impact on value.

In a single-metal open pit, Lane's theories and methods permit analytical optimisation of the cut-off policy for a given mining sequence. There are no simple analytical techniques available to optimise concurrently both the schedule and the cut-off policy.[23] These comments may also prove true underground, but only for some mining methods; for example, sublevel caving.

For such issues as multiple mines, ore sources or ore types; polymetallic orebodies; and complex stockpiling requirements, analytical techniques may be inadequate. Exhaustive search techniques or specialist software may be required to optimise the overall plan. This will be the key focus of the following chapter. In many cases, complex issues can be dealt with by the iterative processes used to determine an optimum cut-off policy. In the simplest analyses, only the opportunity cost is varied from iteration to iteration. In reality,

---

23. The key words here are *simple* and *analytical*, implying analyses that could be done by a competent engineer using spreadsheet software, for example. More complex techniques that could potentially optimise concurrently both the schedule and the cut-off policy, such as dynamic programming and linear and mixed integer programming, will be discussed in Part 2.

every component accounted for may change from one iteration to another, with values in a later iteration being derived from the outcomes of the preceding iteration.

Lane's methodology represents a major advance in cut-off theory. The insights it provides help to explain many of the sometimes counter-intuitive outcomes of a full strategy optimisation. In particular, the concept of opportunity cost – often referred to in discussions of strategy optimisation – is rigorously accounted for to indicate to what extent future production can be deferred to immediately treat additional material as ore. In addition, the concept of the balancing cut-off is introduced in cut-off theory. This shows that it is feasible for the optimum cut-off to be purely a function of the physical features of the resource and the operation's infrastructure – the grade distribution and the capacities of the various stages of the mining and processing operations – and be unaffected by price and cost changes.

CHAPTER 6

Mine Strategy Optimisation – Accounting for Everything

INTRODUCTORY COMMENTS

In the last three chapters, the focus has been on determining the cut-off grade only. We have identified how progressively more of the assumed givens may be accounted for in the analysis, with potentially different cut-offs resulting from both the differences between the cut-off derivation processes and the contrasting goals in the processes. We have considered:

- *Break-even analysis* – a one-dimensional (1D) process, with the implicit goal of ensuring that every tonne classed as ore pays for itself, whatever that might be taken to mean.

- *Application of Mortimer's Definition* – a two-dimensional (2D) process, accounting for everything in the break-even analysis plus the nature of the mineralisation as described by the tonnage and grade versus cut-off relationships. The explicit goals ensure that every tonne classed as ore pays for itself and the average grade of ore treated delivers a minimum profit per tonne.

- *(Simple) Lane's methodology* – a three-dimensional (3D) process, accounting for everything in Mortimer's Definition plus the capacities of the components of the overall production process handling rock, ore and product. It has the explicit goal of maximising NPV.

For all of these methodologies, there are two situations that are common:

1. An implicit assumption that everything other than the cut-off is fixed. The mining plan, which includes sequences, pushback stages, mining rates and treatment rates, has been specified or previously developed. It is then necessary to specify what

material will be classed as ore and classed as waste. The cut-off is determined to effect this discrimination. (For this discussion, stockpiled material may be seen as ore for which treatment has been deferred.)

2. A mining method is assumed and, given the expected costs of mining by that method, a break-even cut-off is derived. This is applied to delineate the boundaries of the ore, for which mine designs and schedules are then prepared, taking account of mining and processing limits that are arbitrarily specified.

The first is more common for an open pit operation, while the second is typical of an underground operation. Regardless of the operation, the cut-off derivation is usually separate from other mine planning, design and scheduling processes. This stems from a misconception that the cut-off is intrinsic to the operation or the orebody, or is a different class of decision from which we select mining rates, treatment rates, mining sequences and so on. This is borne out by the common practice in the mining industry worldwide to recalculate the cut-off using latest cost information and price forecasts, typically using break-even calculation methods. This is usually done at least annually to prepare the end-of-year Ore Reserves report. The cut-off is perceived as having to be accepted, driven by outside forces (in particular, product prices), over which the company has no control.

This is not saying that cut-offs should not be reviewed on a regular basis – they should be. What is being challenged is the common belief that cut-offs are outside the company's control. This leads to the situation where, rather than actively selecting the best cut-offs to weather the storms of change in economic conditions, companies allow themselves to be blown about uncontrollably by passive acceptance of inappropriate cut-off policies, with potentially dire results.

As noted in Chapter 2, this almost unthinking application of break-even cut-offs confuses the definition of a cut-off with how it is derived. The cut-off is a value that can be selected by the company. The discussion of Lane's methodology has already indicated that the optimum cut-off may be a balancing cut-off that is unaffected by price or cost changes. The cut-off decision is just as important as decisions regarding mining rates, treatment rates, mining sequences and so on. As will be seen, the best combination of all decisions can only be made by optimising all strategic decisions simultaneously.

The cut-off, therefore, is not calculated and specified first, after which we optimise the strategy. Rather, an optimised cut-off policy is only one of the outcomes of an overall strategy optimisation study.

The mine strategy optimisation process can therefore be thought of as a multidimensional analysis that is accounting for everything, or at least, anything of importance. It seeks to identify the best combination simultaneously of cut-off grades and other design and strategic decision parameters – the options within the set of value drivers – in order to achieve the corporate goal.

THE GOAL OF STRATEGY OPTIMISATION

It is essential that the corporate goal is specified at the beginning of a strategy optimisation evaluation. As noted in Chapter 2, without the goal being specified, the goal implicit in the cut-off or strategy derivation process will become the de facto corporate goal, whether recognised as such or not. *Goal* is deliberately singular here. Ultimately, there is only one mine plan implemented, and that one implementation can only hit one target – unless by coincidence the same strategy delivers two potential targets simultaneously.

It is not this book's intention to specify what the goal or goals of any company should be. Yet it is worth noting that when a deposit is depleted and the gates closed on a fully rehabilitated site, the benefit is the net cash generated. This cash may be distributed to different stakeholders in various ways, but it is still essentially cash. If we are to maximise the operation's benefits to the stakeholders, maximising the cash generated must be a primary goal. For decision-making ahead of time, we must make predictions about the future outcomes of decisions that must be made now. Arguably the best single number surrogate for cash generated over a period of time – accounting for the pattern of cash flows and the time value of money – is the NPV.

Maximising NPV is a common corporate goal, but it is by no means the only one considered in practice – many companies will specify other goals or multiple goals. This book does not posit a preference for the relative merits of NPV or real options value (ROV), both of which are measures of cash flow. Each is simply a value measure, the maximisation of which could be included in the list of goals dealt with in the evaluation. Chapter 10 discusses briefly certain aspects of a number of common goals, while Chapter 11 describes several valuation techniques, including ROV.

What's important to note is that the goal of the optimisation process must be clearly specified. The goal should be related to some parameter that can be quantified as a measure of value. The optimisation process will then focus on finding the strategy that maximises (or, less commonly, minimises) the value of the measure of value. Again, maximisation of cash generated over the life of the operation, measured by NPV, is a common primary goal.

THE REAL DRIVERS OF VALUE IN A MINING OPERATION

There is a lot of emphasis in the mining industry on improvements in productivity, efficiency, unit costs and the like. These are important issues for any operation, and may be thought of as 'doing things right'. The more fundamental issue is: are we doing the right things? The ultimate aim, of course, should be 'doing the right things right', but strategically the first concern must be 'doing the right things'.[1]

Any operation deals with three big-picture parameters: the things that are given, the things that can be changed (the strategic decisions or options) and the things that flow out of a combination of the first two parameters.

The givens

There are perhaps two key givens with any mining operation. One is obviously the *resource* with which it is endowed. The company has no control over this: it is what it is. (What is known about the resource is a separate issue that will be discussed in the next subsection.)

The other given might be termed the *environment*, which is here taken to be an all-encompassing term that describes the social, political, economic, climatic and ecological framework within which the company finds itself operating. Some of these may be able to be changed as a result of corporate decisions (for better or worse), while others

---

1. The two parts to this distinction are often referred to as efficiency and effectiveness. An operation may be efficient at doing what it is doing, but if it is not going in the right direction, it is not being effective in fulfilling its purpose – we could perhaps describe it as being efficiently ineffective.

may be outside the company's control. For the sake of this discussion they are all taken to be uncontrollable. Some may be predictable or measurable, while others may be intrinsically unknowable, such as future product prices. Some might be considered scenario parameters, for which the effects of different values or forecasts may need to be considered in the evaluation.

Key strategic decision parameters

The key parameters that drive the strategic direction of any mining operation over which the company has control – its options – are as follows:

- *Size of mine* – for an open pit, this is the size of the planned ultimate pit. For an underground operation, the mine size is defined by the cut-offs that determine the size and shape of the orebodies or mining blocks to be mined. This is discussed shortly.
- *Mining methods* – as well as the decision as to whether a mine is to be open pit, underground or both, there are a range of mining methods available depending on the size and shape of the ore zones and the geotechnical conditions. Specifying access and haulage methods and layout topologies could also be included in this category.
- *Production rates* – as noted in Chapter 1, the term mining operation is used to refer to the total production process, from delineating the geology through to selling the final product at the point of sale. Production rates or capacities therefore apply to all stages of the process, and may also include specifying the mix of various products.
- *Sequencing* – for example, sequencing of mines within a region with multiple deposits; sequencing of individual orebodies, mining blocks or pushback stages within a mine; and generally specifying starting points and the direction of mining advances within an orebody.
- *Knowledge of resource* – while the resource itself is a given, our knowledge of it is not. Decisions regarding how quickly a resource is delineated will have a major impact on the types of decisions made. Options involving large capital expenditure may require more certainty about the resource or more tonnage (implying more years of mine life) than low-capital options. Financing covenants can also force an operation to delineate more reserves than might be economically viable while project development loans are being drawn down and repaid.

These will be discussed in more detail throughout the book. This brief description is to set the scene for the initial discussion of the key features of overall strategy optimisation.

Consequential parameters

Consequential parameters flow out of the combination of givens and strategic decisions. They include detailed mine designs, process flow sheets and detailed schedules of physical activities and associated financial outcomes, such as:

- development metres, potentially for several categories of development
- waste tonnages or volumes
- drilled and blasted metres
- ore tonnages and grades
- haulage and hoisting tonnages or tonne-kilometres
- product quantities and qualities

- revenue streams
- operating costs
- sustained capital costs
- project capital expenditure for removing bottlenecks or enhancing products
- taxes.

It is worth noting that these matters typically take up a lot of management time and technical effort at operating mines. Of course, it's important that this be done, but as noted, these parameters are largely associated with efficiency issues. For example, costs must be controlled but costs cannot be selected; if they could, we would naturally select zero costs. Rather, the general level of costs is established as a result of strategic decisions, which lead to specifying certain types of equipment with associated capital costs, productivities and, hence, ranges of operating costs.

The key distinction, then, between strategic decision parameters (or options) and the consequential parameters is the answer to the question: can we make a decision about this parameter in principle? If the answer is yes, we are usually dealing with a strategic option. If the answer is no, because it is set by other decisions we have made, it is a consequential parameter to be controlled within the operation, but must also be accounted for in our optimisation process.

SPECIFYING THE SIZE OF THE MINE

The initial discussion noted that a key decision involves specifying the size of the mine. There are significant differences between underground and open pit mines in this regard, and since it is a fundamental consideration, this will be addressed now before moving on to a discussion of strategy optimisation in general.

Open pit

For an open pit, the size of the mine is essentially the same as the size of the planned ultimate pit. The optimisation task is to select one of a number of alternative pits to use as the ultimate pit limit. These alternative pits will be of increasing volume and surface extent and/or depth, and will typically have a set of nested pit shells derived by using commercial pit optimisation software applying well-established algorithms. Figure 6.1 illustrates the situation as a vertical section.

It should be noted that each pit in such a set will represent the pit with the maximum value for a given volume. The value measure used by commonly available software is undiscounted net cash flow. The term pit optimisation typically used in the open pit mine planning context is not the same as overall strategy optimisation, this chapter's topic. There is no guarantee that the pit shell identified as optimum using common industry practices is the ultimate pit that best satisfies corporate goals.[2] Selecting this is one of the primary tasks of the strategy optimisation process.

---

2. The pit shells derived by pit optimisation software will rarely be suitable as final pit designs. It will be necessary to define pushback stages and locations of temporary and permanent haulage ramps in more detail than is feasible within the pit optimisation software. The final designed pit may then be significantly different from the pit shell from which it has been derived. The open pit planner will have to decide whether the pit optimisation needs to be rerun incorporating the overall effects of the detailed designs into the parameters provided as inputs to the optimisation software.

FIGURE 6.1
Nested pits and alternative cut-off-defined orebodies.

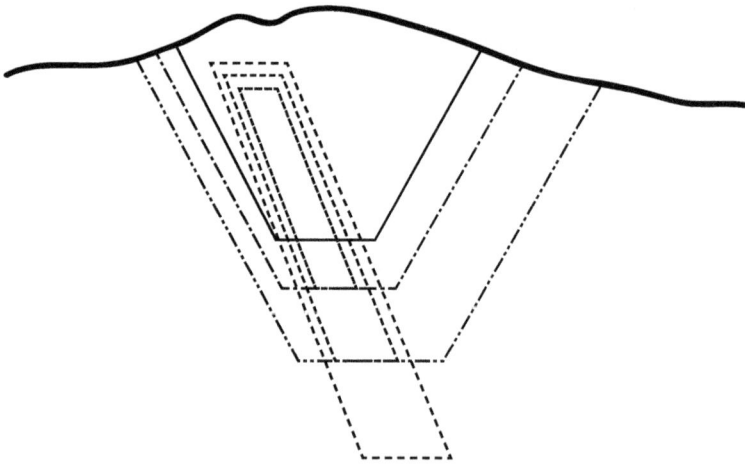

Cut-off selection is often a secondary decision for open pit strategy optimisation, though all the strategic decisions, including cut-off policy, should be made concurrently. The capabilities of different types of software to carry out the required analysis will be discussed briefly in Chapter 13, where it will be seen that computing limitations constrain our ability to achieve real concurrent optimisation in practice. It is theoretically possible to optimise cut-off policy, pit size and all other strategic decisions simultaneously. In practice, some form of iterative process will be required. In this case, specifying a number of alternative ultimate pit limits or pit shells will be necessary as a starting point, but it should also be recognised that the cut-offs selected may drive the overall shapes of the nested pits.

Underground

For an underground operation, the size of the mine is very much defined by the cut-offs selected, which determine the size and shape of the orebody or orebodies to be mined. These parameters then set bounds on such items as potential mining methods, equipment sizes and capabilities and realistic ranges of production rates that may be achievable.

In the same way that we identify nested pits from which we select the best size of open pit, so may we recognise several nested orebodies or stopes to form the basis of an underground optimisation. One of these will, in essence, be selected as the orebody that best delivers the corporate goals, and the sequence of cut-offs that defines the best orebody over time will be the optimum cut-off policy. Figure 6.2 illustrates the type of situation that could apply in a cross-section.

The selection of a range of cut-offs to evaluate, and generating mineable orebodies for each of them, is typically the starting point for an underground strategy optimisation.

DEVELOPING THE BASIC RATIONALE OF THE STRATEGY OPTIMISATION PROCESS

Let us consider two of the key decision parameters: production rate target and cut-off. The following is based on an underground mine but the rationale applies to all types of

FIGURE 6.2
Nested cut-off-defined orebodies for an underground mine.

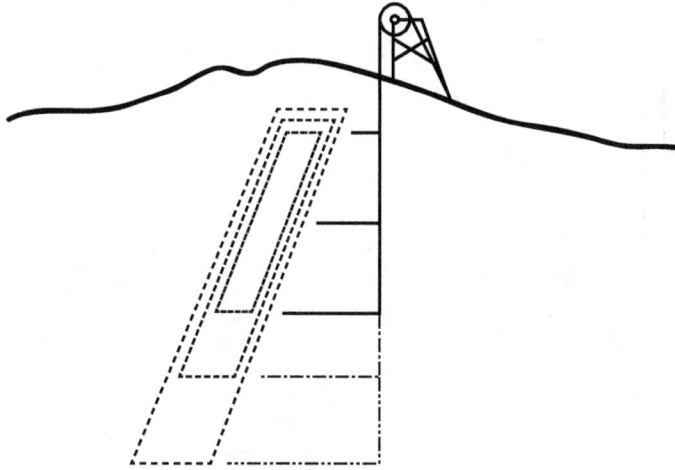

mines. Subsequent discussion will consider issues more specific to open pits and more complex issues for both classes of operations. The discussion at this stage is focused on developing principles common to all approaches, rather than the different optimisation methods that could be employed.

Optimising one strategic decision

If we were to plot value as a function of cut-off, we would obtain a curve similar to that shown in Figure 6.3. As the cut-off increases from a low value, the operation's value will increase, reach a peak and then decline. At lower cut-offs, material that destroys value is included in the reserve, while at higher cut-offs, value-adding material is excluded from the reserve. There is an optimum cut-off grade that maximises value. Note that the vertical axis is simply *value*, and could be any measure that quantifies outcomes of different plans so that they can be ranked, with one selected for what best achieves the company goals.

This concept should not be surprising.[3] Generically this curve is the same as that shown in Figure 5.10, representing the combination of the lowest of the three curves generated for each of the three production stages identified in Lane's methodology. In that case, the maximum value occurs at G_{mhk}, which may be a balancing rather than limiting or break-even cut-off. In practice, the curve may not be as smooth or symmetrical as in this example.

The curve in Figure 6.3 is for a given mining method, capital plan, production rate, etc. This is a 2D plot of value versus cut-off; the other parameters represent additional dimensions. Let us consider another dimension – production rate target – but again as a 2D plot of value versus production rate, as shown in Figure 6.4.

At a coarse scale, there is a relationship of value growing then reducing with an increasing production rate, as for cut-off, but the detail is complex. Production rate or production

---

3. This may be a new concept to some readers, but there is nothing new in it. The author has a sketch similar to Figure 6.3 in his notes made in university mineral economics lectures in the early 1970s. Kelly and Bell (1992) describe a study generating this type of curve for operations at Broken Hill in the mid-1980s.

FIGURE 6.3

Generic value versus cut-off curve.

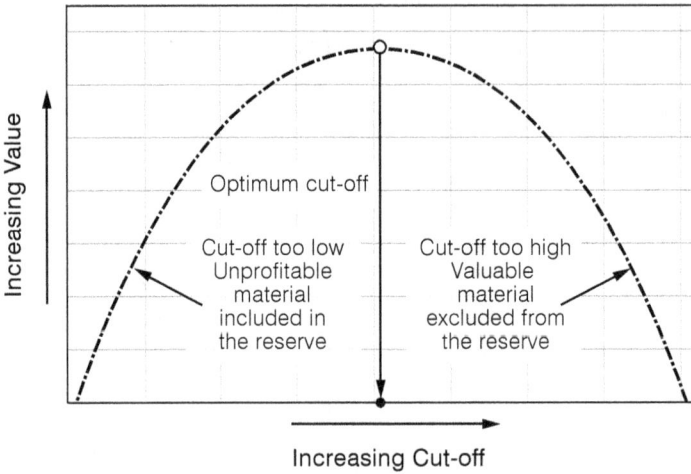

Increasing Value

Optimum cut-off

Cut-off too low
Unprofitable
material
included in
the reserve

Cut-off too high
Valuable
material
excluded from
the reserve

Increasing Cut-off

FIGURE 6.4

Generic value versus production target curve.

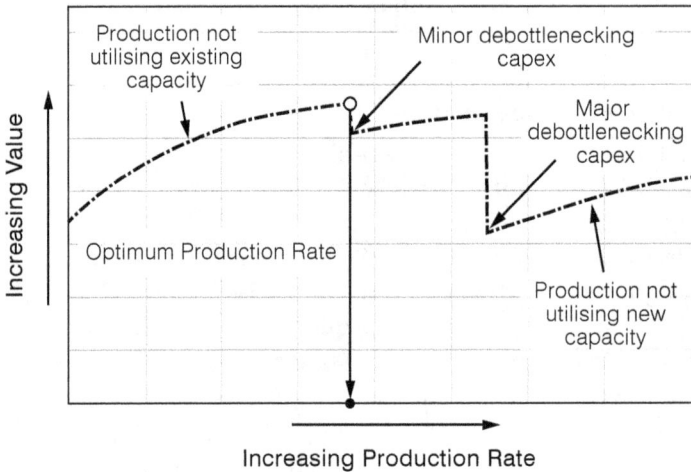

Increasing Value

Production not
utilising existing
capacity

Minor debottlenecking
capex

Major
debottlenecking
capex

Optimum Production Rate

Production not
utilising new
capacity

Increasing Production Rate

rate target is impacted by production limits imposed by the existing or planned facilities. If there is no change to the facilities, value will grow as the production rate increases up to the limits of the production facilities. If the production target were to increase beyond this, the actual production could not because of the physical constraints, and the curve would then plateau for higher targets at the value associated with the production capacity. The stepped curve in the figure is assuming that, as capacities of various parts of the production system are reached with increasing production rate targets, appropriate debottlenecking capital expenditure is made. If capital is spent, it will deliver a step change in capability. But if the production rate is increased only incrementally, the value will reduce by the value of the capital spent – the new capacity must be utilised to justify its provision.

Figure 6.4, therefore, effectively shows a resultant combined curve for three production capacities over a continuum of production rate targets, including the effects of two capital injections to permit the range of production targets to be achieved. Assuming that the range of production rates on the horizontal axis is a practical, realistic range for the orebody or mine under consideration, we see that there is an optimum production rate. This will typically be at a rate that is fully utilising one component of the production system, the implication being that further debottlenecking to allow increased production will be uneconomic.

This example is combining several strategic decisions – each of which could be represented as another dimension – into one dimension to be represented as the production rate target. The capacities of each part of the production system are separate decisions, as is the overall production rate target. The actual production achieved may be yet another variable to be considered – either as an outcome of the capacities provided or, if there are more complex interrelationships than can be modelled, as a separate variable. For simplicity, this discussion assumes that the actual rate is the same as the target and that the appropriate capital is expended to make the full range of target rates feasible.

Finding and climbing the hill of value

The initial discussion has shown how each of two strategic decision parameters – cut-off and production rate – may be optimised independently, all other decisions effectively assumed to be fixed; however, both cut-off and production rate are independent decisions (though, the cut-off selected may impose practical limits on the production rate achievable from the mine). To obtain the best possible value for the project, they must be optimised together. Figure 6.5 shows pictorially how value might respond to simultaneous changes in the values selected for both of these decision parameters. It can be seen that for any selected production rate target, the value grows with increasing cut-off to a maximum, then falls, as shown in Figure 6.3. Similarly, for any given cut-off, the value rises with an increasing production rate, then falls, then begins to rise again, similar in principle to what is shown in Figure 6.4 but with one large capital expenditure only.[4] Figure 6.5 is taken from a real study conducted by the author, and is typical of the outcomes of many underground mine situations where the value measure is NPV or some similar financial or economic measure.

Because we live in 3D space, we are only able to comprehend 3D outcomes graphically or pictorially. The hill of value (HoV) illustrated is a 3D surface showing the behaviour of one dependent variable – value – as a function of two independent variables – in this case, cut-off and production rate. Clearly, there are many strategic decisions to be made, and each of these will be another independent variable to be represented on an axis in another dimension. Similarly, there may be multiple value measures to be evaluated and these will also be additional dimensions in the analysis. To establish the underlying principles of the process, we will continue the discussion of a 3D case with one value dimension and two option dimensions.

---

4. This and other HoV illustrations have been created using the 3D surface chart capability in Microsoft Excel™. This facility cannot handle irregularly spaced data values on the axes; all data points are spaced equally, regardless of their values. Hence, Figure 6.5 does not show the step changes in value seen in Figure 6.4, but the values at all calculated points (the 42 combinations of seven cut-offs and six production rates, shown as a grid on the floor of the chart and the surface itself) do account for such step changes.

FIGURE 6.5

A typical underground hill of value for value versus cut-off and production target.

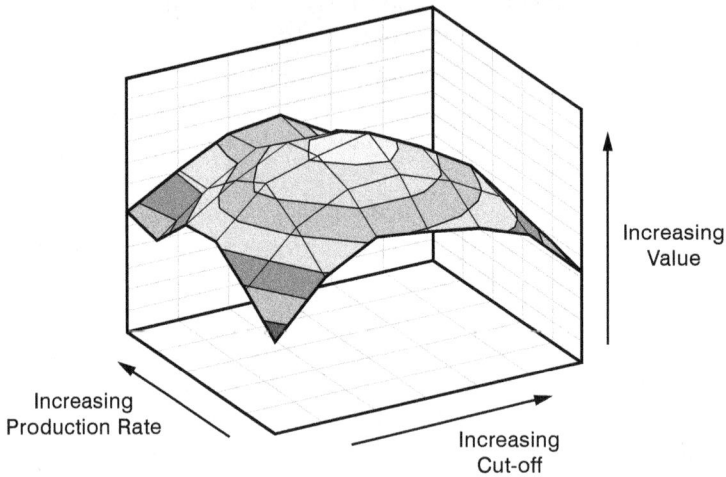

Figure 6.6 shows the typical location on the HoV surface for an operation whose strategy has not been optimised. Typical industry planning processes do not usually deliver a plan that is anywhere near optimal (in terms of maximising NPV).

A common proposal to improve the economics of an operation is to increase the production rate. The idea is to spread the fixed costs of the operation over more tonnage, thereby reducing the unit costs and increasing the average margin. This is illustrated in Figure 6.7.

There are some possible problems with this. Firstly, it is to be hoped that the system capacities will have been identified so that the planned production increase is to the top of the ridge of the value surface; however, there is no guarantee that this will be done, and what may be a logical increase in some areas of the operation may require debottlenecking capital expenditure in other parts. Furthermore, potential gains from

FIGURE 6.6

A typical underground hill of value showing a typical unoptimised strategy.

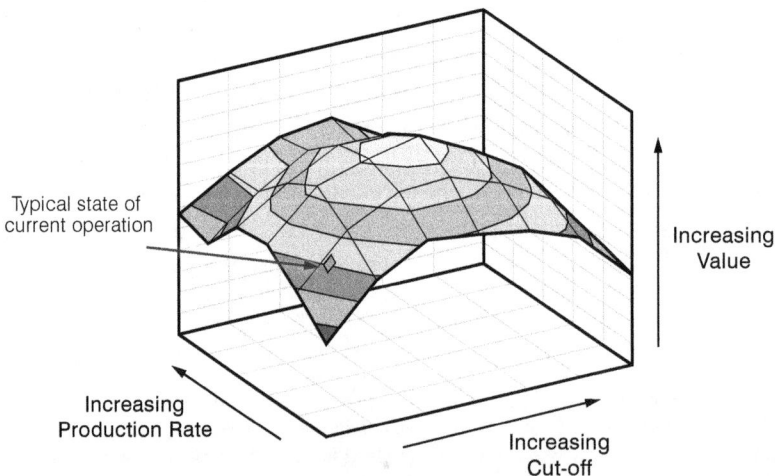

FIGURE 6.7

A typical underground hill of value showing a typical improvement strategy.

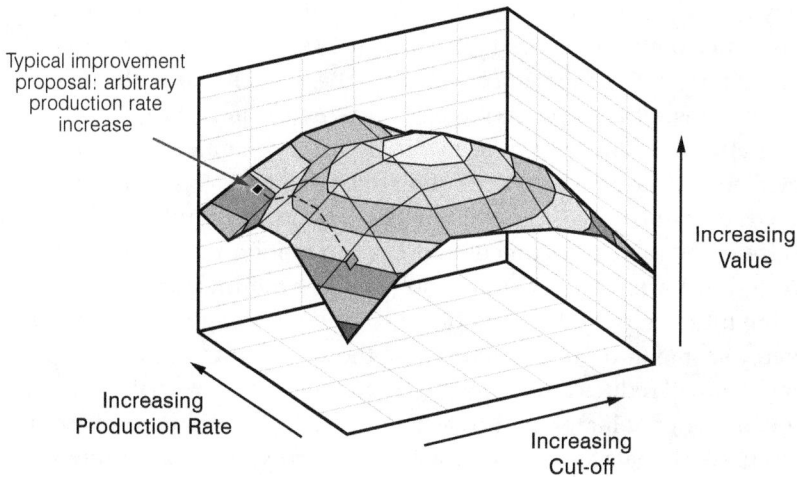

the arbitrary production increase proposed may be more than consumed by the capital requirements. This can be seen in Figure 6.7 where the proposed higher production rate produces less value than the initial rate.

It is implicitly assumed that production can be increased at the same head grade, in which case the margin per unit of product will increase as intended. If this is not feasible – at least in the short- to medium-term – it may be necessary to lower the cut-off to increase the tonnage produced. And although the average unit cost per tonne of ore may reduce, the unit cost per unit of product may not, or at least not to the extent expected. It is also common in these circumstances for it to be proposed that, since the unit cost will have reduced, the cut-off can also be permanently reduced, thereby increasing the reserve. This is a natural consequence of the erroneous assumption that cut-off and break-even are synonymous. A moment's consideration of the HoV's shape will show that this is not a good plan.[5] Reducing the cut-off reduces value.

Figure 6.8 indicates the optimum strategy to maximise value. It is the combination of cut-off and production rate that puts the operation at the top of the HoV. For the case shown, this will require both an increase in cut-off and in production rate. In this case the underlying analysis has taken account of the production capabilities of the orebodies at different cut-offs, so that the production strategy at each point on the value surface is achievable in practice.

It was seen that cut-off defines the size of mine for an underground operation, whereas in an open pit it is the size of the ultimate pit. All the preceding discussion regarding cut-offs in an underground operation can in general be translated into the open pit situation by replacing references to 'increasing cut-offs' with 'reducing pit sizes' and 'reducing cut-offs' with 'increasing pit sizes'. Where cut-off is shown as increasing from left to right

---

5. This conclusion assumes that the key value measure behaves as plotted. This HoV is typical of the behaviour of NPV and long-term profit measures of value. Of course, if maximising the mine life or the reserve tonnage is the key goal, the shape of the HoV will be completely different, falling as cut-off increases. Minimising the cut-off then does become an appropriate strategy.

across the various value versus cut-off plots, the plots are equally valid for open pits with pit size increasing from right to left on the same axis.

The strategy optimisation process is simple in concept, as has been illustrated. Still, the practical application of the underlying principles with several option/scenario parameters to be accounted for – each of which becomes another dimension in the analysis – plus a time dimension (most of the option/scenario parameters may vary over time) may not be easy. Even after the results have been successfully generated, deciding which mine plan to implement will be an even more contentious and time-consuming task. One of the initial problems encountered at an operation that has not been optimised previously will be that the size of the mine as planned before the operation will be too large; that is, for an underground operation, the existing cut-off will be too low (as shown in Figure 6.8), or for a pit, the ultimate pit will be too big and/or the mining rate too low. Other goals will then tend to come into play, as increasing the cut-off underground or reducing the size of an open pit will reduce the ore tonnage and metal in the reserve, which may not be acceptable even though it increases the cash generated. Chapter 10 discusses parameters that are often used as goals and value measures. For now, we note simply that some are inversely correlated with cash generation. Some may be applied as constraints rather than being optimisation goals themselves – the main value measure should be maximised with the proviso that the outcome for another measure must be greater than or less than some specified value.

FIGURE 6.8

A typical underground hill of value showing the optimum strategy.

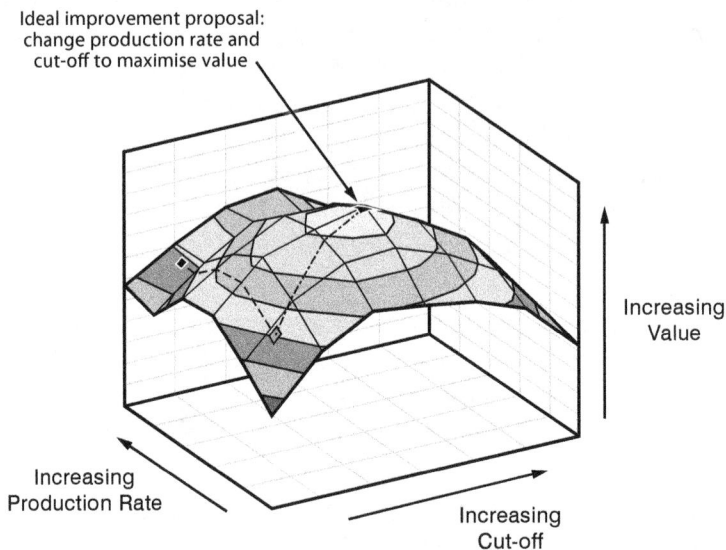

Dealing with multiple goals in an underground mine

If an optimisation model has been developed to derive primary cash-related value measures such as NPV, any other measure that might be of interest should be available in the analysis. Considering similar plots for other value measures of interest permits informed trade-offs to be made, so that the best strategy overall, taking account of both quantitative and qualitative issues, can be specified.

Figure 6.9 shows what is ostensibly a 2D plot of the values of several value measures for a range of cut-offs. This chart is taken from a real study conducted by the author for an underground gold mine. There are five value measures shown, so in the context of the HoV concept discussed, this could be considered to be collapsing six dimensions – one independent option variable, cut-off, and five dependent value variables – into a 2D representation.[6] Alternatively, the discount rate could be thought of as a second independent axis, so it is a 5D analysis with two independent axes, cut-off and discount rate, and three dependent value variables.

FIGURE 6.9

Underground example – various parameters versus cut-off grade.

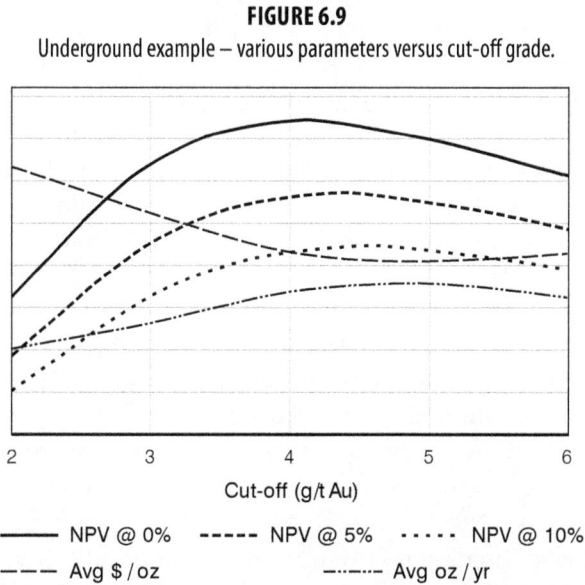

The company's major value measure was NPV at a ten per cent discount rate, but recognising that discounting tends to favour strategies that add value in the near-term, the company was also interested in the total undiscounted net cash flow – that is, the NPV at zero per cent discount rate – and the NPV at an intermediate discount rate of five per cent. Of further interest were the annual gold output, which had to be as high as possible but not less than a specified target, and the unit cash cost of production, which of course was to be as low as possible. Whether these are rational or appropriate goals is not the issue here – the intention is to indicate the type of problem that companies using multiple goals have.

The company at the time was using a cut-off of 3.0 g/t, and the calculated break-even grade was of the order of 2.5 g/t.

It can be seen from the plot that the cut-off that maximises NPV at ten per cent discount rate is approximately 4.5 g/t, while that which maximises net undiscounted cash flow is approximately 4.2 g/t. The NPV at the intermediate discount rate of five per cent is maximised by a cut-off between 4.2 and 4.5. Clearly, different goals require different strategies. Which to choose? Further inspection of the plots suggests that, near the top of the HoVs for these cash flow value measures, the curves are relatively flat, and the

---

6. Other option decisions, such as production rate, are fixed here, though they could vary in a 3D perpendicular to the values versus cut-off plane shown.

practical reality, given also the various uncertainties in the data, is that any cut-off in the range from, say, 4.0 to 4.5 g/t will satisfy all these cash flow maximisation goals.

What about the gold production and unit cost? Interestingly, these measures are both optimised at a cut-off of approximately 5.0 g/t, which is significantly higher than the cut-offs maximising the cash flow measures; however, selection of a cut-off in the 4.0–4.5 g/t range will still deliver outputs and unit costs that are close to optimum.

It becomes evident, therefore, that if there are multiple goals to be satisfied, the optimum strategy may not optimise any individual goal but rather provide a close-to-optimum satisfaction of a number of potentially conflicting goals. Identifying the appropriate trade-offs may be a qualitative judgement, rather than a purely mechanical result of the quantitative processes that generate the numbers for each value measure.

It is also interesting to note that the cut-off that maximises net cash flow (4.2 g/t) is significantly higher than the break-even grade (2.5 g/t) stated by the company. The net cash flow at that break-even cut-off is significantly lower than the maximum possible, which indicates that at the break-even there is material included in the reserve that is consuming more cash in input costs than the revenue it is generating. This in turn implies that the break-even calculation is not adequately incorporating all costs that should be included. This is a common outcome and typically results from excluding sustaining capital costs[7] (and sometimes some or all of the fixed costs) from break-even calculations.

Not shown in this chart are plots of mine life and tonnes and gold ounces in the reserve. These, of course, all fall continuously (not necessarily as straight lines) from left to right across the chart as cut-off increases.[8] When these types of measures become important, finding a strategy that well satisfies both maximisation of cash flow or profitability and maximisation of life or reserves measures may not be easy.

There is often a lot of discussion within companies around the appropriate value of the discount rate to be used. It is interesting to note that the magnitude of the maximum NPV is, not unexpectedly, strongly influenced by the discount rate. Clearly getting the discount rate right is important if the aim of the study is to identify the desirable purchase or sale price for the operation. If the aim is to identify the optimum operating strategy to maximise cash returns, in this particular case, and in a number of other studies conducted by the author, the discount rate used, perhaps unexpectedly, has little impact. It is important to identify the purpose of the study.

In the project illustrated by the HoV in Figure 6.8, the company's senior management identified that the mine life – delivered by the cut-off and production rate strategy that maximised NPV – was shorter than what was accepted as a given by both investors and the nearby townspeople, which included most of the workers at the operation. Management's assessment was that the scale of the reduction in life would result in a major reduction in the company's share price, and cause problems with the workforce and town. Yet after identifying a reduced but acceptable mine life – with no significant reduction in share price and adequate time for the workforce and town to prepare for the mine's closure – it was found that the resulting strategy, with a somewhat lower cut-off

7. The distinction between capital and operating costs is to a large extent irrelevant for cut-off determination. Often the distinction between project capital, usually excluded from break-even calculations, and sustaining capital, which should usually be included, is not appreciated, and all capital is often excluded from break-even calculations.

8. Mine life will also decrease with increasing production rate as well as increasing cut-off.

and production rate, was still close to the flat portion at the top of the HoV. Most of the upside in value (from the current position on the front-left corner of the HoV) could be achieved without the angst of a serious reduction in mine life. A number of value versus strategy surfaces were used to make an informed decision about the best strategy (which did not optimise any one measure of value), taking account of a number of interrelated issues, some of which were qualitative judgements.

These case studies suggest that the ultimate aim of a strategy optimisation study is not just to mechanically identify the optimum, if that is in fact possible, but to provide senior decision-makers with information from which informed decisions can be made. Decision-makers can be made aware of the benefits and dangers of strategies and, applying their collective knowledge of bigger-picture issues not necessarily captured by the optimisation evaluations, make the best possible decisions. Chapter 12 describes how these techniques are applied to quantify the trade-offs between the upside of correct assumptions, such as future metal prices, and the downside of wrong assumptions.

Dealing with multiple goals in an open pit

Figure 6.10 illustrates results from an open pit case study. The HoV is shown as a contour plot, indicating the maximum value occurring slightly above-right to the centre of the plot. The horizontal option axis represents the run-of-mine (ROM) cut-off – the cut-off applied to the rock mined from the pit to determine what is ore – which is to be sent to the plant for processing. The vertical option axis is the rate of mining rock, expressed as a percentage of the rate in the current mine plan (at the time of conducting the study). It can be seen that there is a ridge of value running from low rate / low cut-off to high rate / high cut-off within the range of option combinations evaluated. The base case – the then current mining plan – is identified by the circular marker on the '100 per cent of base case mining rate' grid line.

Open pit mining plans generally automatically optimise the ROM cut-off, given a mining rate. There will be short-term variability, but overall the ROM cut-off will be the

FIGURE 6.10

Open pit example – net present value versus run-of-mine cut-off and mining rate.

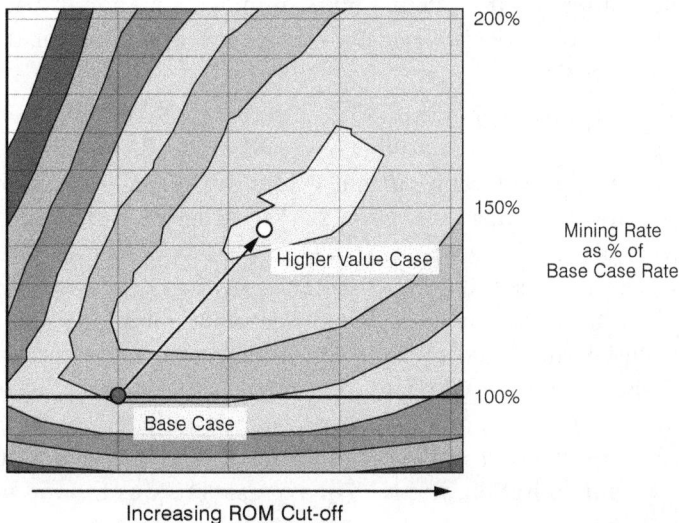

Lane mining–treatment balancing cut-off, subject to its not being less than the mining-limited break-even cut-off, and this will be the overall optimum cut-off. With short-term grade distribution variations, there will be some variation in the ROM cut-off to suit. This maximises value at the planned mining rates, which are set to deliver the planned (plant capacity) ore tonnes at a planned ROM cut-off; this in turn will be calculated as a break-even. By selecting the mining rate to balance with the ore treatment rate at that cut-off, the ROM cut-off is optimised at those rates automatically. There is a circular argument here. What starts as a break-even cut-off that may or may not be optimal becomes optimal because capacities are established so that a physical balance is achieved at that cut-off.

However, it can be seen that value can be increased, up to a point, by increasing the mining rate and the ROM cut-off. For a given ore treatment rate, faster mining generates more material of all grades, so the cut-off can be increased naturally to keep the mill full of ore at a higher grade. Up to a point, the higher grade and extra revenue will cover the additional mining costs of the faster mining rate, but eventually the distribution of grade in the mineralisation will be such that this no longer occurs. Additionally, while mining is in progress and fixed costs are incurred regardless of what is classed as ore, the marginal break-even below which material cannot be economically treated is Lane's mining-limited cut-off; however, once mining is completed and treatment of stockpiles only is progressing, fixed costs must be included in the marginal break-even, which therefore rises. Low-grade material that could have added value if treated as marginal material at low mining rates becomes unprofitable if it has been stockpiled for treatment after mining is completed because of higher mining rates. The value therefore also drops at higher mining rates because of the overall cut-off applied to the material within the defined ultimate pit.

This should not be seen as an argument in favour of a low mining rate and lower overall cut-off and increased reserve. In reality the decision to mine at a slow rate has resulted in an increased mine life, and although the fixed costs are committed and irrelevant to the cut-off decision once the mining rate has been specified, they are, in fact, a real cost of that marginal material. There is a direct relationship between the mining rate and the cut-off; once we specify one, and given a pre-specified treatment rate, the other is almost automatically determined. So a decision to reduce the mining rate and increase life is effectively a decision to reduce the cut-off and increase the life, so the cut-off situation is a treatment-limited situation rather than a mining-limited situation. It becomes mining-limited by virtue of a policy decision to mine at a certain rate, rather than a genuine physical mining rate constraint.

In the case presented here, the current ROM cut-off may be optimal given the current mining rate, but taken together, the mining rate and the ROM cut-off are not optimal. Both need to be increased from their current values to become optimal for the goal of maximising NPV. However, this subsection is discussing multiple goals in the open pit context. So what other goal/goals are evident on the NPV versus ROM cut-off and mining rate plot? The answer is: mining costs. These will be proportional to the mining rate. Management at this operation were being pressured to minimise annual costs, and costs will obviously be minimised by reducing mining rates. The option selected was therefore to minimise the mining rate. This in turn required reducing the ROM cut-off to maintain ore feed to the mill. The net effect was to significantly reduce the NPV achieved by moving down and to the left in the contour plot shown, rather than up and to the right; however, cost minimisation, not NPV maximisation, was the measure by which management was being judged.

Whether this was the corporate goal or not is another issue – it probably wasn't. What this case study highlights is how goals for mine staff that are thought to be aligned with the corporate goals, but are not, can drive counter-productive decision-making. Selecting strategies based on measures that are truly aligned with corporate goals is obvious, but this frequently does not occur. Explicit management goals such as cost minimisation neglect to account for situations where cost reductions reduce the revenue more than the costs. This is what occurred in this case study. Implicit goals, such as 'every tonne pays for itself' for break-even cut-offs, can also result in suboptimal strategies as has been shown here. Optimising the mining rate and adopting the corresponding optimum cut-off will generate a higher net cash flow, even though costs are higher and the cut-off is not a break-even grade as commonly calculated.

STRATEGY OPTIMISATION VERSUS TYPICAL STRATEGY SELECTION

How does a strategy optimisation study differ from long-term strategy specification practices in the mining industry?

Typical evaluation processes handle many relationships in separate, independent processes to create a limited number of preset strategies. Technical staff produce alternative plans and schedules for the operation being evaluated. Mine plans are generated using industry-standard mine planning, design and scheduling software, often to a high level of detail. By virtue of their training and experience, technical staff (geologists, mining engineers, geomechanics specialists, metallurgists) tend to ensure that all considerations within their fields of expertise are accounted for. The capabilities of modern software tools also help. If anything, a false sense of accuracy is attained – just because the scheduling package is capable of working in daily or hourly time slices for decades into the future, doesn't mean that this will happen. Most technical staff would recognise this, yet the capabilities of the software almost drive us to generate plans for much longer time frames than are warranted by the accuracy of the underlying data, not to mention the unplanned events that will undoubtedly cause changes to plans in the future.[9]

Because of time constraints and the limited availability of technical staff,[10] it is common for only a base case and a few alternative options to be generated at this level of detail and intellectual rigour. The mining plans thus developed will then be used to estimate the capital and labour requirements, metallurgical performance and such items as environmental and closure plans. The schedules of various physical activities and costs derived by each of these processes will then be consolidated in a relatively simple model, which in reality will be little more than an arithmetic calculator to derive revenues and costs from imported, unchangeable physical quantities, though there may be complex

---

9. One-year budgets are often obsolete by the start of the budget year to which they pertain; how much more so for five-, ten- or 20-year plans? Yet the increasing power of modern software is leading to an expectation that a long-term plan cannot be relied upon if it does not have this level of precision across the whole time frame being evaluated.

10. Consider the reason for these constraints. Apart from the numbers of people available with degrees of experience and qualifications, the topics on which these people are working do not necessarily align with the importance of the work. It is all too common to find that the *urgent* displaces the *important*. Chapter 7 discusses a number of mine planning process issues. It is evident that urgent issues that displace work on the important issues are often the result of not having done the important tasks properly in the first place, if at all! It can become a vicious circle, and it needs breaking.

calculations in some areas, such as the derivations of revenue, fleet replacement schedules, stock movements, working capital and accounting statements. Figure 6.11 shows schematically the study activities and flows of data and information.

FIGURE 6.11
The typical strategy selection process.

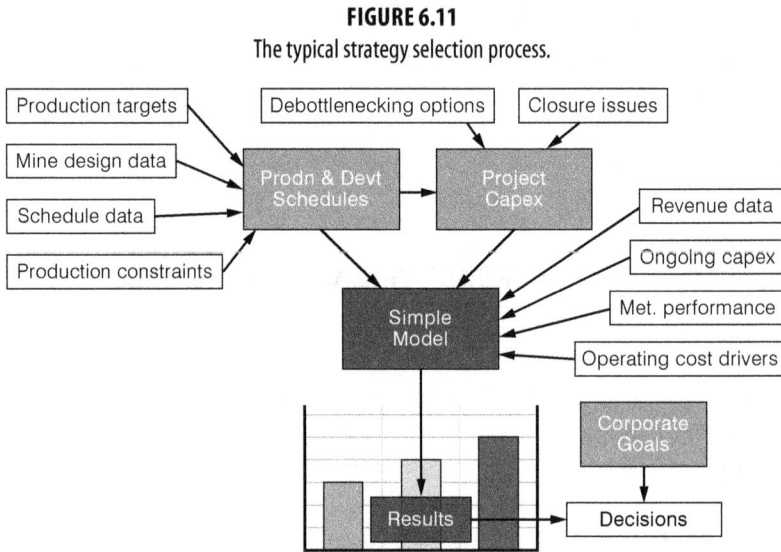

The *simple model* will often be used to flex things that can be easily changed within such a model. These will be such items as product prices, exchange rates, operating and capital costs and perhaps ore grade and metallurgical recoveries, and will usually be scenario parameters. These will be all flexed independently, up and down by specified percentages of the base case values, regardless of whether the likely range of variations in each of these inputs is greater or less than these arbitrary variations. As expected, on plotting the outcomes of these sensitivity analyses (often referred to as a spider diagram), we will find that if prices go up or costs go down, the value of the plan evaluated will increase, and if prices go down or costs go up, the value of the plan will decrease. It is also usual to find that value is most sensitive to variations in revenue-related parameters, such as price, exchange rate, grade and recovery.

What we have done in this type of analysis has identified how the value of a plan varies in response to external parameters over which we have little or no control. We have not in any way assessed how the plan itself might actually change in response to these external issues. The mining schedules and other underlying physicals will not change in the model; they are fixed by the detailed work done externally, with values only imported into the evaluation model. The absolute value of the project is not so important when it comes to strategic decision-making. Of more importance is identifying changes in circumstances that would cause us to change our plan. The typical strategy selection process only addresses this issue at a superficial level by evaluating a small number of alternative operating plans, none of which is likely to be optimal.

A strategy optimisation study, on the other hand, must consider the same issues as the strategy selection process. It must also identify the optimum operating strategy (the best options) for the base case economic assumptions, such as prices and exchange and inflation rates, and establish how changes in various scenarios might result in changes to the optimum plan. The process must, theoretically, be able to evaluate all the options to ensure the best plan can be found. For strategy selection, engineering resources are used to develop a few detailed plans. For strategy optimisation, these resources help to identify the relationships that experienced planners are using automatically on the basis of their experience as the detailed plans are developed, so that the relationships rather than specific plans are incorporated into the evaluation model. Instead of several separate, independent processes to generate plans and schedules and then determine their associated revenue and cost streams, the optimisation model handles all the physical scheduling and resulting cost, revenue and value measure relationships internally. This enables a very large number of options to be evaluated. Various inputs can be flexed simultaneously, and the best strategy, including cut-off policy, will be identified.

Figure 6.12 shows the strategy optimisation process schematically. Note that all the study process inputs in boxes around the outer edges of the figure are the same as in Figure 6.11 for strategy selection. However, as indicated in the preceding discussion, the information flowing into the model will be both data values and relationships that enable schedules to be developed by the model, rather than predetermined unchangeable schedules.

There are other subtle differences between the two figures for strategy selection and strategy optimisation. Earlier chapters have identified the importance of specifying corporate goals to drive the evaluation process, and how, if this is not done, the goal implicit in the evaluation process will become the de facto goal of the plan developed. Note that in Figure 6.11, the corporate goals are not linked to the evaluation processes but are only applied to the results of those processes. In reality this is common for strategy selection. The few operating plans that are developed are not focused on achieving any particular goal. Rather, their outcomes after they are developed are compared with each other in light of the goals to decide which plan best delivers the goals. If none of the plans deliver satisfactory results, attempts are made to improve the best of them; for example, by reducing costs (often unrealistically).

FIGURE 6.12
The typical strategy optimisation process.

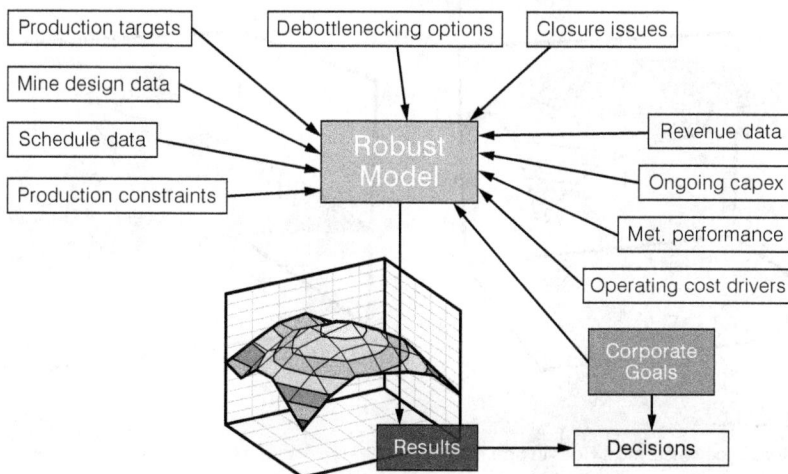

In Figure 6.12, however, the corporate goals are used not only to assess outcomes, but also feed into the evaluation process. Because the optimisation process identifies the best plan out of a large number of options, the goals, the measures that quantify them and the ways they are to be accounted for within the process must be key parts of the evaluation processes. The goals are not just used to compare options after the evaluations, as is done with typical strategy selection.

The other major difference at this stage is the difference between the sensitivity analyses in each process. Typical sensitivity analyses flex a number of variables, such as costs and prices, which are usually scenario parameters outside the control of the company; then assess the impact of these changes, independently, on the value of a specified operating plan. Strategy optimisation, on the other hand, is primarily flexing decisions about option parameters. The HoV is in fact a sensitivity plot of value versus decisions.

Knowing the behaviour of the value versus decisions surface is a crucial part of strategy optimisation, indicating how much flexibility there is, as well as the optimum strategy. Is the HoV shaped like a church steeple, so that there is a significant loss of value if one deviates a little from the identified optimum? Or is it a broad plateau with a range of options, indicated by little variation in value across the upper surface? In this case, deviations from the precise optimum may result in insignificant value reductions, as long as the limits defined by steeper sides around the flat upper surface are not breached. Figure 6.13 illustrates the types of situations that could occur.[11]

While noting that three dimensions are all that we can graphically conceive, there is no reason why we should not consider higher numbers of dimensions in the analysis and include both options (which we can make decisions about) and scenarios (which may vary but are outside of our control) as axes of a multidimensional HoV. To continue the geographical analogy, consider a change in predicted prices. For strategy optimisation, we are not so much concerned to know how much the hill rises (the value of any operating plan options will increase), but whether the pattern of value increases is such

FIGURE 6.13
The importance of the shape of the hill of value.

HoV with "flat" top
Precise strategy
specification not critical

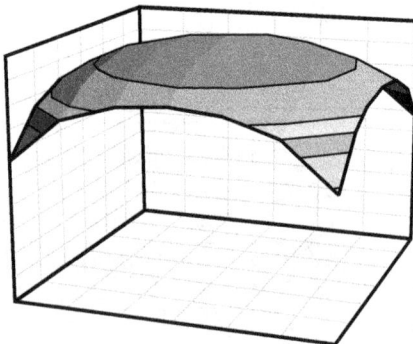

HoV with "pointed" top
Precise strategy
specification critical

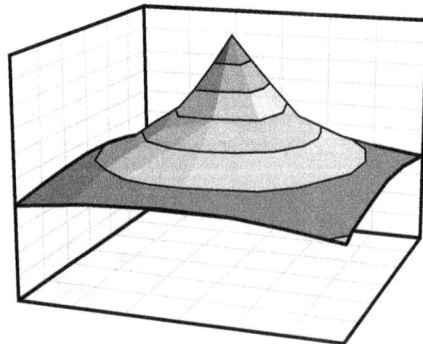

---

11. In the author's experience, most HoVs in optimisation studies are flat near the optimum but become steeper with distance from the optimum, like the first of the two hills in Figure 6.13.

that the HoV also moves sideways (that is, the optimum strategy changes) as a result of the changes in external parameters. If the axes of the HoV can include both option and scenario parameters, the spider diagram types of sensitivity analyses can be seen to be a subset of HoV-type outputs, depending on what parameters are selected for the axes of various plots that can be understood graphically in 3D space.

THE STRATEGY OPTIMISATION PROCESS

Preceding chapters discussing the ways to determine cut-offs describe the underlying theory and processes to derive a number as the cut-off. Simple yet rigorous formulas are used to obtain the desired result. Unfortunately, there is no formally accepted recipe for strategy optimisation. This chapter has described the principles underlying an optimisation evaluation, of which finding the optimum cut-off policy will be a component. Although there are similarities between operations, each has its own peculiarities, and some model customisation will be required to fit the circumstances of each case.

There are a number of methodologies that can be applied to determine the optimum strategy, and the details of the processes and data requirements will depend on the methodology employed. The processes for a strategy optimisation generally involve generating data and relationships in order to model the influence of constraints and decisions related to geology, mining, metallurgy, operating cost, capital costs, end-of-life and social issues and financial and economic forecasts. Data requirements and relationships are discussed in more detail in Chapter 8. Later chapters will consider features of various optimisation methodologies and cover some aspects of the processes in more detail.

OPTIMISING ONE PROJECT VERSUS OPTIMISING ALL PROJECTS

Many companies own more than one operation. Each operation is often optimised in isolation, without considering the impact on other operations. The task of senior company executives will be to maximise the value of the company as a whole, not just one of its operations.

For an individual operation, value will typically be maximised with a relatively large project. If there is no limit on the amount of capital that the company can spend across its enterprises, it may be appropriate to spend whatever is needed so that each operation is optimised individually. However, if there are limits on what can be spent, each operation must be considered as part of the total portfolio of projects. Each potential capital expansion, though it may be economic and add value to that project, must be assessed in conjunction with other potential projects at the other operations, so that scarce capital can be allocated to projects that generate the greatest NPV for the total capital invested.

Although it is important for a team evaluating an operation to identify what strategy maximises the value of that operation, in principle it is insufficient on its own. Results should also indicate the maximum incremental NPVs derived from sequential capital injections. Corporate decision-makers can then identify the optimum sizes for all projects considered together, rather than in isolation. The overall optimum corporate strategy may well be to have a number of smaller high-return operations instead of one or a few larger but low-return projects. Identifying the values of expansion projects at each operation allows the optimum allocation of scarce capital to maximise the total value of the company.

Decision-makers should also consider the impact of sequencing various deposits within one operation. Previous discussions have highlighted the need to flex uncertain items to determine whether they influence current decisions. These decisions may be significantly impacted by whether or not additional resources are mined and treated and, if so, when. Figure 6.14 gives a simplistic representation of the type of situation that may be found.

FIGURE 6.14

Values of one and several sequentially mined deposits.

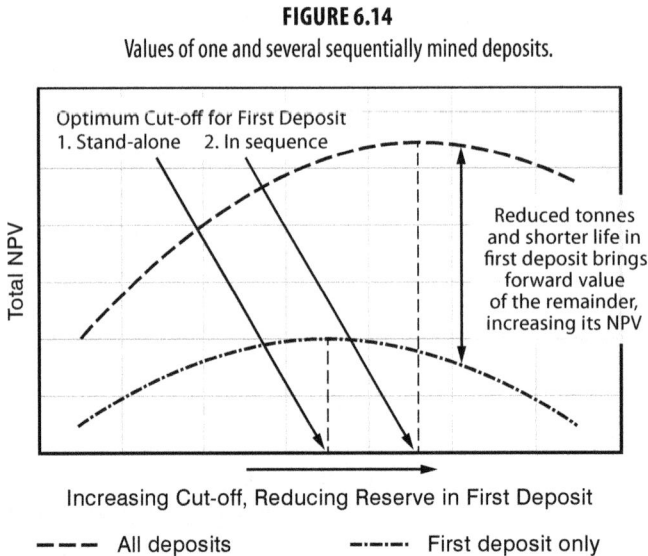

The horizontal axis is the cut-off to be applied to the first deposit or mine area in the sequence of possibly many deposits/mine areas. This is the initial decision to be made: what should the cut-off be for this resource? The lower curve shows the HoV value versus cut-off curve for the first deposit. The optimum cut-off is clear. But if the other resources are included in the analysis, the cut-off for the first deposit that maximises the total value of the operation – as shown by the upper curve – is significantly higher. The reason for this is easily explained: the loss of NPV resulting from increasing the cut-off and reducing the reserve and mine life for the first deposit, judged on a standalone basis, is more than compensated for by the increase in NPV resulting from the consequent bringing forward of the cash flows of the subsequent deposits.

Decisions to mine the other deposits are not being made now but will be made at various times in the future. There is therefore no guarantee that the other deposits will be mined in the same way and with the same values, as indicated now, or even that they will be mined at all.

CHAPTER SUMMARY

In the previous three chapters, the main focus has been on determining the cut-off grade, which is just as important as the decisions regarding mining rates, treatment rates, mining sequences and so on. The best combination of decisions can only be made by optimising all strategic decisions simultaneously. The mine strategy optimisation process can therefore be thought of as a multidimensional analysis that is accounting for everything. It is seeking to identify the best combination of cut-off grades and other design and strategic decision parameters – the options within the set of value drivers – in order to achieve the corporate goals. The corporate goal/s must therefore be clearly specified. The strategic policies that deliver one set of goals will probably differ from the policies that deliver other sets of goals.

Any operation deals with three types of big picture parameters: the things that are given; the things that can be changed (the strategic decisions or options); and the things that flow out of a combination of the first two parameters. These must be considered in the context of external parameters, such as metal prices and inflation rates, that the company has no control over (the scenario parameters).

There are two key givens associated with any mining operation: the *resource* with which it is endowed, and the *environment*, which is taken to be an all-encompassing term that describes the social, political, economic, climatic and ecological framework within which the company finds itself operating.

The key parameters over which the company has control that drive the strategic direction of any mining operation are typically the size of the mine, mining methods, production rates, sequencing and knowledge of the resource. Consequential parameters are what flows out of a combination of the givens and strategic decisions, including detailed mine designs, process flow sheets and schedules of physical activities and their associated financial outcomes.

Each of the strategic options can be optimised on its own, but to optimise value the interactions between all options need to be considered. Relationships need to be developed so that in the strategy optimisation model – in whatever software package and with whichever technology is employed – the effects of changes made to the values of one or more option parameters will be accurately represented throughout the model and, of course, in its outputs.

Although powerful analytical techniques can be employed to find the optimum, it is important to know how value responds to changes in the options settings about the optimum.

Although formulas and defined processes exist for determining break-even, Mortimer-style and Lane-style cut-offs, there is no such formal, well-accepted recipe for strategy optimisation. The processes for a strategy optimisation generally involve generating data and relationships in order to model the influence of constraints and decisions related to geology, mining, metallurgy, operating cost, capital costs, end-of-life and social issues and financial and economic forecasts.

There is nothing among these processes for a strategy optimisation that would not form part of a well-conducted prefeasibility or feasibility study. The typical strategy selection process flexes scenario items such as metal prices and costs for a limited number of pre-specified mining and treatment plans. Strategy optimisation involves flexing decisions. The evaluation model must therefore be able to generate the mining and treatment plans in response to changes in strategic decisions. The key distinction of a strategy optimisation is that multiple data sets may be required and relationships developed so that the impact of flexing decisions is fully accounted for in the model, not in various external design and scheduling processes. Final selection of the best option will come from evaluating the trade-offs between relevant measures of value in light of all the knowledge available.

CHAPTER 7

Cut-off Derivation Methodologies and Their Place in the Overall Mine Planning Process

INTRODUCTORY COMMENTS

Several ways of deriving cut-offs and the optimum mine plan have been discussed to this point. The reader may now wonder when or in what circumstances each should be used. Determining cut-offs and strategy optimisation are part of the overall planning process at any mine.

The author considers it a privilege to have spent most of the formative years of his career in companies with a strong planning culture, where best practice planning is absorbed as part of ongoing professional development. Unfortunately, many in the mining industry – in management, operating and technical roles – do not know what constitutes a good planning process and the benefits of it. In many cases, they often do not know that they do not know.[1] As an industry, we (quite rightly) award kudos to those who apply their management skills to manoeuvre their operations out of difficult problems; however, many of these operational problems could have been foreseen and avoided with good planning. It is easy to see the costs of the technical team in the cost reports. The benefits

---

1. Possibly a result of the operations-centric culture in many companies as well as the worldwide shortage of technical professionals, resulting in rapid staff turnover, lack of continuity and lack of mentoring. References to staffing shortages in this book take into account the industry's boom-and-bust cycle. During boom times there is a shortage of qualified people available, but during downturns, while there may be plenty of qualified people available, companies are not employing them and are running with reduced staffing levels. So, for the work being done within companies, there is an effective shortage of staff, leaving those who are employed overworked and without the time for thinking about long-term value-maximising strategies.

are less easily quantified but will often be orders of magnitude greater than the costs of the planning – they are the cost savings and revenue gains from better plans and, perhaps more importantly, from the mistakes that did not get made.

This chapter discusses firstly a recommended overall mine planning process in order to establish a framework of good process, and then indicates how the various cut-off and strategy optimisation procedures can be included in the process. The processes that will be described have long been well-recognised in the industry.[2] While primarily intended to give context to the cut-off and strategy optimisation techniques in this book, the discussion may also give some insight to those who do not know (or do not know that they do not know) what good planning processes are.

THE MINE PLANNING PROCESS

Mine plans and strategies should have a clearly defined goal (or a set of multiple goals). As noted previously, if plans are not deliberately created to deliver specified corporate goals, the de facto goal of the corporation will become the goal that is implicit in the derivation of the plan (such as ensuring that every tonne pays for itself if using a break-even cut-off), regardless of official pronouncements stating the company's goals.

For a mining company, the primary goal is to maximise value for shareholders. Ultimately, this is achieved by maximising the net cash generated by the operations. Taking into account the time value of money, the net present value (NPV) is the measure often used to represent the cash returns. It is for each company to specify its corporate goals, and how value is to be measured. There may be a number of measures, and the best strategy may satisfy a number of these, rather than optimising any one particular goal. Various common value measures and goals are discussed in more detail in Chapter 10. The discussion in the rest of this chapter will be in terms of one goal only.

Much of the traditional economic theory regarding maximising the value of a company has been developed in the manufacturing industry where the factory is the primary asset. In the simplest economic models, there are no constraints on inputs or outputs. Successive time periods are essentially independent – if product is not made and sold in any time period, the opportunity is lost forever. If the capacity of the factory is not changed, value is optimised by maximising the value of each time period independently by simply maximising outputs and minimising unit costs.

Mining, in general, is different. The primary asset is not the factory (the mining facilities and metallurgical treatment plant). Rather it is the mineral resource. With a finite and depleting resource, decisions made in any time period affect the resource remaining at the end of it, hence affecting what is possible in all subsequent time periods. Value cannot therefore be maximised by considering each time period independently. Plans for all time periods must be optimised together.

In a long-life operation where there are significant lag times between exploration, planning, development or waste stripping and finally production, plans must be developed in a hierarchy from long term to short term. Shorter-term plans should be

---

2. Many authors have written on these aspects, but for simplicity, readers are referred to the *Cost Estimation Handbook* published by the AusIMM, particularly Chapter 1, which deals with preproduction studies (Dewhirst, 2012). Several relevant papers are listed references to that chapter. (That so many have felt the need to write such papers supports the author's contention that many in the industry 'do not know that they do not know' what good planning procedures are.) Steffen (1997) discusses the planning processes at an operating mine.

consistent with, and provide more detail than, the longer-term plans from which they are derived. It is the long-term or life-of-mine plan (the LOMP) that ultimately generates value. Once the optimum LOMP has been identified and approved, all shorter-term plans must be developed to deliver that LOMP.

Key performance indicators (KPIs) should cascade down through the organisational structure to ensure the ultimate delivery of the LOMP.[3] Almost by definition, KPIs for senior operating managers should focus not on short-term measures but on those related to activities that help realise the LOMP. Focusing on short-term plans and one-year budget KPIs is almost guaranteed to result in long-term outcomes that are at variance with the optimum LOMP. Operations in the long term will then be suboptimal and value-destroying. Long-term plans will be forever reactive to previous years' short-term decisions, rather than setting the overall direction to be facilitated by the detailed short-term plan.

Since these discussions refer to planning processes internal to each company, terms such as *ore* and *reserves* are used in their colloquial sense and not according to the strict definitions for public reporting in international reporting codes. Some of the studies may result in publicly reported information, but that is beyond this book's scope.

PREPRODUCTION STUDIES

Preproduction studies advance through a number of stages with increasing levels of detail. These stages go by various names, one set of which has been used here. Recommendations for incorporating cut-off determination and optimisation processes are included in the discussion for each stage. For all preproduction study stages, the cut-offs derived will almost certainly be planning cut-offs, as defined in Chapter 2.

Scoping or conceptual study

Typically stated to be at a high-level order-of-magnitude accuracy (±30 per cent), the purpose of this level of study has been to demonstrate early in the resource discovery and evaluation process that further detailed work is justified. Theoretically it is therefore only necessary to find one case that proves this. An underlying purpose will be to avoid relinquishing a prospect too early in the evaluation process that has the potential to add value. So scoping studies may quite legitimately be optimistic in their assumptions.

Strategies adopted at later stages of evaluation may be different from those used in the scoping study, without implying that the scoping study was incorrect. Anecdotally, however, there seem to be many instances where the opposite is true, and either the credibility of those conducting the scoping study is called into question, or subsequent studies are constrained to the strategy used in the conceptual study; worse, they are expected to deliver similar financial outcomes.[4] In many cases, the scoping study is becoming akin to a high-level prefeasibility study. It is common to find projects that move straight from the scoping study to a full feasibility study without a formal

---

3. Targets for metal or ore production are often used as value measures, but they must be recognised as imperfect surrogates for value – maximising metal output does not necessarily maximise value. Metal targets may be a useful way to align an operation's plans with the corporate strategy, but only after the value-optimising plan has been identified so that optimum (not necessarily maximised) metal targets can be reliably specified.

4. As noted, it is legitimate for a scoping study to be optimistic, but that would not be appropriate for later, more detailed stages of investigation. Somewhat poorer outcomes in later studies should not be unexpected.

prefeasibility study, which is seen to save time and costs for getting a good project into production. Both of these study types are discussed in more detail below, but the implication is that, whether it is expressed in the scope of work for the scoping study or not, there is a growing expectation that it will identify the preferred option for a feasibility study and will be implemented. The scoping study is therefore becoming the de facto decision point for operating strategies for a growing number of projects.

If there is a requirement – express or implied – that the outcomes of a scoping study are as close as possible to the optimum at a later stage of investigation, it may be a necessity to conduct a high-level strategy optimisation (which will include cut-off optimisation) at this stage of evaluation. The quality of the data available is usually, almost by definition, inadequate for this to be reliable, yet it is an increasingly common requirement.

If, on the other hand, it is only necessary to demonstrate that further investigations are warranted, some form of full-cost break-even cut-off specification may be adequate. If there is a reasonable amount of tonnage and grade data, a Mortimer-style analysis might be a better option, particularly if the deposit is low-grade. If the case selected for evaluation does not then demonstrate viability and a better case must be found, a high-level optimisation style of study might be necessary, investigating a few major value drivers such as pit size, cut-off, mining method and production rate.

If a strategy optimisation is conducted at this level, it is essential to include caveats allowing for major changes in strategy without loss of credibility as more information becomes available, even if the study is expected to identify the optimum plan.

Prefeasibility study

The prefeasibility study (PFS) should have a two-fold purpose:

1. demonstrate to a higher level of confidence (typically ±20–25 per cent) that the project is technically and economically viable
2. evaluate a wide range of options to ensure that the best is selected for further detailed study and potential implementation.

The reality is that the case selected for subsequent stages of study will be the one built and brought into operation if it is proven to be feasible. Obviously it is desirable that the most valuable option is the selected, investigated further and built.

In practice, many studies will evaluate a limited number of options for various design parameters, but only considering variations from the base case for one parameter at a time, rather than concurrently; it is deemed too costly and time-consuming to evaluate all combinations of strategic options. This need not be a burden – most of the information and relationships will be generated anyway – and the financial benefits of optimising the strategy with the evaluation at the PFS stage can be far greater than the extra cost.

A full strategy optimisation study should be seen as an important and integral part of the prefeasibility study. Some prefer to interpose an options study between the scoping study and PFS. With this approach, the latter becomes a study of a single option or limited number of options, but less detailed and accurate than a full feasibility study. For the purposes of this discussion, the semantics are not important. What is important is that a comprehensive evaluation is conducted at the PFS level prior to committing to the option to be evaluated in the feasibility study.

Feasibility study

The feasibility study (FS) is intended to demonstrate the technical and economic viability of the project at a level of accuracy (typically ±10–15 per cent) sufficient to proceed with the project.

So-called optimisation studies may be done after a major FS, but typically these merely investigate ways of reducing costs or improving the efficiencies of the selected strategy, rather than look for better strategies. Additionally, if the study results are positive, the identified plan will be implemented quickly; delays for studies of other options are normally unacceptable. From a public relations point-of-view, there will have been much corporate 'face' invested in the published outcomes of the prefeasibility and feasibility studies. Announcing a major change of strategy soon after the publication of such outcomes will be politically unacceptable, destroying the company's credibility. If the best strategy has not been identified by the PFS and becomes the focus of the FS, it is unlikely to be found during the FS.

The policies identified as best in the PFS, such as production rates and mining and treatment methods, will normally carry forward into the FS, only being changed if more detailed investigations expose previously unidentified problems. Yet in many studies, the cut-off will have been determined as a break-even, and this may well alter if price and cost projections have changed since the previous study. Depending on how much the orebodies change – both because of cut-off changes and additional geological information obtained – it may become necessary to completely rework other strategic decisions. If cut-offs have been determined by some sort of optimisation methodology (such as Lane's simple methodology or a full strategy optimisation evaluation) instead of break-even, and there are significant changes in price and cost forecasts, an early part of the FS should be an update of the PFS optimisation. This update should use the latest data and forecasts to assess whether the major strategic decisions going forward into the FS should be changed. It may need to be repeated as forecasts change and information becomes available during the course of the study; the FS may even need to be halted and the wide-ranging PFS redone using the new information.

As noted in the discussion on scoping studies, there is a developing trend for projects in some circumstances to go straight from a scoping study to a final FS without a PFS. In such circumstances, a strategy optimisation study should form the initial stages of the FS to ensure that the FS is evaluating the best option, though, as indicated, this is rarely done. The author has seen cases where the strategy optimisation is conducted towards the end of the FS, the strategies of which have already been publicly announced. This is apparently done simply to tick a box to say that it has been done, with little chance that the outcomes of the optimisation study will change the direction of the FS.

The feasibility study becomes the official LOMP until the planning processes for the operational phase of the project start generating new LOMPs on a regular basis.

Detailed engineering

While the feasibility study will include mine, plant and infrastructure designs to a level of practicality and detail, its primary purpose as the name implies is to verify the technical feasibility of the proposed designs and processes, and derive quantities to support the predicted financial outcomes. It will usually not produce plans that can be immediately implemented for mining and construction. Detailed engineering and design are required to convert the principles in the FS into construction and mining plans for practical implementation.

By this stage of the development process, mining and plant capacities should be well defined. With improving knowledge of the mineral deposits and updated price and cost forecasts, it may be necessary to revise cut-offs at the detailed design phase. Long-term mining plans may require updating using optimisation techniques. Simpler cut-off models may be appropriate for shorter-term plans developed within the context of the longer-term plans. If the nature of the deposits and mining methods permit, a simple Lane-style cut-off optimisation may be appropriate. If not, a Mortimer-style cut-off specification will always be preferable to a simple break-even analysis. Lane-style balancing cut-offs should also be identified – in the light of production capacities being installed, these may be more appropriate if use of Mortimer-style or break-even cut-offs were to result in only one stream of the processes operating at capacity. A balanced operation, with two or more parts of the system at capacity, will often generate better values. A Lane-style opportunity cost should also be considered in break-even formulas so that, if the ore-handling and treatment processes are operating at capacity, marginal ore adds value over the life of the operation. If prices have increased greatly, the cut-off should not be depressed in a break-even manner to the extent that ore grades fall too low and product targets cannot be met – a balancing cut-off may again be more appropriate.

In practice, there is a focus on only the short-term requirements to commission the mine so that medium- and long-term operating plans are not developed until the mine is in its production phase. This may lead to reduced output early in the mine life after the initial ore stocks are depleted; the feasibility study's long-term plans will have become inappropriate as development, production and knowledge of the orebody progress and improve, but the long-term planning processes are not yet in place to handle this.

All aspects of the ongoing short-, medium- and long-term planning processes must evolve during the construction and commissioning phase so that the transition to production is seamless. From a planning point-of-view, the preproduction construction and development phase should be treated as if the mine and plant are already operating. The following sections recommend how planning should proceed at an existing operation and how the cut-off derivation and strategy optimisation processes fit into this.

A RECOMMENDED PLANNING PROCESS AT OPERATING MINES

In the following discussion, *plans* and *planning* refer to all technical activities supporting the operation and the outputs generated by these activities. They include such items as geological data gathering, geological interpretation and models, mine designs and schedules. The following integrated planning process has been found to deliver good outcomes at a number of mines where similar processes have been implemented. The key components of the recommended planning process are:

- strategy options analysis (SOA)
- life-of-mine plan (LOMP)
- five-year plan (5YP)
- two-year rolling plan (2YP)
- annual budget (budget)
- shorter-term detailed operational plans.

The process works in a hierarchy from long to short term. Long-term plans set the overall strategic direction. Short-term plans provide progressively more detail and accuracy at the front end of the long-term plans. Approvals processes also work from long- to short-term and are an integral part of the process.

The importance of focusing planning activities on the corporate goal has already been indicated. Long-term plans must focus on delivering that goal in the long run. Short-term plans must indicate in detail how operations in the near future will contribute to achieving long-term plans; if this is not the case, the plan will fail to deliver the corporate goals long term, and in the worst case may preclude attaining them.

Mining schedules should allow for indirect operating functions impacting on the schedule such as backfilling, and major planning and other non-operating activities in the critical path such as diamond drilling, assaying, geological interpretation and plan preparation. As well as the operations, it is also necessary to plan the planning. Planning activities and formal approval processes are often in the critical path of establishing a new mining area, but not considered when mining schedules are being developed, with potentially serious impact on achieving the mining plans.[5]

Formally documenting the planning processes at all levels of the organisation is essential. This is important given the rapid turnover of staff at many operations. Documentation may seem to be unproductive, but such processes should not depend on the knowledge or capabilities of individual people during times of high staff turnover. Investing in an effective planning process should not be squandered because of lack of continuity of staff in key roles. Formalising documentation standards and storage and archival of plans is also essential so that plans and supporting information are available to all concerned, both while they are current and in the future.

In the following discussions of different stages in the planning process, the lengths of the plans, the time periods reported in each and the frequencies at which they are carried out are suggested. Before the advent of computerised scheduling packages, schedules would have been created manually at this frequency and level of granularity in well-planned operations. Modern computerised systems permit both a much finer level of detail in the internal workings of the software and a full integration of short-, medium- and long-term schedules in the one working file. As always with software applications, users should be aware of the distinction between precision (relating to the internal calculations) and accuracy (the overall effects of data uncertainties and the simplification of relationships built into the calculations). The computer's precision will be the same at all points in time, but the accuracy of any schedule produced will be expected to decrease with time into the future.

---

5. The author has from time to time worked with planning staff at underground operations without formalised planning processes as recommended in this chapter. To emphasise the lead times required, a skeleton schedule of the critical path activities is developed, working from the earliest planning activities to the first blast in a new mining area. The schedule must include planning and operating activities, and allow for two or three detailed cycles of obtaining geological information (mapping, samples, diamond drilling, etc); obtaining survey and assay information; updating geological databases; geological interpretation and model updates; designing and scheduling development, development; and finally, designing the stope drilling pattern, drilling, designing the charging pattern, charging and blasting. The lead time from the first geological data gathering to first blast, depending on the operation and mineralisation is four to six years (hence the recommendation later in this chapter of a 5YP). This lead time is often a revelation to improperly trained planners, who are attempting to do all this in a fraction of the time and wondering why their plans and schedules fail to work.

The author suggests that the frequency and level of granularity recommended, though carried down from pre-computerised times, are still appropriate for official plans and schedules distributed outside the confines of the planning office. This will avoid the precision of the software's calculations creating a false impression of the accuracy of the outputs. Chiefly, it will help establish a disciplined and prepared planning process.

How the cut-off derivation and strategy optimisation processes fit into the planning process is discussed after describing the recommended process.

Strategy options analysis

An SOA is the first step in identifying the optimum LOMP to be adopted and followed by all detailed levels of the planning process. The SOA is the strategy optimisation evaluation that is the subject of Chapter 6. It evaluates, at a high level, the impacts on value, or a number of different measures of value, of all the strategic decisions that the company can make, separately and together.

Key issues to be evaluated include, but are not limited to:

- whether various resources are included in or excluded from the mining plan
- mining and treatment capabilities/capacities and the timing of potential upgrades or changes to methods and processes
- sizes of ultimate pits (open pits) and cut-off grades (underground and open pits)
- mining sequences[6]
- the impact of changes in external or uncontrollable factors such as metal prices and the existence of additional resources that may be discovered by exploration.

A full SOA is not normally done on an annual basis at an operating mine. It should be reviewed annually by qualified and experienced expert planners as part of the planning cycle, so that it remains a regular component of the process and ensures that the approved LOMP continues to be the best long-term plan. It should not be an irregular *ad hoc* process that has to be reinvented each time it is deemed necessary. It may be useful to rerun the previous optimisation model each year with updated price and cost forecasts (and perhaps updated tonnage and grade information for mining blocks) but without time-consuming changes to the model logic, to check if previously identified optimum strategies have changed. Major updates and reworking of the SOA would be expected perhaps every three to five years, triggered by, for example:

- significant changes in metal price forecasts
- changes in corporate aspirations or strategic goals
- discovering substantial new resources
- identifying new options for processes, markets, mining strategies, etc.

Life-of-mine plan

The LOMP is the formally approved long-term plan for the mine or business. Normally it would be selected after conducting an SOA as the plan best delivering the corporate goal.

---

6. Sequences and schedules are related terms, often used synonymously. In general in this book, *sequence* refers to high-level ordering of mining activities without specific reference to time frames, and *schedule* refers to a more constrained and detailed tabulation of activity quantities over time, resulting from the application of sequencing and rate decisions. There is no clear boundary between the terms and some degree of overlap will often exist.

If an SOA has not been done or is out of date, a formal LOMP is still required to establish the framework within which all other short-term plans are developed.

The LOMP should at least be reviewed annually as part of the planning cycle. This should be done by qualified and experienced planners who can recognise factors that should trigger a complete reworking of the plan. This would take account of constraints identified in short-term plans or resulting from actual events and changes identified in the SOA. Unless a major change in strategic direction has been signalled by an update or reworking of the SOA, changes in successive LOMPs should be minor.

Depending on the level of detail in the SOA and when it was last completed, the LOMP may be extracted directly from the SOA, or may require some additional work to finalise it. Either way, it should be seen as distinct from the SOA – an outcome from the SOA, not the SOA itself.

The LOMP establishes the planning framework over time for such items as:

- cut-off grades (underground and open pits) and sizes of ultimate pits (open pits)
- mining rates and capacities including timing of changes
- mining sequences and schedules, especially specifying when new mines, mining areas, orebodies, pushback stages and the like are to commence
- treatment rates or capacities for all plants, including timing of changes
- sources and destinations of all products and intermediate products (for example, if the company has its own smelting and refining facilities, the concentrates to be smelted in-house or sold)
- major infrastructure additions such as mine shafts and mill expansions
- project and sustaining capital requirements to implement the plan
- operating costs and associated resources, such as mobile equipment fleets and labour.

Five-year plan

The 5YP is an essential part of the planning process at operations with more than a few years of life remaining. It forms a critical medium-term link between the high-level strategies in the LOMP and the detailed short-term implementation plans. A five-year time frame will provide sufficient lead time for all or most long-lead items to be adequately planned, but in some cases the medium-term plan may need to be longer – up to ten years in exceptional circumstances – or somewhat shorter. The key issue for the time frame of this plan is establishing a reliable link between the LOMP and short-term plans. For this discussion, five years is assumed to be typical.

The author recommends that the 5YP should be recognised as an essential component of the planning process, with formal approval of the 5YP being part of the annual planning cycle occurring between approvals of the LOMP from which it is derived, and the short-term plans that emanate from it. The 5YP would be generated annually, with quarterly periods reported, though longer periods may be suitable in the later years of the plan. Monthly time periods can give a false sense of the level of accuracy of plans in later years. Annual values, while satisfactory for high-level summaries, may be too coarse for more detailed consideration of component activities, especially earlier on.

The appearance of a new mining area in the last year of a new 5YP would be the flag for designing or initiating such elements as:

- underground
  - block infrastructure and access development with major ventilation circuits

- stope boundaries, general layouts of stope development and stoping sequences
- open pits
  - for new pits: environmental studies, negotiations with other stakeholders and approvals processes
  - for new pits and new pushbacks in existing pits: overall sequencing and location of temporary and permanent access ramps
- underground and open pits
  - additional diamond drilling, metallurgical test work, geotechnical data gathering and obtaining information that is required to carry out more detailed designs as production from these areas draws nearer.

When new mining areas have advanced to the third and fourth years of the 5YP, major design and scheduling parameters should be well established. Shorter-term development and waste stripping schedules should be taking account of the 5YP production schedules to include the initial access for these new areas, so that they are available to produce when scheduled.

Production scheduled in the 5YP should be consistent with optimised schedules identified in the SOA and formalised in the approved LOMP.

Two-year rolling plan

The 2YP, like the 5YP, is a critical link between the long-term plans and short-term implementation plans and schedules. It provides a higher level of detail supported by detailed engineering work at the front end of the 5YP. The author recommends that the 2YP is formally updated quarterly and reports activities on a monthly basis.

The deliberate inclusion of *rolling* in the name indicates that it is an ongoing part of the short-term planning processes and is not just done once a year as part of a budgeting cycle. Technical staff would recognise that updating short-term plans and schedules on an annual basis is too infrequent – too much can change. Most operations would have more frequent updates of plans and schedules when situations change, but often done on an *ad hoc* basis. Formalising regular updates requires both operators and planners to regularly look at the effects of current operating and planning issues up to two years ahead, thereby of avoiding actions that may be expedient in the short term but create major problems in the longer term.[7]

The appearance of a new stope or bench, for example, in the last quarter of a new 2YP would be the flag for commencing the design of:

- underground
  - stope development
  - ventilation networks for development and production
  - general layouts of drilling designs and blasting sequences
- open pits

---

7. In some parts of the world, staff turnover rates are such that operating managers and planning staff are unlikely to remain in the same roles for as much as two years. The organisation must therefore ensure that short-term expediency at the expense of long-term performance is not rewarded. Having in place formalised planning processes like those recommended here may assist with this.

- sequencing of operations on benches to maximise flexibility and the number of working places available, and to avoid where possible operations on one bench preventing operations on another
- timing of the creation and removal of temporary ramps to maintain access and flexibility
- although not specifically related to milestone events in the pit, open pit plans must also plan for changes or additions to surface access roads, waste dumps, stockpiles, infrastructure construction, surface drainage and pit dewatering
- underground and open pits
  - additional diamond drilling, mapping, geotechnical and geometallurgical data gathering, etc if required.

When the start of these activities has advanced to being 15 to 18 months away, all design and scheduling parameters should have been decided. Mining schedules in at least the first half of the 2YP should allow for:

- grade variations within stopes or benches based on planned blasting sequences
- underground: backfilling and associated curing times required in adjacent stopes before final stope preparation and blasting can commence
- open pits: detailed planning and sequencing of operations on each bench
- planning and other non-operational activities in the critical path, such as surveying and drilling design after stope development or a bench blast is complete.

Each 2YP should be approved as realistic and achievable by management, planning staff and operators responsible for implementing it. All potential problems with the schedule should be identified and solutions found at least 12 to 18 months ahead of time. It is counter-productive to approve a plan with known problems that make it unachievable.

As with the 5YP, the duration of the 2YP could be longer or shorter to suit the situation. The well-settled period at the start of the plan should be of the order of 12 months, but a shorter time may be acceptable in some operations.[8] There are perhaps two considerations when defining this duration.

The first is operational, to ensure that there is adequate time for planning once a new stope or mining level appears in the last quarter of the schedule, so that designs are locked in and schedules well settled for a suitable time period into the future. This is to ensure that the mine can operate with as few unpleasant surprises as possible, implying that detailed designs and schedules need to be generated well ahead in order to identify problems when they are far enough in the future to be fixable without causing undue stress on the schedules and personnel.

Of course, not all possible problems can be foreseen and avoided. Geological and geotechnical knowledge will always be incomplete and shortages of technical staff will continue to ensure that plans developed are not as good as they could be. However, if the operation is continually lurching from crisis to crisis, it is a sure sign that the planning (in the broad sense) is inadequate and should be improved. Some do not realise that continual crisis management does not have to be the norm in mining operations.[9]

---

8. If a good planning process has not existed in the past, early efforts to establish a process will start with short time frames with the aim of extending these over time. We must plan the planning as well as plan the operations.

9. The author suggests, without proof, that, since mining tends to be a cost-focused industry and the benefits of good planning are often unrecognised and difficult to quantify, there will often be a tendency to under-resource the planning functions in many organisations, regardless of whether there are staff shortages or not.

The other consideration is organisational, but has similar rationale. The following subsection discussing the budget suggests that the budget schedule should simply be the activities in the 12-month budget period within one of the 2YPs generated shortly before the start of the budget year. The 2YPs should ideally allow sufficient time for designs to be completed for all stopes appearing in the budget year in the 2YP from which the budget is extracted. This may effectively extend the well-settled schedule out to 15 or 18 months. This may not be practical in some situations, but, while recognising that budgets are frequently obsolete before the start of the budget year anyway, the ideal would nevertheless be to have the best possible information regarding mining activities in the budget. On this basis, the minimum duration of a 2YP would be 18 months, but targeting a minimum two-year time frame is often desirable.

The annual budget

If the SOA/LOMP/5YP/2YP process is in place and plans are formally approved sequentially from long term to short term, then plans for all time frames are continually evolving, but are also realistic, achievable and continuously approved. The budget plan then very simply becomes the plan for the budget year in a version of the 2YP created three to six months before the start of the budget year. This should generate the budget schedule with no additional work, since the rolling process of plan and schedule generation and associated approvals ensures that it is realistic, achievable and aligned with the corporate goal. There is no undue focus on the budget process for weeks or months to distract planners and operators from their primary tasks of planning and operating effectively and efficiently.

Physical quantities in the realistic and corporate-goal-oriented budget plan should drive the budget costs via appropriate cost models. In the same way as physical plans and schedules have increasing detail and accuracy as the time frame reduces, so too should cost driver models. The cost models developed for the SOA will typically provide the broad framework for what is relevant in the budget at its level of detail. The discussion in Chapters 2 and 6 has given some indications of how costs behave, and more detailed discussions in Chapter 8 expand on a number of issues.

Given that there is often pressure at budget time to reduce costs, it is highly desirable that models of both physical and cost behaviours should enable identifying how reductions can be achieved. Ideally, budget costs will be derived from production physicals by a two-stage process. As noted in Chapter 6, focusing on and rewarding arbitrary cost reductions without considering the impact on revenue can at times result in revenue reductions larger than costs saved. Cost savings therefore can result in a reduction in value or profit, rather than the intended improvement.

Physical quantities of inputs should be derived from activities by consumption rates. These process inputs will include such items as labour hours, drilling consumables, explosives, fuel, power, crushing and grinding consumables, reagents and so on. Costs should be derived from these process inputs and the unit costs of each input. It is only by using such a form of zero-based budgeting that requests to reduce budget costs can be reliably dealt with. Cost reductions can only be achieved by:

- *Reducing the unit costs of inputs* – this is usually a purchasing and supply issue beyond the control of operating and planning staff; however, as with so many things, you get what you pay for. Minimising the purchased unit cost of supplies and consumables is counter-productive if low price results in low quality, so that the total cost of providing a particular capability in the operation is increased because of greater consumption rates. In the extreme case, the impact is not just on the cost of providing a service but also leads to reduced availability of production equipment, and hence lost production and lost revenue.

- *Reducing consumption rates* – if these have been explicitly identified, there may be some opportunity to reduce consumption rates, but in each case there will be a fully efficient, underlying consumption rate that cannot be improved upon. Any further attempt at reduction, for example, by artificially or arbitrarily restricting supply, can only result in a reduction in the associated activity.

- *Reducing the physical activities scheduled* – if the plans in the hierarchy from long term to short term have been developed to achieve the corporate goal and been previously approved, any reduction in activities must result in deviating from the plan that delivers the corporate goal. Planning staff should have the capability to identify the effects of reductions in physical activities in the budget year on the long-term strategies, so that decision-makers requesting cost reductions can be made aware of their impact on achieving the corporate goal and make trade-off decisions.

- *Deferring sustaining capital* – if the physical assets of the mine are not maintained in the state required to continue to meet operational targets, diminished capacity in future years (for example, by reduced availability) will result in reduced levels of physical activities.

Without linkages of integrated models in place – both physical and financial – it is impossible for staff to assess the impacts of short-term goals on the long-term goal, and provide information to decision-makers. Without integrated schedules, the impact of reducing near-term activities on later years cannot be assessed. Similarly, without integrating the costs and revenues, the financial impacts cannot be assessed. The intention is to avoid making short-term knee-jerk-response plans that create problems by failing to look to the long-term health of the company.

There will be a need for some formal budget development and approval processes specific to the budget, but if the LOMP/5YP/2YP/budget process is implemented as recommended, the additional work required at budget time can be significantly reduced, and more reliable plans, both physical and financial, will result as a matter of course.

Shorter-term operational plans

Detailed operating plans, consistent with the 2YP and budget, should be developed to ensure that all activities required to implement the approved plans are scheduled, with all necessary inputs available as required.

These plans would typically be for three-monthly periods that are updated monthly, with time frames of weeks and days, or even shifts. The schedules should identify individual items of mining equipment, crews, work locations and other relevant details.

THE IMPORTANCE OF CORPORATE PLANNING POLICIES

To derive all the benefits of a good overall planning process, it is essential that policies and procedures are supported and driven by the company's most senior management, and faithfully implemented at all levels. The LOMP should be specified at a high level to deliver the corporate goal and an SOA will have defined the LOMP that best achieves the goal. The 5YP, 2YP and budget each provide more detailed and better engineered plans at the front end of the LOMP. Each ensures that all planned activities are focused on achieving the goal of the next longer-term plan, and thus ultimately the LOMP.

Without that top-level commitment from the longer to the shorter term, everything else is a waste of time. Many of the planning activities will have to take place so that the operations can continue with a degree of future planning. However, without the framework of the formalised process and a LOMP focused on the corporate goal to guide short-term plans, it is unlikely that the plans will actually achieve the stated goal. Rather than top-down, the annual planning process will more likely become a bottom-up process. Each department within an operation or each operation in a corporate group will typically build its own plans without any clear guidance of what the company requires of it other than generic 'reduce costs and increase output' directives.

The corporate long-term plan will then be a rolled-up summation of all the individual operations' LOMPs. When these are aggregated, the capital requirements will be deemed too high and arbitrary cuts or deferrals will be decreed. Unfortunately, there is typically no time available for plans to be reworked to account for the reduced capital spending, so the benefits of capital expenditure are built into the forecasts without the capital having been spent to achieve them. Such plans are likely to fail, and as the old adage says, failing to plan is planning to fail.

Similarly, without a good model linking physical mining activities, required physical inputs and resulting operating costs, any operating cost reductions can only be arbitrary. If budgeted unit costs are based on past history, there is no way to see how realistic any reduction might be. If unrealistic arbitrary reductions are made to satisfy demands from upper management, there are two extremes of after-effects during the budget year:

1. if the primary focus is on production activities, costs will be over budget
2. if the primary focus is on costs, physical activities will be below budget, with the dangers of
   - subverting the long-term plans (if there is a real plan to be subverted)
   - reducing the revenue generated by more than the costs saved.

Is it any wonder that many corporate long-term plans are unrealistic and not respected or referred to by the people who are supposed to be implementing them at the operations?[10] There must be a top-level commitment to a planning process that cascades from the longer to the shorter term, with all plans focused on achieving the corporate goal.

INCLUDING CUT-OFF DETERMINATION AND STRATEGY OPTIMISATION IN THE PLANNING PROCESS

Strategic options analysis

A strategy optimisation evaluation (SOA) is not normally done annually as part of the planning cycle at an operating mine. Major updates and reworking would occur every three to five years, triggered by anything that makes the previous plan obsolete such as:

- significant changes in metal price forecasts
- changes in corporate aspirations or strategic goals

---

10. The author suggests that 'challenging but achievable' targets have a less than 50 per cent chance of being met.

- discovering substantial new resources
- identifying new options for processes, markets, mining strategies, etc.

By definition, a full cut-off and strategy optimisation is both the key component and purpose of conducting the SOA.

From life-of-mine plan to budget

Once the strategy optimisation is completed, the optimum mine plan will be formalised as the LOMP. The LOMP will identify not only the immediate cut-offs to be applied but also important inputs for ongoing cut-off derivation such as pit limits and pushback staging in open pits, mining methods underground, mining and treatment rates, products and the like. These will become the overall strategic decisions implemented between successive strategy optimisations, which may be several years apart. Nevertheless, cut-off specification or optimisation is an ongoing process in the more detailed shorter-term planning processes.

The cut-offs in the LOMP will be planning cut-offs, as defined in Chapter 2. As noted there, normal variations in mining would be expected to result in times when the specified planning cut-off is temporarily inappropriate and a short-term operational cut-off should be applied. Referring back to the freeway analogy in Chapter 2, the LOMP cut-off policy represents the freeway to get from where we are to where we want to be, whereas operational cut-offs in the shorter-term plans may be thought of as changing lanes on the freeway from time to time to pass slow-moving traffic and negotiate blockages along the way. So how should cut-offs be derived at an operating mine within the context of an existing LOMP?

A simple break-even cut-off will rarely be optimal. A marginal break-even cut-off may be appropriate as an operational cut-off to account for short-term fluctuations in availability of ore to feed the treatment plant.[11] If no other material is available, any mineralisation whose value pays for its downstream treatment and product-related costs may be added to the ore stream and will add value. Short- and medium-term plans should be focused on returning the operation to the point where it can operate at the planning cut-off identified by the LOMP. The following subsections identify processes to use at different stages for both open pit and underground mines.

Open pit shorter-term plans

For an open pit operation, the LOMP sets pit limits, pit staging and overall mining sequences, as well as mining and treatment capacities. The 5YP, 2YP, budget and shorter-term operational plans will refine the detail of these plans. It is likely that, with each increasing level of detail and shorter period of time considered, there will be variations from overall LOMP plans. The top-down process of planning time frames and approvals will set the overall direction to be implemented by the short-term plans. In particular, it is implicit for short-term plans that capital will not be spent to increase mining, treatment or marketing (that is, rock, ore and product-handling) capacities. Mining capacities could, however, be reduced by mothballing some of the mining fleet. This does not preclude the possibility that capital expenditure could be justified on the basis of benefits obtained

---

11. In Lane parlance, this is the mining-limited break-even cut-off. There are, however, situations where this is the optimum long-term planning cut-off. The marginal break-even should not become the planning cut-off through a failure to identify what the optimum longer-term cut-off should be.

within the shorter-term plan being considered, without notionally considering the longer-term plans. In that case, however, the effect of the capacity change on the overall SOA, and hence LOMP, may be significant enough to affect the short-term plan under consideration. The *prima facie* case is that short-term cut-off evaluations should assume the existing capacities as upper limits and not allow for possible expansion capital expenditure.

There is no reason not to conduct simple Lane-style cut-off optimisation evaluations for all short-term open pit plans once the pit limits, mining sequences and capacities have been defined by the long-term plans. In practice, particularly with the budget and shorter-term plans, the process may be simplified as more parameters become fixed by longer-term plans, but the full simple Lane methodology should remain the underlying reference point for any such simplifications. Any simplifications should not be allowed to become arbitrary rules in their own right – they may become inappropriate as conditions change in the future – but should always have firm theoretical support. Chapter 5 indicates how Lane's simple methodology may be applied in short-term situations:

- current values for each cost component of Lane's limiting cut-offs can be derived from the cost records
- the opportunity cost can be estimated from the latest LOMP
- the latest price forecasts will be available
- capacities for each process stage are known
- metallurgical recovery relationships are known
- tonnage and grade versus cut-off relationships can be formed for the rock to be mined
- the optimum cut-off for the next increment of rock or period of time can be determined.

In principle, the time frame is irrelevant – it may be days, weeks, months or quarters. All schedule durations in plans with time frames shorter than the LOMP can be catered for in a simple Lane-style analysis. Therefore, assuming that the style of mineralisation is amenable,[12] the *prima facie* argument is that a Lane-style analysis should be done for all open pit plans shorter than the LOMP.

The optimum cut-off will often be a balancing cut-off,[13] frequently the mining–treatment (rock–ore) balancing cut-off. If the mining operations are working through a low-grade portion of the deposit, the operation may become mining-limited. When mining rock at the maximum rate, it is not possible to keep the treatment plant operating at full capacity. In this case, the cut-off can be lowered to Lane's mining-limited break-even cut-off, but no lower.

If the mining operations are working through a high-grade portion of the deposit, the balancing cut-off will increase and higher-grade ore can be treated. Yet at many operations, rather than increasing the cut-off to account for the short-term variability in the orebody, the cut-off is kept constant and the number of trucks operating is reduced,

---

12. Refer to Chapter 5 for a discussion of situations where a Lane-style analysis may not be appropriate.

13. If one process is limiting for an extended period of time, it may have been over-resourced, or another under-resourced; however, this may legitimately be the case when the mine is highly constrained during early mining stages or when running down towards closure, or if the cost to increase the capacity in one process is not justified by the additional value generated from the new optimum mine plan.

thereby reducing the mining capacity to maintain a rock–ore balance at the specified cut-off. This may have the apparent advantage of reducing costs. Planners should consider carefully, however, whether the cost savings from reducing the mining rate are greater than the revenue gain from treating higher-grade ore using the higher balancing cut-off. Reducing costs is not the best tactic if it reduces revenue by more than the cost saving.

In the rare situations where Lane's methodology is inapplicable, a Mortimer-style cut-off specification is preferable to a break-even analysis. A Lane-style opportunity cost should also be included in break-even formulas so that, if the ore-handling and treatment processes are operating at capacity, additional ore adds value over the life of the operation. Lane-style balancing cut-offs should be identified as accurately as possible. If prices have increased somewhat, the cut-off should not be depressed in a break-even manner to the extent that ore grades fall too low and product targets cannot be met; one of the balancing cut-offs will be more appropriate.

Underground shorter-term plans

As noted in Chapter 5, applying Lane's simple methodology to underground operations can be difficult. The methodology assumes that the mining sequence is specified, and cut-offs are derived for that sequence using the formulas for limiting cut-offs and the processes for identifying balancing cut-offs. But in an underground mine, cut-offs define the size and shape of the orebodies, and in many cases, the mining sequence will depend on the nature of the orebodies defined; a Lane-style cut-off optimisation, which depends on sequence, therefore depends on the cut-offs. The process has become circular and may be impractical for a simple Lane-style analysis; however, as indicated, some cut-off-dependent inputs to the analysis may be accounted for by the iterative process. Chapter 5 also indicates that there are some styles of mineralisation and mining methods amenable to a Lane-style cut-off optimisation. In particular, this occurs where there is a single production front moving in one direction, with no returning to extract additional ore from parts of the resource that the production front has already moved through, such as with sublevel caving.

Where the mining method allows it, a Lane-style analysis should be applied to optimise cut-offs for shorter-term plans where the overall development rates and ore handling and treatment capacities have been established in the LOMP and long-term plans. The rationale described for open pits is directly applicable for underground operations. There are, however, some additional considerations. As noted in Chapter 5, *mining* in a Lane sense relates only to major access development, while most underground operations deal with ore and are therefore *treatment* activities. Particular care must be taken to ensure that costs and capacities are attributed to the correct Lane-style process stages.

In the more common underground situations where Lane's simple methodology is not easily applied, a Mortimer-style cut-off specification will be preferable to a simple break-even analysis. A Lane-style opportunity cost should also be considered in break-even formulas so that, if the ore handling and treatment processes are operating at capacity, additional ore adds value over the life of the operation. As with open pits, Lane-style balancing cut-offs should be identified. If prices have increased significantly, the cut-off should not be depressed in a break-even manner to the extent that ore grades fall too low and product targets cannot be met – a balancing cut-off will often be more appropriate.

The discussion in Chapter 3 regarding the time required to change a cut-off in an underground operation, particularly upwards, should also be noted. Short-term responses to fluctuations in prices, costs or other inputs may not be practical.

OTHER ISSUES AFFECTING CUT-OFF SPECIFICATION

Planning and operational cut-offs at operating mines

As indicated in Chapter 2, there is a danger that, having identified that it is easier to fill the mill at a lower operational cut-off, the longer-term planning cut-off is changed to this lower cut-off. It is perfectly valid to use an operational cut-off that is different from the longer-term planning cut-off to account for short-term variability in operating conditions. The shorter-term planning processes must generate realistic short-term plans and also return the operation to the optimum long-term strategy using the planning cut-off identified in the longer-term plans.

Elevated and depressed cut-offs

Often one hears comments that a particular operation is using an elevated (or depressed) cut-off. The implication of the terminology is that the cut-off ought to be a particular number, but for some reason – which is explainable but not considered justifiable – another value is being used. In such cases there is usually a rational explanation for why the cut-off in use has been specified. It will often revolve around the fact that a break-even calculation has been conducted but the mine plan resulting from using that break-even as the cut-off is unsatisfactory for some reason. Typically:

- The mill cannot be filled with ore, so the cut-off is lowered until it is. The cut-off thereby established will be either a Lane-style mining–treatment (rock–ore) balancing cut-off or, at the limits, a Lane-style mining-limited (rock-limited) cut-off.
- The product targets cannot be achieved, so the cut-off is raised until they are. The cut-off thereby established will be a Lane-style balancing cut-off, either mining–marketing (rock–product) or treatment–marketing (ore–product).
- The resulting head grade is too low to make an acceptable profit, so the cut-off is raised until the desired outcome is achieved. The cut-off thereby established will be a Mortimer-style minimum profit cut-off.

The underlying fallacy is that the cut-off grade ought to be a break-even grade. Because the outcomes from such an assumption are unacceptable, there is an implicit recognition that the break-even model for cut-off derivation is inadequate. The response is to invoke other cut-off models discussed in this book, but without recognising that they exist or understanding their rationale. The resulting elevated or depressed cut-offs are perceived to be aberrations. Additionally, because the models being implicitly applied are not understood, there is the danger that they are misapplied and the outcomes are suboptimal. Iterative trial-and-error evaluations will often be conducted until an acceptable outcome is obtained. Correct application of the appropriate cut-off derivation model could have achieved a better outcome with more rigour and less work.

In one sense, if the cut-off is the number that distinguishes between ore and waste, as described in Chapter 2, there is no such thing as an elevated or a depressed cut-off. The cut-off being used is the cut-off. It is neither elevated nor depressed: it is what it is.

If, in light of Chapters 3 to 6, we suggest that the cut-off ought to be the value that best delivers the corporate goal, it is proper to use adjectives such as elevated or depressed if a different cut-off is used; however, taking this view will often result in the opposite usage of what is encountered. Assuming that the goal of most companies is value maximisation rather than every-tonne-pays-for-itself, cut-offs in general ought not to be break-evens. If the cut-off being used has been changed from a break-even starting point to achieve a mine plan that does deliver the corporate goal, a so-called elevated cut-off is actually the right cut-off, and the original break-even is a depressed cut-off.

Using a so-called elevated cut-off is often referred to pejoratively as high-grading, with the sense that it is somehow improper. If this is a result of achieving the corporate goal, it should be seen as right-grading; traditional break-even cut-offs could then be described as low-grading. To be consistent, this should be just as pejorative a term as high-grading.

In some jurisdictions, cut-off determination is mandated by law. In such places, the cut-off ought to be what is decreed, and any variations will be elevated or depressed, leading to high- or low-grading. Whether the latter terms are pejorative will depend on whether the deviations are legally permitted and the extent to which they may frustrate the underlying intent of the law.

Cut-offs and design rules

For the optimisation studies, there must be consistency in the processes used for all options. Inconsistencies in generating mining reserves data between, for example, cases with different cut-offs, can lead to incorrect option ranking. Design rules in the optimisation process must then be carried forward into the ongoing planning and design to ensure that the operational plan using the latest and most detailed information honours the selected strategic plan identified using older, less detailed information.

Figure 7.1 illustrates the problem that may arise applying cut-offs to mineralisation with irregular ore boundaries in an underground situation.

The figure shows an extreme hypothetical situation where the grade distribution in two dimensions is circular, with grades highest in the centre and reducing with distance from the centre. Three grade boundaries are shown. It is assumed for practical mining reasons that square stopes must be mined. An optimisation evaluation (an SOA) has determined that the square stope shown is the optimum. We wish to specify the boundary cut-off, so that future planners will design the optimum stopes after the block model has been updated with better geological information. What is the design cut-off to be specified?

FIGURE 7.1

Cut-offs and design rules with irregular ore boundaries.

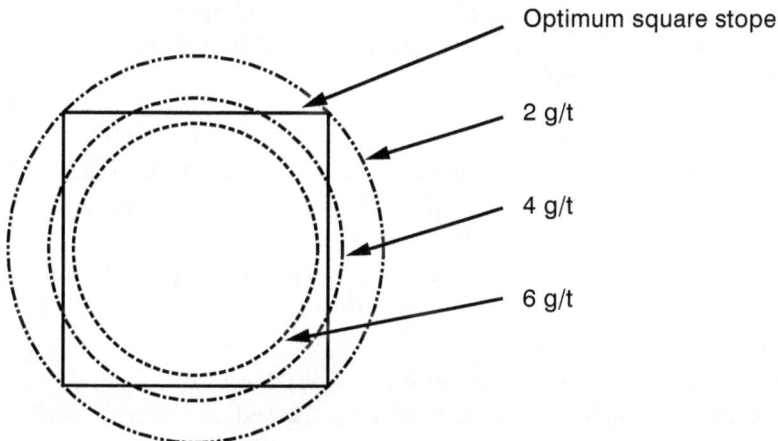

Optimum square stope

2 g/t

4 g/t

6 g/t

It could be any of the following:

- 2 g/t, designing to extract the maximum tonnage of above-cut-off material but with no sub-cut-off material included
- 4 g/t, designing to balance the tonnages of sub-cut-off material included and above-cut-off excluded
- 6 g/t, designing to extract all the above-cut-off material, and including the minimum quantity of sub-cut-off material necessary to achieve that.

If the outcomes of the SOA were viewed as hills of value, there would be different hills for each design rule. Each rule that described the same optimum stope would of necessity have the same maximum value; however, the maximum values would be located at different cut-offs across the range from 2 g/t to 6 g/t.

What if the design rule used to generate the reserves for different cut-offs in the SOA were not passed on to the detailed design engineers?

If the 'no sub-cut-off material included' rule had been used for the SOA, the optimum cut-off would be 2 g/t. If the detailed design engineers lacked guidance with the design rules, they might decide to use a 'no above-cut-off material excluded' rule. If this were applied to the 2 g/t grade boundary – since the cut-off has now been specified numerically as that – a significantly larger and lower-grade stope than the optimum would be generated. Figure 7.2 illustrates this situation.

FIGURE 7.2

Cut-off applied using an inappropriate design rule 1.

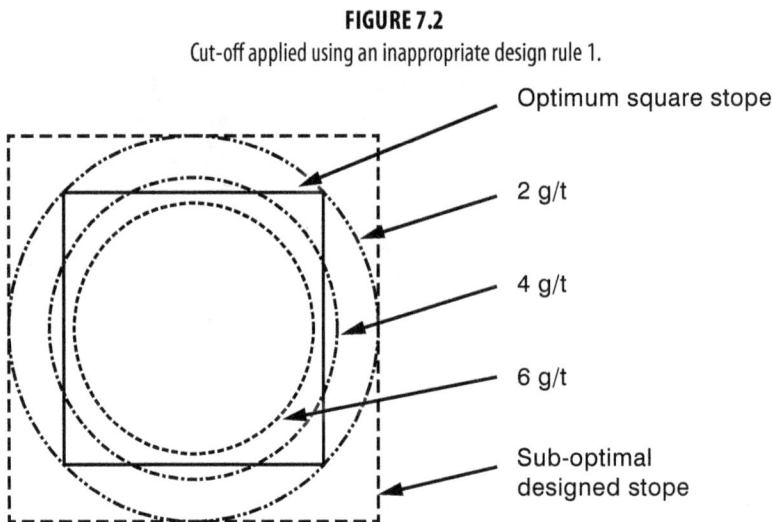

Optimum square stope

2 g/t

4 g/t

6 g/t

Sub-optimal designed stope

Similarly, if the 'no above-cut-off material excluded' rule had been used for the SOA, the optimum cut-off would be 6 g/t. If the detailed design engineers had no guidance regarding the design rules, they might use a 'no sub-cut-off material included' rule. If this were applied to the 6 g/t grade boundary, as the cut-off has now been specified as that, a significantly smaller and higher-grade stope than optimum would be generated. Figure 7.3 illustrates this situation.

Clearly, design rules used in the optimisation evaluation must carry over into the later ongoing mine planning and design processes so that the operational mine plan honours the strategy selected as a result of the SOA and incorporated in the LOMP. How critical this

Cut-off applied using an inappropriate design rule 2.

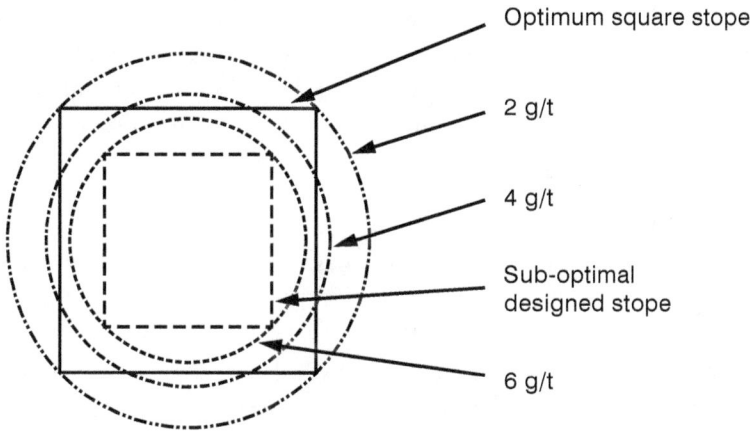

might be will depend on how irregular grade boundaries within the mineralisation are and how easy or difficult it is to define stope boundaries that include all above-cut-off material and exclude all sub-cut-off material. As soon as there is the practical necessity for the stope design to exclude ore and include waste, the problem potentially exists.

The increasing availability of software that automatically defines stope boundaries is making it easier to record design rules for underground operations. Rather than being a verbal description that may be easily lost or forgotten, numerical values can now be specified for each of the input settings in the software. These can be recorded and reused at all stages of the planning process. If the mine is already operating and various settings are in common use, these should be adopted in any SOA undertaken, unless there is a good reason for doing otherwise. If the mine is not already operating, the settings used in the SOA should flow on to the detailed design engineers. These may be found to be unsuitable for the style of mineralisation encountered in practice. It will then be necessary, initially, to compare the tonnages and grades generated by the software for a range of cut-offs for both the SOA and detailed planning software settings. This is so that any necessary adjustments can be made to the cut-off used in the mine to ensure that mine plans generated using the new design rules, as reflected in the software settings, continue to honour the strategy in the LOMP. Eventually, it will become necessary for the SOA to be redone using the new stope design settings.

There should be consistency across all processes that generate information used as input data for options evaluated in the SOA. Inconsistent design processes could produce erroneous results. It should go without saying that if there are a number of design engineers generating information, perhaps at different cut-offs, all of them should be using the same design rules or software settings for the same orebodies to ensure consistency of designs.

To illustrate the importance of this, consider the opposite – that each engineer chooses their own design rules. Suppose the engineers working with 2 g/t and 6 g/t cut-offs chose to use the rules that generated the same optimum stope illustrated in Figure 7.1. The same tonnage and grade would be generated for both of these widely separated cut-offs. Suppose the engineer doing the designs for 4 g/t cut-off also used one or other of these two rules. We would now have a data set with tonnages and grades at 2 g/t and 6 g/t cut-offs being the same, while those for a 4 g/t cut-off would be either higher or lower.

At best, the inconsistency will be identified when tonnage–grade relationships are initially produced by combining the data from each engineer. A potentially costly and time-consuming reworking of at least some of the input data sets would become necessary. At worst, the inconsistencies between the data sets would remain undiscovered, leading to unreliable outcomes of any study based on them.

Different cut-offs in different mine areas

Different cut-offs can be applied to different areas within a mine to increase value (Hall and Stewart, 2004; Horsley, 2005). Figure 7.4 shows schematically how different cut-offs in different areas could be optimal in an underground mine.

In the upper part of the figure, two areas with similar mineralisation, mining methods and costs structures, and therefore the same break-even grade, used as cut-off, are mined. Different lives for the two areas result in an unprofitable low-production-rate tail that will be truncated by mine closure if fixed costs cannot be sufficiently reduced to allow it to

FIGURE 7.4

Value-adding by different cut-offs in similar areas.

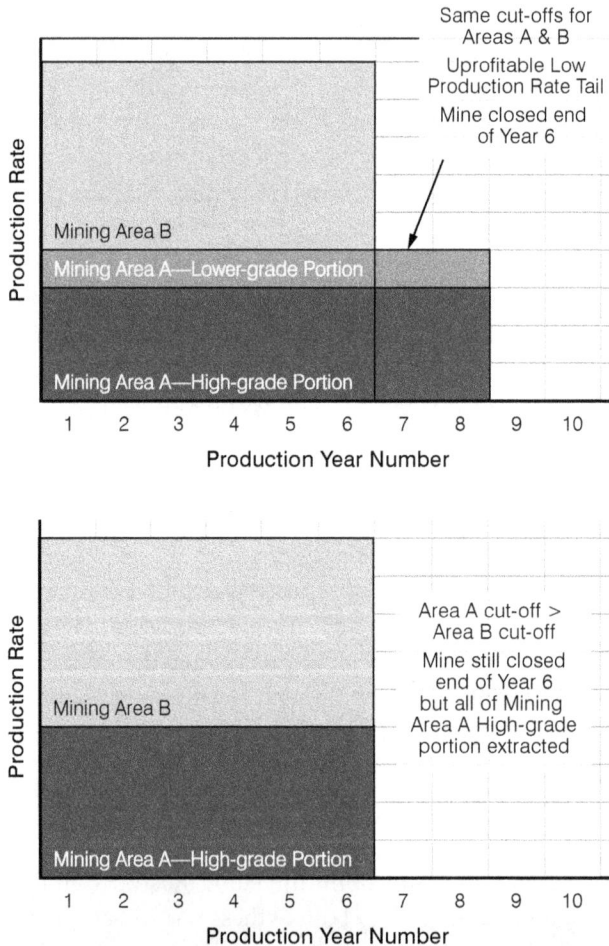

be mined profitably. During the life of the mine, production from the longer-life area will contain material that is above the common cut-off for both areas, but which is of lower grade than some of the material that remains unmined when the mine closes. The lower part of the figure shows how the cut-off in this area can be increased so that, at the time the mine does close, the best possible grades have been mined from both areas.

In this simple example, the cut-off policies to maximise value would be easily determined. In a real operation, predecessor/successor dependencies between mining areas and other scheduling and sequencing constraints and options may make selecting the optimal cut-off policy for all areas over time non-intuitive and non-trivial. Sophisticated optimisation techniques, some of which are described briefly in Chapter 13, may be required to select the best strategy to bring into the LOMP.

Applying individual stope break-even cut-offs

Some underground operations, however, take this 'different cut-offs in different areas' principle to the extreme, effectively applying a cut-off to each stope. The process makes use of a stope value calculation that takes account of the stope-specific, rather than orebody-average, physical quantities and costs to determine whether an individual stope should be included in the reserve. There is nothing wrong with this in principle for short-term planning within the context of approved long-term plans, but the way in which it is implemented at many operations is not aligned with the longer-term plans and corporate goal. The process is applying a break-even volume cut-off to each stope. If the long-term plan makes use of a break-even cut-off – that is, the corporate goal is to ensure that every tonne of ore pays for itself – the break-even stope valuation process will be appropriate. However, if the long-term cut-off is an optimum rather than simple break-even,[14] the individual stope valuation process has the potential to subvert the long-term plan by using short-term tactics not aligned with the long-term strategy.

One possible way to align such a stope valuation process with the long-term strategy is to identify the difference between the optimum cut-off and the overall break-even cut-off and include this difference as an additional cost to be borne by each stope. Alternatively, the Lane-style opportunity cost could be included in the costs to be accounted for in the break-even style of analysis. These options may be appropriate if the optimum cut-off has been identified as some form of break-even, such as a Lane-style limiting cut-off. If the optimum cut-off is, for example, a Lane-style balancing cut-off, break-even calculations applied to individual stopes or blocks of mineralisation may be completely inappropriate. The only justification for such a process would be to exclude any block that did not generate sufficient revenue to cover its direct variable costs.

The preceding discussion is in the context of, for example, open stopes that are designed by planning and design engineers some months or years before the stope is developed, drilled and extracted. Similar rationale applies to all other underground mining methods. When ore and waste are identified immediately prior to ore extraction, greater care may be required to ensure that short-term considerations do not prevent achieving the long-term strategy. Examples include caving methods, where the grade of the broken material in the drawpoints may be reassessed on a regular basis, and cut-and-fill, where geologists will often mark up the ore–waste boundary in operating

---

14. It is, of course, possible that a simple break-even is the optimum cut-off; however, this will rarely be the case, so in this statement, optimum and break-even should be seen as different.

stopes to direct the mining operations. In both situations, there is often the temptation to classify as ore anything that is above its own marginal break-even grade – the rationale being that it is profitable and, if not extracted now, will be lost forever. However, if the treatment stream (as defined by Lane – the ore handling and treatment processes) are operating at capacity, any additional material classed as ore must cover the costs included in Lane's treatment-limited break-even cut-off – ore and product-related variable costs, fixed costs and Lane's opportunity cost.

Assuming that these break-even processes are appropriate for short-term planning and the costs have been correctly identified, care must be taken to ensure that costs are allocated to the correct blocks of mineralisation. The author has, for example, seen cases where pillar stopes with uncemented (therefore lower cost) fill are allocated lower marginal break-even cut-offs than primary stopes with cemented fill and higher costs. The reality is that the cemented fill has been placed in the primary stopes so that the pillar stopes, or at least some proportion of them, can be extracted. By this rationale, the cement costs should be allocated to the pillar stopes, not the primaries, and pillar stopes should have the higher cut-off, all other things being equal.

The fact that both positions could be argued with some degree of logic indicates that both are spurious. The real situation is that having specified that the mining method is to be by a combination of primary and pillar stopes, there is a commitment to add an overall cement content to all fill placed across primaries and pillars. The average cost of fill for all stope types is a cost of the mining method used for the whole block. The cost of fill in individual stopes is irrelevant. The average fill cost should therefore be attributed to all stopes, not differential costs based on what type of stope each happens to be. Having made the decision that a certain amount of cement, with its associated costs, is to be included in the total fill placed in the mining block, it is irrational to then say that higher grades should be left unmined in some areas and lower grades extracted as ore in others simply because of the distribution of cement in the whole mining block. If two stopes, a primary and a pillar, were both being extracted from the same mining block, the lowest grade mined from each stope should be the same. The ore–waste decision was made when the mining method was decided upon and the mining block limits defined, before stopes were designed and allocated primary or pillar designations. The filling decision regarding the total amount of cement has already been made.[15]

However, nothing in this discussion should be taken to preclude the possibility of extracting additional low-grade stopes that will be unfilled, or filled with uncemented fill, on the periphery of the ore boundaries defined by higher-level evaluations. This will only be appropriate if all extra costs incurred, including the opportunity cost of displacing higher-grade material in the ore stream, are covered.

Similar care should be taken with access and stope development. These are designed to make the overall mining method work, and to the extent that various mine openings provide access or other benefits for more than one stope, they should be seen as method

---

15. Similar logic applies in an open pit. Classifying material as ore and waste is made at the pit rim. That some material may have come from near the surface and some from the bottom of the pit is irrelevant. The cut-off does not depend on the depth of mining. Shallower material does not have a lower cut-off than deeper material because of different haulage costs. The haulage cost at the point where the ore–waste decision is made – the pit rim – is a sunk cost and does not come into future decisions. In Lane parlance, it is a mining variable cost, and we know from Chapter 5 that these do not come into Lane's limiting cut-offs. This assumes that the mining schedule has already been determined. Depth-related costs will, however, affect the optimum mining schedule.

or block costs, distributed among all stopes, not just to the one within whose boundaries they happen to lie.

Grade control and sampling

All the cut-off models we have considered assume the cut-off grade specified is a true grade. The grade available for decision-making will often be a sample grade.

It is well known that samples are frequently biased. Therefore, it is necessary to know the relationships between the sample grade and the true grade of the rock in question so as to make an informed cut-off decision within the operating mine.

Grade control and sampling are complex specialist topics in their own right and this book does not purport to deal with them in any way, other than to highlight their potential importance when specifying cut-off policies.

CHAPTER SUMMARY

This chapter has discussed recommended mine planning processes and indicated how the various cut-off and optimisation processes described in previous chapters might be included. Mine plans and strategies must have a clearly defined goal. If plans are not intentionally derived to deliver a specified corporate goal, the de facto goal of the corporation will become the goal that is implicit in deriving the plan.

In a long-life operation, where there are major lag times between exploration, planning, development or waste stripping and production, plans must be developed in a hierarchy from long- to short-term. Short-term plans should be consistent with but provide more detail than the long-term plans from which they are derived. The long-term or life-of-mine plan (LOMP) is set to achieve the corporate goal and ultimately generates value. Once the optimum LOMP is approved, all short-term plans must be developed to deliver that LOMP.

Preproduction studies for a project go through several stages with increasing levels of detail, and potentially different ways of determining cut-offs and other strategic options.

A scoping or conceptual study is typically at an order-of-magnitude accuracy (±30 per cent). Its major purpose is to demonstrate early in the resource discovery that further work is justified. It is therefore only necessary to find one case that proves this. A cut-off must be selected to derive ore tonnages and grades, but there may be insufficient information to rigorously derive one by any of the methods discussed in previous chapters. Some form of full-cost break-even is perhaps the most appropriate method. If there is a reasonable amount of tonnage and grade data, a Mortimer-style analysis might be the preferred option, particularly if the deposit is low-grade. Strategy optimisation is generally unsuitable at this level of study but, if the cases evaluated are not viable and a better case must be found, or if the study's purpose is to determine what the optimum strategy might be, a high-level optimisation study might need to be conducted.

A prefeasibility study (PFS) should demonstrate to a higher level of confidence (±20–25 per cent) that the project is technically and economically viable. It should evaluate a range of options to ensure that the best is selected for further study. Usually the case selected at this stage will be the one developed if it is proven to be feasible. A full strategy optimisation study should therefore be seen as an integral part of the prefeasibility study.

The feasibility study (FS) is intended to demonstrate technical and economic viability of the project at a level of accuracy sufficient to justify proceeding with the project (typically ±10–15 per cent). The policies identified as best in the PFS, such as production rates and mining and treatment methods, will normally carry forward into the FS, and will only change if more investigations expose previously unidentified problems. However, cut-offs will be recalculated if price and cost projections have changed since the PFS was done. Depending on how much the resulting orebodies change, it may be necessary to completely rework other major strategic decisions. The first part of the FS may need to be an update of the PFS optimisation using the latest data and forecasts to assess whether the major decisions going forward into the FS should be changed.

Detailed engineering is required to convert the principles in the FS into construction and mining plans for practical implementation. With improving knowledge of the mineralisation and updated price and cost forecasts, it may be necessary to revise cut-offs at this stage. With many parameters such as mining and treatment capacities now locked in, simpler cut-off models may be better. If the nature of the deposits and mining methods permit, a Lane-style cut-off optimisation may suit. If not, a Mortimer-style cut-off specification is preferable to a simple break-even analysis. Opportunity cost should also be considered in break-even formulas. It would also be desirable to consider the capacities of the now-defined production stages. Lane-style balancing cut-offs should therefore be identified. If prices have increased significantly, the cut-off should not be depressed in a break-even manner to the extent that ore grades fall too low and product targets cannot be met – a balancing cut-off may be more appropriate.

At an operating mine an integrated planning process is recommended. The process works in a hierarchy from long term to short term. Longer-term plans set the overall strategic direction. Progressively shorter-term plans provide more detail at the front end of long-term plans. Approvals processes also work from long term to short term. Long-term plans must focus on delivering the corporate goal. Short-term plans must then detail how operations in the near future will help achieve long-term plans.

The strategy options analysis (SOA) evaluates, at a high level, the impacts on value of all strategic decisions that the company can make, both separately and together. A full SOA is not normally done annually but should be reviewed annually as part of the planning cycle to ensure that the approved LOMP continues to be the best long-term plan. Major updates and reworking the SOA should occur approximately every three to five years.

The LOMP is the formal long-term plan for the mine or business. It is selected after conducting an SOA, and is the plan identified as best delivering the corporate goal. It establishes the framework within which all other short-term plans are developed. The LOMP should be reviewed and updated annually as part of the planning cycle, taking account of constraints identified in shorter-term plans or resulting from actual events, and changes in the SOA.

The five-year plan (5YP) forms a critical medium-term link between the high-level strategies in the LOMP and the detailed short-term implementation plans. The 5YP would usually be generated annually, with quarterly time periods reported. It provides sufficient look-ahead time for long lead items to be identified and planned for. Formal approval of the 5YP should be part of the annual planning cycle.

The two-year rolling plan (2YP) provides a higher level of detail and is supported by detailed engineering work at the front end of the 5YP. It would be formally updated quarterly and report activities on a monthly basis. It is a regular ongoing part of the short-term planning process and is not just done once a year as part of the budgeting cycle. Formalising these regular updates requires both operators and planners to regularly look at the effects of current issues up to two years ahead, avoiding actions that may be expedient in the short term but create major problems long term. The appearance of an activity at the end of the 2YP would be the flag for detailed design work to commence. All activities occurring in the first 12–18 months of each 2YP should have been planned in detail, with solutions found for any problems identified.

The annual budget plan is simply the plan for the 12 months of the financial year to be budgeted in the version of 2YP that is created three to six months before the start of the budget year. Its preparation should not require any special attention, since the rolling process of plan and schedule generation and associated approvals ensures that it is realistic, achievable and aligned with the corporate goal. Physical quantities in the realistic and corporate-goal-oriented budget plan should drive the budget costs via cost models, whose detail will have evolved in the same way as the physical plans.

Short-term detailed operational plans, consistent with the 2YP and, when appropriate, the budget, should be developed to ensure that all activities required to implement the plans are scheduled, with all necessary inputs available as required. These would be for three-monthly periods, updated monthly, with time frames of weeks, days or shifts.

Cut-off determination and strategy optimisation would be incorporated in the planning process as follows. The SOA will identify the optimum mine plan. This would be formalised as the LOMP, which will identify not only the immediate cut-offs but important inputs for ongoing cut-off derivation, such as pit limits and pushback staging in open pits, ore boundaries and mining methods underground, mining and treatment rates, products and the like. These will be the strategic decisions to be implemented between successive strategy optimisations, which may be several years apart.

For an open pit, once mining sequences and capacities of various stages of the production process have been set by the LOMP, most of the conditions required for a successful Lane-style cut-off optimisation have been satisfied. For an open pit, a Lane-style analysis should be used for short-term cut-off optimisation for all time frames – from the 5YP to short-term operational plans. If for some reason Lane's methodology is not easily applied, a Mortimer-style cut-off specification is preferable to a simple break-even analysis, but a Lane-style opportunity cost should also be considered in break-even formulas. Lane-style balancing cut-offs should be identified as a balancing cut-off will often be more appropriate than a break-even.

Underground, cut-offs define the size and shape of the orebodies, and in many cases, the mining sequence will depend on the nature of the orebodies defined. A Lane-style cut-off optimisation, which in its simple form depends on sequence, depends on the cut-offs. It may therefore be impractical, though, in some circumstances, the iterative nature of a Lane-style optimisation will account for this and other cut-off-dependent inputs. Lane-style cut-off optimisation is most easily applied underground when there is a single production front moving in one direction, with no returning to extract additional ore from parts of the resource that the production front has already moved through. In such cases, a Lane-style analysis should be applied to optimise cut-offs for short-term plans where the overall development rates, ore handling and treatment capacities have been established in the LOMP and long-term plans. Care must be taken to ensure that costs and capacities are correctly attributed to the Lane-style production stages.

In the more common situations where Lane's simple methodology is not easily applicable underground, a Mortimer-style cut-off will be preferred to a simple break-even analysis for short-term plans. A Lane-style opportunity cost should also be included in break-even formulas. Lane-style balancing cut-offs should be identified as a balancing cut-off will often be more appropriate than a break-even. The time required to change a cut-off in an underground operation, particularly upwards, should also be considered. Short-term responses to fluctuations in prices, costs or other inputs may not be practical.

For both open pit and underground operations, a simple marginal break-even cut-off will only be suitable as an operational cut-off to account for very short-term fluctuations in availability of ore to feed the treatment plant.

PART 2

IMPORTANT CONSIDERATIONS FOR STRATEGY OPTIMISATION

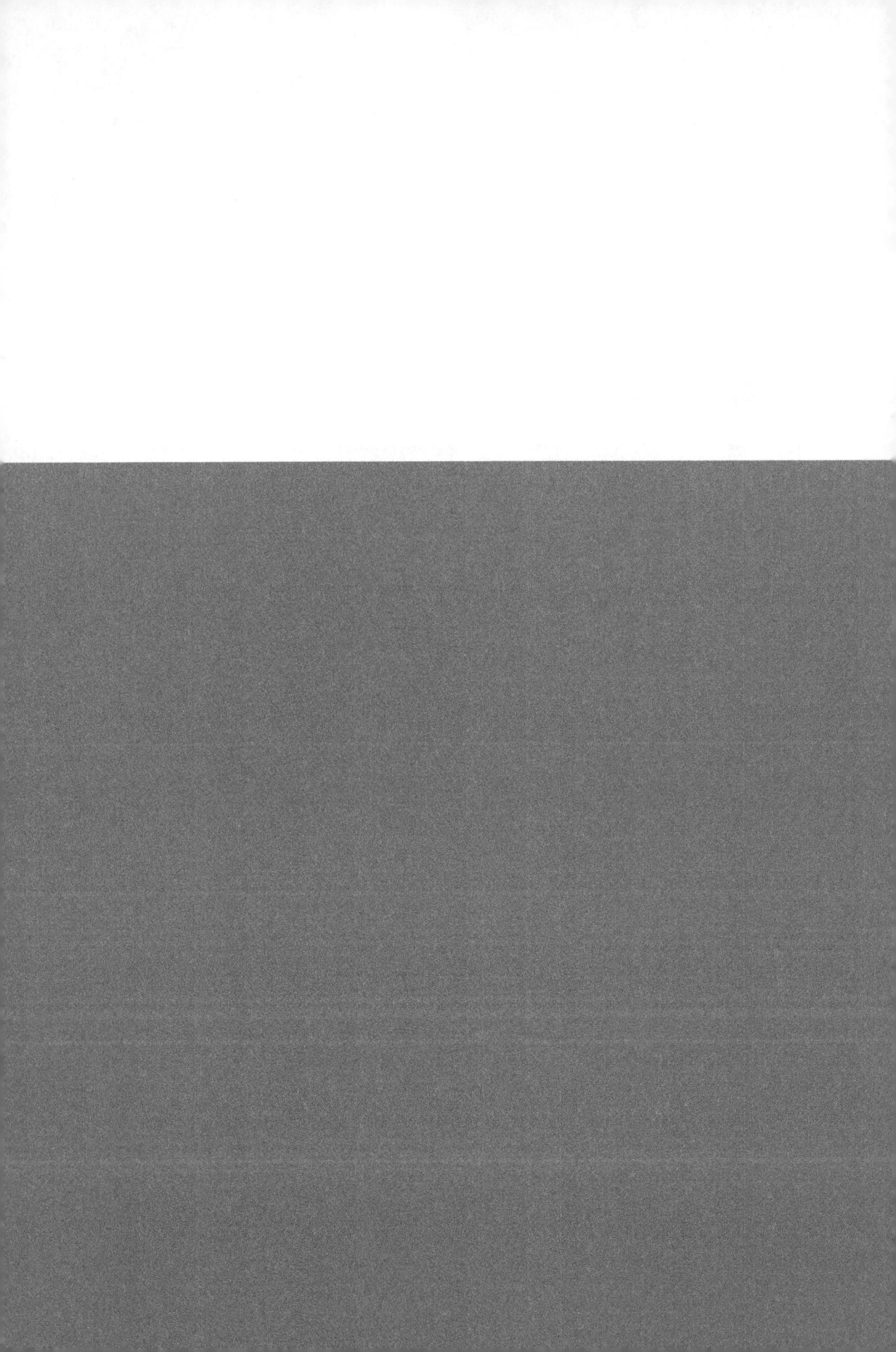

CHAPTER 8

The Mine Strategy Optimisation Process

INTRODUCTION

So far, we've discussed ways of determining cut-offs (break-evens, Mortimer's Definition and Lane's (simple) methodology). We've looked at the underlying theory and processes used to derive a number as the cut-off for a predefined mining plan. The concepts and rationale underlying a strategy optimisation evaluation, of which the optimum cut-off policy will be a part of the outcomes, has been covered, as well as the general data requirements and procedures of performing a strategy optimisation. This chapter, the first in Part 2, describes these requirements and procedures in more detail.

As noted, there are several methodologies that can be applied to determine the optimum strategy, and the process details will depend to some extent on the methodology employed. This chapter discusses common features of these. It is not possible to fully optimise the mining strategy without considering the metallurgical treatment processes and vice versa. But often, one or the other is attempted by teams within the operating departments, assuming that the other process remains unchanged. A full strategy optimisation requires an integrated evaluation of all parts of the production process across the whole life of the operation. That this is done is an implicit assumption in the following discussion.

IDENTIFYING THE OPTIONS AND SCENARIOS

The corporate goal must be borne in mind at all stages of the optimisation process. What do we need to do to derive value measures associated with that goal, and what processes and associated data do we require to achieve this? As well as the goal, it is important to identify the options that will be evaluated. Data and relationships generated must support the full range of options, some of which will represent different values for what may be thought of as a continuous parameter, such as the optimum production rate within a specified range. Others may represent discrete ways of achieving certain outcomes, such as the types of trucks/loaders; whether haulage is to be by truck, conveyor, or shaft; and whether stoping is to be by primary/secondary or continuous advance sequencing. If certain options are not to be investigated, the model need not handle these. But if certain options are to be evaluated and compared, the data must enable options to be distinguished, with the model handling all the relationships required to do this.

As a general rule, it is impossible to evaluate something that has not been included in the modelling. It might be feasible to use a surrogate in the model, but there is no guarantee of a reliable outcome.[1] It is therefore important at the start of the evaluation to identify what needs to be taken into account during the studies. Items not deemed essential at first may be easy to incorporate during the early model building, or can be excluded from the initial stages of the project with the model structured to allow for their easy addition later if necessary. Items that are to be added at later that have not been anticipated in the model design may be difficult to retrofit into an existing model. At worst, the model's structure may make such additions impossible, necessitating a complete rebuild. Project sponsors should ensure that the optimisation team is aware from the outset of everything that might be of interest, and should not presuppose what may be difficult to model; experienced teams may have developed processes that simplify tasks that others perceive are difficult.

It is imperative that decision-makers at as high a level as possible in the company have input into the study-scoping phase. It is not uncommon for staff at one level in an organisation to be unaware of higher-level concerns. Similarly, matters that are perceived to be unchangeable and not needing investigation at one level may be seen as crucial matters by senior staff. The involvement of management in these early stages is therefore essential, or the study may be constrained from the outset and unable to deliver what is expected.

UNCERTAIN PARAMETERS AND RELATIONSHIPS

There will be levels of uncertainty associated with all parameters of the strategy evaluation process. In many cases, it will be such that the study outcomes may be considered unreliable. Sometimes there is a propensity to leave such things out of the analysis so that only reasonably certain parameters are included. This is thought to enhance the rigour and the reliability of the analysis.

The author suggests this is a fallacious argument. For a strategy optimisation study, it is vital to take account of the influences of uncertain parameters. Consider that the most uncertain parameter in any mining evaluation will usually be the product price. We would never contemplate a study that ignored prices 'because we don't have good

---

1. As a simple example, if the model has not been constructed to allow for different metallurgical recoveries, one could model the revenue effects by flexing the product price. If the cost change resulting from the different volume of product were insignificant, using price as a surrogate for recovery might be satisfactory. However, if the change in product-related costs resulting from a change in recovery and product volume had a significant impact on total costs, the overall result might not be sufficiently reliable for the decisions needed.

information'. Rather, we'd estimate the most likely scenario and flex the price inputs over a range. As has been noted, a change in the price will shift the value of any operating plan. The purpose of an optimisation study is not so much to determine value *per se*, but to identify the optimum plan and how it might change if uncertain parameters change. So the issue is not how variable the value might be because of price uncertainty; we wish to identify how changing price might cause the optimum plan to vary, and therefore how it might change the decisions made.

The same rationale applies to any important but uncertain parameter. It might be the numerical value to be used, such as a price or cost, or the tonnage and grade of an Inferred Resource or exploration target. Or the uncertain parameter might be a relationship, such as how metallurgical recovery varies with feed grade. It is critical that every important (or thought to be important) but uncertain parameter is recognised as a scenario parameter to be flexed in the analysis over the likely, though as yet uncertain, range to identify what causes the strategic decisions to change. If the decision-changing situation is found to be outside the likely range for an uncertain parameter, we will be able to state that it is not vital to defining the optimum strategy. If the decision-changing point is found to lie near the most likely value or relationship, we know that we must either take steps to reduce the level of uncertainty or consider the trade-offs between the benefits and costs of making decisions based on correct or incorrect assumptions.

This risk–reward trade-off will be discussed more in Chapter 12. The issue of levels of uncertainty is introduced here to raise awareness of its potential impact on the process. Every data input parameter should be reviewed early in the process to identify whether it is certain enough to be a fixed input value or relationship, or whether it should be a scenario parameter that may influence strategic decisions. Disagreements about whether a parameter is important or not can ultimately only be resolved by including its effects in the analysis and investigating it quantitatively. In practice, many input parameters that have what might be termed *set and forget* fixed values will also have flexing factors applied for conducting a *sensitivity analysis*, as described in Chapter 6. Such factors facilitate including those parameters in the scenario analyses, and will often be the simplest way to investigate the impact of other uncertain parameters.

GEOLOGICAL INPUTS

A geological model is the basis of any mine planning activity. For strategy optimisation, the model must be reliable over the range of cut-offs that may be investigated. In practice there may be a trade-off between geological rigour and the time required to develop such a model. It is common to find that an existing model is deemed by the geologists to be reliable for only a portion of the range of cut-offs, leaving two alternatives. One is to defer any further analysis until a model has been developed that is deemed reliable across the full range of cut-offs. This is rarely an acceptable option – the time frame for the study will already have been set and the geological staff will have too many other tasks to attend to anyway. The other course of action – often the one adopted – is to make use of the existing model, recognising that it may have inadequacies[2] and taking account of these.

---

2. In addition, it is not uncommon to find that the existing geological model is out of date and in the process of being updated, but the new model is expected to be delivered just after the results of the optimisation study!

This may not be as bad as it sounds. Rigorous analyses can still be conducted with the existing model. If the optimum strategy is found to lie within the range where the model is reliable, the optimum will be reliable, and additional work to improve the model would be unnecessary. If, however, the optimum strategy lies outside the reliable range, the analysis will indicate the range of cut-offs over which the geological model must be reliable, and whether creating such a model is worthwhile. The analysis using the existing model will indicate the value of the optimum strategy and the value of a strategy using the cut-off at the limit of the model's reliability. The difference in value will indicate the maximum cost that could be incurred for the additional reliability.

The extra cost might simply be for reprocessing the existing data. Alternatively, there could be significant costs for such items as additional diamond drilling at a closer spacing. If the extra cost is less than the extra value, it's worth generating a new model. If the cost is greater than the potential value gain, it is not worth doing – the hill of value (HoV) will rise but, before it reaches a peak, there will be a cliff at the cut-off at the limit of the model's reliability, which will then be the optimum cut-off.

It is necessary to specify the level of reliability of geological information required, and this may be better indicated by establishing what is not required. It is not necessary, for instance, that the model locates every potential ore lens in 3D space so that final designs for access and stope development can be generated. That is not the purpose of a strategy optimisation study.

What is necessary is that the model represents, at a level of accuracy consistent with the study, the tonnages and grades at various cut-offs, as well as the general size, shape and location of orebodies defined at those cut-offs. The required modelling outcomes for optimisation are that the shape characteristics are sufficiently accurate for mining method options and related parameters to be reliably estimated. Underground, this may include aspects such as sublevel interval, types of equipment and development metres. The locations of the orebodies need only be sufficiently well established to generate, for example, access development quantities. For an open pit, the orebody size and shape may be important for identifying options if the changing nature of the orebody with cut-off were to result in, for example, different mining equipment or drill and blast patterns. There is an assumption that the nature of the geology and grade distribution is such that, when final designs are needed in the future, it will have been possible to obtain the additional information to generate detailed designs. This is, of course, part of the normal planning cycle with phases of increasingly detailed geological data gathering and generating more accurate plans, as described in Chapter 7.

The tonnage and grade versus cut-off relationships will change as the amount of information increases. With closer-spaced drilling, grade variability will be better identified and there will be more high-grade tonnage with greater knowledge, even if the total resource tonnage is unchanged. This means that the tonnes and grades that exist above certain cut-offs will be greater than predicted by early-stage geological modelling. If there is adequate history to derive relationships, it may be appropriate to apply factors to tonnages and grades generated at various cut-offs in preliminary studies to better estimate the real situation. Depending on the nature of the mineralisation and the cut-offs applied, the early models could be either underestimating or overestimating the potential value to be obtained from the resource.

With a geological model that is sufficient for the study, it is now necessary to generate orebodies[3] at each cut-off of interest. From these, we need to identify geological domains that allow different mining methods, scheduling constraints, metallurgical treatment, etc. For this discussion, *geology* also includes geomechanical information such as stresses, rock strengths, structures impacting designs and the like, again at a level of accuracy commensurate with the overall study.

Exploration potential is often a critical input. This may have several categories, such as extensions of existing orebodies, new zones within the existing operations and other deposits in the region that could be combined with the ore feed to the existing plant. The nature of other deposits in the region could also lead to constructing another treatment plant or similar. These external deposits could belong to the company conducting the evaluation or a third party, with toll treatment, ore purchase or outright purchase of the properties as options to be evaluated. By its very nature, exploration potential is uncertain data, though there are degrees of uncertainty. Some may be almost to the stage of formal reporting as resources, while others may be as-yet untested targets.

Uncertainty, as has been noted, is a poor reason to exclude exploration from the investigation. Optimum strategies could be found both with and without the exploration potential, and scenarios including it could consider optimistic, pessimistic and mid-range estimates. An outcome of this type of analysis could be whether the optimum strategies with and without the exploration potential were similar or significantly different. If similar, then, although the exploration potential is uncertain, it would not affect operating strategy so is not a concern in that regard. If, however, strategies are very different, particularly in the short term, decision-makers can then decide how quickly to advance the exploration to obtain good information, or make decisions that maximise the upside or minimise the downside of potential additional resources.[4]

MINING INPUTS

The preparatory geological modelling will have identified the size and shape of the orebodies at a range of cut-offs. The first task of the mining-related processes is to identify which of the potentially suitable methods for the orebodies at each cut-off will be considered in the evaluation. The big-picture mining method options are open pit and underground, and within them exist various methods depending on size, shape, geomechanical conditions and so on.

It was identified earlier that the primary mining-related decision may be thought of as the size of the mine, though, in reality all decisions should be optimised concurrently. Size of mine is, however, the primary decision driving how resource data will be generated for the underlying database for the optimisation model.

For an open pit mine, sizes can be defined by a number of outlines, such as nested pit shells formed by pit optimisation software.[5] For an underground mine, nested stopes or

---

3. As previously indicated, in this book terms such as *ore* and *orebody* refer to material above a cut-off being evaluated. They are not used in the strict sense of the codes for public reporting of resources and reserves.

4. This risk–reward trade-off is discussed in more detail in Chapter 12.

5. To ensure an adequate data set, these may range from the smallest pit that could realistically be mined with pessimistic economic forecasts of prices and costs and including only mineralisation identified with a high level of confidence, to the largest pit that might be mined with optimistic economic assumptions and including low-confidence or hypothetical extensions of the geological resource.

nested orebodies need to be generated for cut-offs spanning the range within which the optimum is likely to lie.[6] Until recently, this was a labour-intensive task, and in some cases, several designers had to be used to generate the stope shapes and reserves data at the cut-offs required in a reasonable time frame. This had the potential for workers to produce the information differently, so that the data sets at different cut-offs were inconsistent. The recent development of software to generate realistic stope shapes at specified cut-offs has reduced the time and effort required to create the underlying database and improved the consistency of the data across the range of cut-offs considered.

Ideally, we would feed the whole geological model into the strategy optimisation software, along with all the operating characteristics and relationships, operating and capital costs and revenue relationships and the like, and the solution would emerge from the process. However, the limits of computer power, now and in the foreseeable future, requires that small geological model blocks need to be aggregated into a number of larger mining blocks to make the optimisation problem computationally tractable. The mining inputs of an optimisation study form the primary database of material for mining. The geological model will be the main input, but it is the mining considerations that result in the data that the model or software works with.

The mining blocks to be generated for the database will be different for open pit and underground operations.

Open pit mining inventories

For open pits, the aggregated mining blocks will be bounded at the coarsest level by pit shells corresponding to each major pushback stage and by a number of elevations corresponding to pit bench elevations, spaced at some integer multiple of the bench height. For small pits, a block may be the ring defined by a range of pit shells and bench elevations around the full circumference of the pit. For larger pits, these rings may need to be further subdivided – perhaps by specified northings or eastings, or by lithological boundaries, or whatever is logical for the situation. Intermediate pit shells within pushback stages and individual bench elevations may also be used to further subdivide the mining blocks. The optimisation software will then select the blocks to be mined and schedule them in such a way as to generate the optimum pit size and mining sequence.

This description implies that decisions regarding pushback staging of the ultimate pit have already been made before the evaluation starts. Such sequencing and scheduling should ideally be part of the outcome, not an input into it. So while some thought is needed with staging at the data creation stage, it is preferable to allow as much flexibility and as little preconception as possible. More and smaller mining blocks will assist with this, but will also increase the computation time and make the problem potentially less tractable. A more precise outcome may be achieved, but better decisions aren't necessarily produced.

---

6. In the author's experience, the NPV-maximising cut-off for an underground mine is typically 1.3–1.5 times the cut-off derived by typical break-even calculations. In the absence of other information, a range of cut-offs from somewhat less than, to perhaps double the existing cut-off may be a good starting point for the range to be considered. If available, a marginal break-even cut-off (Lane's mining-limited cut-off) with optimistic prices could assist in defining the likely minimum cut-off, and Lane's treatment-limited cut-off with pessimistic prices may in many cases be a useful guide to the upper-bound of the likely range of cut-offs.

The impact and timing of irrevocable decisions must also be considered. For example, although the size of the ultimate pit using assumptions about future conditions will be a result of an optimisation study, it may well be that the decision regarding pit size does not need to be made for many years into the future. Notionally, the decision about the size, mining rate and starting time of each pushback stage, including the last, which creates the final pit walls, only has to be made just before the last stage starts mining; however, each pushback stage must be considered in the context of the whole pit, not independently. The number of mining blocks that can be adequately dealt with will be a function of the capabilities of both the software and the computer hardware on which it operates.

To facilitate cut-off optimisation, it is necessary to have tables of data for tonnages and grades of material versus cut-off for each identified mining block, enhanced by other available data such as geometallurgical information. Further subdivisions of the data tables by such items as rock and ore types and different categories of deleterious materials may also be necessary. Depending on the orebodies defined at different cut-offs, it may be appropriate to have tonnage and grade data for two or more open pit mining methods. For example, we might require data for selective and bulk mining carried out by two distinct mining fleets with different productivities, ore recovery and dilution, and capital and operating costs if this cannot be adequately estimated by applying different factors to the one data set.

If the optimisation study includes the transition from open pit to underground mining, it may be necessary to generate different sets of pit shells and mining blocks, accounting for any mining of higher-grade parts of the resource from underground before the pit is expanded into those areas. Pit slope stability may need to be re-evaluated if stopes have been extracted adjacent to what will later be pit walls.

Underground mining inventories

For underground studies, mining blocks will best be defined after nested stopes or nested orebodies have been produced at different cut-offs, so that the best location of boundaries between mining blocks can take account of the shapes, potential mining methods and similar factors. They will typically be defined by a combination of ore zones, sublevels, northings or eastings, geological structures, changes in the size and shape of the orebodies and boundaries defining geological domains within the rock mass or orebodies. The mining inventory for each cut-off must be as realistic as possible, taking account of both the ore loss and dilution required to generate realistic stope designs.

Outliers of material above any specified cut-off that would be uneconomic to extract should be excluded. A simple economic analysis comparing costs to access and develop an outlier with the extra revenue it generates may be adequate to exclude it. Some areas may be economic with some price forecasts but uneconomic with others. Areas that are economic according to such an analysis but are some distance from the nearest large zone of mineralisation should not be automatically aggregated into those adjacent areas – it may still be preferable to consider them separately to obtain the best plan. For example, it may be that the best allocation of scarce resources to all the tasks for a mine plan to work is such that an apparently economic zone is never allocated mining resources. If there is a limit on the amount of jumbo development that can be done, optimisation software may allocate resources preferentially to sources with the highest net revenue per metre of development. An outlying area that is notionally economic but of low net value may well have a low return per unit of the mining resources required to develop and produce from it. The optimiser may therefore allocate equipment and mining crews to areas where the most money will be generated per unit of the scarce resources and decide not to access and produce from other areas.

Underground evaluations will usually be simpler if boundaries between mining blocks are the same at all cut-offs. If, for some reason, the natural boundary between two mining blocks is a function of cut-off, special care may be needed. It might be essential to ensure that, if it is feasible for the two adjoining blocks to be mined at different cut-offs, there is no gap of unmined material in the data set, though this may be a practical outcome from the optimisation process. It will also be important to ensure that a region that is part of two mining blocks at different cut-offs is not mined twice by the optimisation scheduler. If this is found to be an issue, consider whether the tonnages and grades affected are material and must be accurately modelled, or whether they are insignificant and will not affect the overall optimum strategy. Unnecessary modelling complexity may affect the computational tractability of the problem.

If the optimisation study includes the transition from open pit to underground mining, it may be necessary to generate different sets of underground mining block data for mineralisation that is extracted within different final pit limits and to account for any differences in methods or costs associated with proximity to the pit wall.

Other physical data

In order to model the interrelationships that may be necessary to account for all the identified options to be investigated, other physical data will need to be generated in association with the mining inventories described.

For open pits, it's necessary to define haulage distances for each mining block to both ore and waste dumping points. It is quite common to find geological block models with a single haulage distance ascribed on the basis of a predetermined classification as ore or waste. Strategy optimisation will make this decision. In order to facilitate this, both distances need to be available to the optimiser. More complex information may be required depending on dump locations, rock types and the like. Block models may also record truck hours and fuel consumption, which may need multiple records for combinations of different truck types and destinations if that is an issue. The author has also encountered a situation where the acid-generating potential of the rock necessitated mining additional waste rock from elsewhere for the sole purpose of neutralising the acid generation in waste dumps. This in turn generated additional loader and truck hours if a block were classified as waste rather than ore.

For underground mines, development requirements for each mining block will need to be specified as data, though, in some cases relationships may be specified to allow calculation within the model. Depending on the level of study conducted, development might need to be split into categories such as major access, sublevel development and stope development in ore and waste. How development and stoping schedule interactions are to be handled in the evaluation may also suggest what proportion of each development category needs to be completed before the mining block can start producing. Stope drilling and blasting parameters may need to be specified separately at each cut-off, or, depending on the study's level of accuracy and the materiality of any differences, might be adequately catered for in the stoping cost per tonne of ore. Trucking distances may vary if haulage and hoisting options are to be evaluated. Quantities of different types of backfill may need to be specified, particularly if the mining methods being evaluated have different proportions of the orebody to be filled or are using different fill types.

Additional underground data requirements may also include elements such as access methods or topologies. For example, decline layouts may differ depending on whether the main access from the surface is by shaft or by decline, which in turn could be from the surface or within an overlying open pit. If there are two orebodies relatively close to each other, the decline topology options could include a single decline with accesses to each orebody or twin declines, one adjacent to each orebody.

For open pits, the additional physical information will typically be for activities or parameters that occur in parallel with the mining of the rock, and will be used either to generate costs accurately or to derive activities to which constraints may be applied. For instance, the loading fleet will constrain the tonnes mined, but the trucking fleet will constrain the tonne-kilometres hauled or truck hours worked. Depending on variations in the average depth of mining, the effective constraint on mining rate could vary from one to the other over time, but both may need to be specified and appropriate data and relationships provided.

For underground mines, some of the additional physical information will be for activities or parameters that occur in parallel with ore production, and will be used either to generate costs accurately or derive activities to which constraints may be applied, such as drilling, blasting and haulage. However, some information will be for activities that logically precede or follow the ore production activities, such as development and backfill. If included in the analysis, these will have associated costs and constraints on rates. Their constraints and their predecessor–successor relationships with ore production will often have a major impact on the overall mining schedules and therefore need to be accounted for in the scheduling component of the analysis. Similarly, decline topology could influence access development schedules and hence, depending on what mineralisation has been accessed, the ore production schedules. Twin declines could make available the option of a one-way haulage loop, which could impact on haulage productivity and on mining capacities.

There are no universal specifications for either open pit or underground, though the preceding discussion illustrates some of the more common situations. While it is vital to account for everything impacting on value, it is also important to minimise the computational requirements within the optimisation component of the modelling. This is done by combining items that logically may be combined and which, by doing so, will not affect the accuracy of the results, such as accounting for drilling in the cost per tonne. Such parameters may well need to be calculated and displayed separately for final outputs presented in reports, but in the context of this discussion, that is a post-processing task for the cases presented, and not part of the optimisation process itself.

Scheduling rules and capacities

Having generated data for mining blocks, it is then necessary to describe sequencing and scheduling rules. These must result in schedules that are practical and realistic at the level of accuracy of the study and that are able to account for all the combinations of cut-offs, mining and treatment rates and other options that could be evaluated.

Some rules are simple and obvious, while others are less easy to describe in accounting for all the possibilities that might potentially arise. In open pits, blocks closer to surface are mined before underlying deeper blocks. If the pit shell boundaries are based on pre-identified pushback stages, it is probable that minimum bench widths in pushbacks have been accounted for. If mining blocks are smaller than this, it may be necessary to define minimum widths of each pushback, so that mining blocks can be aggregated by the optimiser's scheduling processes. These thicknesses will be functions of the size of the loaders and trucks in the mining fleet, which in turn might vary with mining rates evaluated.

Mining rate specifications must look at what is possible practically, not just at what is scheduled in the existing mine plan. Very often the current schedule will have been set to deliver the required amount of ore at a pre-specified cut-off, which may not be optimal. This will have the effect of making the current cut-off, however it has been determined, a Lane-style balancing cut-off. Simply applying the currently scheduled rates as constraints will result in the current cut-off appearing to be optimal. If different rates are applied as constraints – perhaps the capacity of the existing mining fleet used to its full potential or by the acquisition of additional equipment – different optimum cut-offs and higher values may well result, which would more than pay for the increased mining rates. (See, for example, Figure 6.10 and the accompanying discussion in Chapter 6.)

Although related, equipment size and mining rate are independent decisions. While options involving high mining rates with small equipment and low mining rates with large equipment can probably be ignored, the author suggests that the optimisation team should avoid prejudging the mining rate at which the size of the mining fleet will change. Discovering that should ideally be an outcome of the evaluation, not a pre-specified input into it. To do that will require the ability to model higher mining rates with small equipment and lower mining rates with large equipment than might be thought to be optimal, with associated data and model logic to correctly evaluate these.

Underground, access development must precede stope development, which must precede stope production. Stope sequencing rules may be more complex. Although rules often specify that one stope must precede another, this is frequently done for convenience or because of preconceived logical sequences. In reality, underground scheduling constraints are often simply that two particular activities cannot occur concurrently. For example, two adjacent stopes cannot be mined simultaneously: one must be mined and filled before the other can produce, but there may be no logical necessity for either to have priority over the other. On the other hand, it might be that the general direction in which the production front is advancing forms part of the description of the mining method. In that case, different predecessor and successor relationships may be needed for each alternative. Depending on the size and shape of the orebodies and mining methods, complex proximity rules may need to be specified with the minimum separations between stoping activities, including filling; these will differ depending on the material between the activities, whether void or solid rock or fill of various types.

Every scheduled activity has a notional maximum rate or capacity. In an open pit, the maximum mining rate (or rock capacity) may be either the loading or truck haulage capacity. As indicated, haulage distances may vary for ore and waste, and if so, the haulage requirement and hence trucking capacity may be a function of cut-off. In hard rock, drilling capacity may be the rock constraint. The maximum ore mining rate in an open pit is sometimes limited by the turn-around time for grade control assay results.

Underground, we have seen that the main access development rate, which is exposing new mineralisation that can be then classified as ore or waste, is equivalent to the open pit rate of mining rock. Most of the processes in an underground mine are dealing with ore.

The mining data for an underground optimisation study will typically need to identify at the least the maximum access development rate and the maximum ore production rate within the mine. Depending on the problem's complexity, it may be necessary to specify capacities for a number of components of the ore production system in the mine, such as stope development (which may be the total development capacity less what is committed for access development for future production), and loading and truck haulage capacities (similar to the open pit situation). The overall production rate may be the result of complex interactions between the size of the orebody, stope layouts and adjacency factors, and predecessor–successor relationships between development, production and filling operations in various stopes.

The ventilation system may be the ultimate constraint on an underground operation, but this is often not easy to bring into the modelling process. The ventilation requirement may be a function of the depth of mining and the needs for different types of equipment that could be used. The overall constraint may therefore be a function of the total ventilation capacity, the average depth of scheduled mining and the equipment fleet selected, as well as the usual constraints previously described. The overall ore capacity will be the lowest of the maximum rates of all ore-related processes in the mine and treatment plant.

It was noted in Chapter 6 that, for an optimisation study, it is important to ascertain relationships between parameters so that the effects of making changes in one will be reflected in the reported results. With regard to system capacities, linear relationships between inputs and outcomes may be inaccurate. Figure 8.1 illustrates a common situation with trucking, particularly in an underground mine where there may be a limited number of places where trucks travelling in opposite directions can pass.

If there are no constraints on truck movements, no queuing, no delays for passing and so on, capacity will increase linearly with the number of trucks, as shown by the 'theoretical linear' line in the figure; however, when there is interference between trucks, such as waiting to pass a truck travelling in the opposite direction, the incremental capacity of each extra truck is progressively less, leading to a curvilinear relationship between capacity and number of trucks, as shown by the 'with interference' line in the figure. Further complexity is added when the trucks interact with a fixed capacity system, such as the 'downstream' line in the figure. The overall response will now be a complex function of the breakdown and repair characteristics of every component, their individual capacities and the amount of surge capacity or storage between successive parts of the overall system. The loading capacity will impose a similar flat line to the downstream capacity on this two-dimensional plot, but this may also be affected by a curvilinear relationship in a third dimension, representing numbers of loaders in the loading fleet, increasing the complexity of the interactions significantly.

A simulation of the physical system may be needed to account for these interacting components to estimate the overall production capability. The importance of these non-linear system responses will depend on the study's level of accuracy. As noted when discussing Lane's methodology, the optimum solution will often be found to occur when two components of the system are both operating at capacity. As can be seen in Figure 8.1, this is often where the greatest error will occur if simple relationships, such as the theoretical straight line and the downstream limit, are applied.

Options for the removal of constraints will usually be specified, unless the scope of the project is clearly defined to be limited to optimising the value of the operation

FIGURE 8.1

Production capacity versus number of trucks with various assumptions.

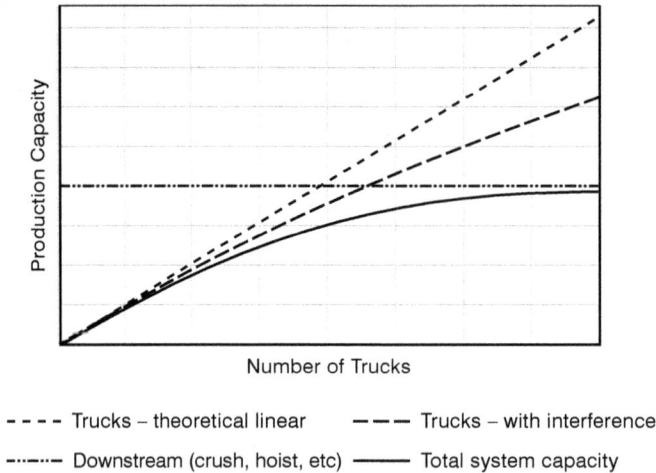

Legend:
- - - - Trucks – theoretical linear
- — — Trucks – with interference
- ·——·— Downstream (crush, hoist, etc)
- ——— Total system capacity

(Y-axis: Production Capacity; X-axis: Number of Trucks)

subject to existing constraints. Often this will simply be the addition of more units of development and production equipment or crews, such as drills, loaders and trucks. Other debottlenecking options could involve, for instance, sinking a shaft or upgrading the ventilation capacity.

It is of course possible that the schedules generated by the optimisation software will be unconventional, but so long as they are practical and realistic, that is what counts. Conventional wisdom is not always wise, and a logical process obeying appropriate rules may well find a better solution to which conventional wisdom has blinded us. The scheduling rules should therefore be as simple as possible to specify genuine constraints, with as little imposition of a planner's prejudgment as possible to define a tractable problem. The more that is locked in, the less opportunity the optimiser has to find better solutions.

It will be rare for a change in strategy to be made instantaneously. There will typically be a period of time at the start of the evaluation that needs to have its physical activities locked in, representing the time taken to conduct the evaluation and make any strategy-changing decisions, plus the time taken to create a plan to effect the transition from old to new. Depending on the corporate culture and the nature of the deposit and mining method, this unchangeable period could easily be as much as two years; in some cases, a reduced amount of locked-in activity could also extend into later periods as transition to the new plan is effected. The scheduling rules will therefore need to make allowance for these locked-in activities at the beginning of the evaluation. Depending on the accuracy of the study being conducted, limitations may have to be placed on the transition from what is effectively the current long-term plan – represented by the locked-in activities – and the new optimised plan generated by the strategy optimisation processes. For example, there might be a constraint on the rate at which production, development or waste stripping rates may be increased for a number of years.

It is important to identify how much of the locked-in activity is a practical reality and how much is related to corporate culture and inertia. It may be appropriate to evaluate a range of locked-in periods, in order to identify the value generated by challenging and overcoming organisational (as opposed to practical) impediments to change.

Concluding comments on mining

There may be practical problems generating the required information. Computer software can produce most of the mining data required for open pits. Depending on the level of study, it may be necessary to do some designs with human engineering input in order to calibrate data generated automatically by software for a large number of cases. Transforming data so that it's suitable to input into optimisation software may present challenges.

Automated software processes are nowhere near as advanced for underground operations as they are for open pits. At the time of writing, stope optimisation software that is capable of generating nested stopes is commercially available, but is not necessarily applicable to all styles of orebodies and mining methods, though developments continue. General purpose software for automatic generation of stope and stope access development is limited to in-house software owned by some companies and applicable to specific methods and styles of mineralisation. Commercially available applications are, however, being developed. Software to optimise the placement of declines[7] has been developed but, to the best of the author's knowledge, is not commercially available. Generating adequate data for an underground strategy therefore requires an amount of human-generated mine design, and availability of technical staff to do this work may be a practical constraint on the overall strategy optimisation process.

Technical staff often produce rigorously engineered designs for every case to be evaluated, yet this might not be necessary. If the optimisation software requires graphical objects created by a mine planning software application as its inputs, each case may need to be designed, and this may not be a trivial task within the constraints of both the design and the optimisation software. If, however, the optimisation software requires only tables of data to define the mining requirements, it may well be that a small number of cases require engineering input, while data for other cases can be interpolated. The guiding principle for input data generation is: maximum information from minimum work. Exactly how this balance is achieved in each case depends on the complexities of the mineralisation and mining methods and on the engineering judgement of the technical staff. If the optimum solution is one for which engineered designs have not been done, there should be sufficient engineering input into other cases from which the data for the optimum case has been interpolated so that the results can be accepted with confidence within the study's level of accuracy.

There is no general rule as to what constitutes sufficient rigour in the data – what is judged to be adequate by one team, or within a particular corporate culture, may be perceived to be inadequate by another. Despite this, if any input items are deemed to not be sufficiently accurate – perhaps because of time or availability of engineering resources constraints – those parameters may be treated as scenario parameters to ascertain whether changes have an impact on the decision to be made. If not, the additional accuracy might be nice to have, but it is not needed to make a sound decision.

---

7. For a simple single orebody, an experienced design engineer will be able to produce a design as good as any computer algorithm, though if multiple designs are needed (for example, to service the different orebodies defined by different cut-offs), an automated process could be useful. For many optimisation studies, knowing the depth of the mine and the gradient of the decline will be sufficient to derive development quantities at an acceptable level of accuracy. For multiple orebodies that could potentially require a complex network of access declines to link them, an access layout optimisation tool may be essential.

METALLURGICAL INPUTS

Physical metallurgical inputs to a study will generally be related to both metallurgical performance (typically recovery and product quality) and processing constraints.

It is essential to specify recovery relationships that are reliable for the head grades resulting from the range of cut-offs to be evaluated. Other parameters such as blends of different rock types may also need to be accounted for. Other relationships, such as the mathematically rigorous four-way relationship between recovery, head grade, concentrate grade and tailings grade,[8] or empirical relationships between recovery and feed rate, may be relevant. Care must be taken to ensure that relationships are valid for the range of head grades encountered. Often, recovery relationships have been derived by regression of results in the current operating range, assuming a polynomial or some form of exponential function of head grade. Plotting recoveries over the wider range of head grades anticipated for the study may reveal unrealistic behaviours. For example, a cubic function may produce recoveries that rise, fall and rise again with increasing head grades, or that fall to zero or negative values with lower head grades that are actually economic to treat.[9] Some metallurgical investigations may be needed to identify better relationships to use for the study, or scenario analysis may be necessary over any uncertainty range in the recovery relationships to identify whether strategic decisions are likely to change if the recovery relationships vary.

Treatment plant production constraints must be identified. As a minimum, these should be for ore feed and each final product. Depending on the complexity of the treatment circuits, it might be necessary to specify constraints for some of the intermediate process streams. Any points in the circuit where it may be feasible to build stockpiles of in-process materials may potentially act as logical breaks into separate processes, each of which might have feed and output constraints, as well as metallurgical performance characteristics.

Ore and product streams must be considered separately, taking account of the real constraints in each independently. The ore constraint is often expressed as a maximum tonnage rate that reduces when the feed grade increases above a specified grade. This typically means that, at the specified grade, the ore and product streams are at full capacity, and any increase in head grade must result in reducing the ore feed rate to avoid overloading the product circuits. For strategy optimisation, this combined relationship should ideally be split out and ore and product stream constraints specified separately. The combined effects of both constraints will be taken into account in the optimisation process, but by specifying them separately, the effects of changes in each can be more easily handled.

---

8. Recovery = $[c.(h - t)] / [h.(c - t)]$, where h = head grade, c = concentrate grade and t = tailings grade.

   This formula is particularly useful for base metals operations, where the tonnage of concentrate is a significant proportion of the ore feed tonnage. For free-milling precious metals, where head grades are low (a few parts per million), the tailings tonnage is the same as the ore tonnage, and the concentrate grade is very high relative to the tailings grade. Then, as an approximation, $c = (c - t)$ and the formula reduces to Recovery = $(h - t) / h$.

9. The author has also encountered relationships that, with high head grades, generated concentrates with contained product grades in excess of 100 per cent. The point of the discussion is that quoted recovery relationships (or any relationships) should not be blindly accepted simply because they are provided by technical experts. The optimisation team must ensure that any relationships used will be valid across the full range of the values that may be encountered, and the relevant technical experts should be intimately involved in this to ensure that relationships are accurate across the range of interest.

Constraints may be more complex than simple ore and product tonnages. It may be useful to consider the constraint on throughput to be the number of operating hours available. Since different rock types in the ore feed may have different milling rates, the ore tonnage capacity of the plant will be dependent on the ore blend. In such circumstances, it may be preferable to apply a tonnes-per-mill-hour factor to the various components of the ore feed, change the primary quantity measure from tonnes to required mill hours and apply the annual operating hours as the constraint. If appropriate, this process could be extended to account for power draw required, with the constraint becoming the available milling power.

As for mining, options for removing constraints in the treatment plant will usually need to be specified, unless the scope of the project is clearly defined to be limited to optimising the operation's value subject to existing constraints. Since a strategy optimisation study will be evaluating a range of cut-offs and hence head grades, some strategies investigated may be high production rate / low-grade cases, which may only encounter ore stream constraints, while others may be low production rate / high-grade cases, encountering only product stream constraints. Metallurgical staff will frequently have a list of sequential upgrades for the plant. Typically these will be a mix of ore stream and product stream upgrades, based on an assumption of increasing ore feed tonnages at the same head grade. To properly account for the distinction between ore and product in the analysis, it's necessary to separate the list of plant upgrades into separate sequential lists for each process stream.

As well as identifying upgrades to be implemented, it will be necessary to specify any changes in metallurgical performance that result from capacity changes, such as recovery and product quality. Operating and sustaining capital cost changes may also need to be accounted for.

The increasing focus on geometallurgical properties of mineral resources is leading to an increase in the amount of information that may be taken into account to optimise operating strategies. The example of different milling rates for different rock types has been available to evaluation teams for many years, but more sophisticated relationships relating various metallurgical performance parameters to mineralogy and other rock properties are being developed. The availability of extra information, however, while enabling extra opportunities to add value, potentially increases the complexity of the analysis that must be undertaken to do so.

OPERATING COST INPUTS

Chapter 2 describes the types of cost behaviours to be accounted for in a strategy optimisation study. As with the physical inputs to the study, the cost models should be as simple as possible to minimise the complexity of the calculations, but sufficiently detailed so that differences between cases brought about by the options and scenarios selected will be described accurately to discriminate between cases.

The simplest cost model is a fully variable single cost-per-tonne model. At many operations, cost reports only report total cost dollars and the resulting unit cost per tonne, without any further attempt to identify what physical quantities are driving the costs. This may be satisfactory for management control purposes, but is usually inadequate for a strategy optimisation evaluation.

As soon as an optimisation study considers changing production rates, a fixed-and-variable model will be necessary since not all costs increase in direct proportion to the quantities involved. Some are related to time only, not physical activity. Similarly, if considering changing cut-offs, ratios of various physical quantities, such as ore tonnes per metre of development and product quantities per tonne of ore, will potentially vary.

For a given production rate and cut-off, the ratios of all the various physical parameters will tend to be constant, and the overall cost per tonne of ore implicitly accounts for the current ratios of physical quantities to each other.

For a more detailed optimisation study, the general principle is that variable costs need to be specified separately for any physical parameter that does not vary in direct proportion to ore tonnes as any of the options evaluated are varied. The reason is that any overall average cost-per-tonne will include the costs associated with all physical parameters, assuming that the ratios of the quantities involved are unchanging. If the physical ratios change with changing strategies, the unit cost will as well. As discussed in Chapter 3, this will introduce a circularity into the analysis that is resolved by relating costs to the physical parameters that actually drive them; however, any physical parameter for which quantity maintains the same ratio to ore tonnes across the range of strategies need not be costed separately – its costs can be included in the variable cost per tonne of ore.

This rationale is based on the common approach of expressing all costs as a unit cost per tonne of ore; however, discussions in previous chapters have indicated that, as a minimum, variable costs also need to be separated into those that are related to rock, ore and product. If this is recognised and implemented, the rationale would be expanded to a principle that variable costs need to be specified separately for any parameter that does not vary in direct proportion to either rock, ore or product quantities. The costs of those activities that do vary in direct proportion to either rock, ore or product quantities can be included in the variable cost per tonne of rock, ore or product as appropriate.[10] If there are separate constraints for these other activities, their physical quantities at least will need to be modelled separately so that appropriate constraints are applied, even if costs are included in the rock, ore or product-related variable unit costs.

Fixed costs will often be best split into those that are related to the mining operation, the treatment operations and general and administration overheads. With open pits, treatment of stockpiles often continues beyond the closure of the mining operations, at which point the incurring of mining-related fixed costs will cease, but plant fixed costs continue.[11] It may be appropriate to identify general and administration fixed costs with and without mining occurring. In some organisations, where the custom of expressing all costs as unit costs per tonne of ore is particularly entrenched, one may find that, while variable costs are, correctly, expressed as a constant unit cost per tonne for all levels of activity, fixed costs are expressed as a cost per tonne that

---

10. If there is a corporate requirement to report the quantities and costs of such activities, the strategy optimisation should be as simple as possible to reduce the amount of computation required. The required outputs can then be derived for final reporting for the final case/s presented after the optimisation has been completed.

11. As identified in the discussion of Lane's methodology, the mining processes are represented in an underground operation by the major access development. It may therefore be appropriate, if the difference is material, to split underground mining fixed costs into those related to development and production activities, as development activities may finish some time before production.

varies with changes in the production rate. If the production rate is to be varied in the strategy optimisation, the fixed cost input has to be expressed as a function of the production rate. This is convoluted logic that can be simplified by recognising that fixed costs are incurred because of the elapse of time, not by the quantities of physical activities that occur in each period. Fixed costs should be expressed as a constant cost per time period.

For open pit mines, variable costs will typically be driven by:

- rock tonnes drilled, blasted and loaded
- truck tonne-kilometres hauled or hours of operation.

In many cases, the haulage cost is accounted for by generating a mining cost per tonne of rock that varies with depth, so that all costs are driven by rock quantities, but the unit cost is a function of the average depth of mining. Either way, total mining costs will be driven by two parameters that must be generated: rock quantities and a distance measure, such as average haulage distance or average depth of mining. Alternatively, the geological block model blocks, and hence the mining blocks used in the strategy analysis, may be allocated a mining cost that accounts for differences in haulage costs. In this case, mining costs are derived directly from parameters in the mining blocks' data, rather than from mining physicals and unit costs specified as input data. In some cases, the underlying truck productivity and fuel consumption rates, based on haulage route specifications, may be included in the block model and hence in the mining block data.

As discussed in previous chapters, there may need to be a base case cost assuming all rock mined is dealt with as waste, and a differential cost, which may be positive or negative, for rock that is to be classified as ore.

For underground mines, variable costs will typically be driven by:

- development metres, perhaps split into several categories
- ore tonnes drilled, blasted, mucked and hoisted
- truck tonne-kilometres hauled or hours of operation
- backfill tonnes, perhaps split into several categories.

In the treatment plant, variable costs will typically be driven by:

- tonnes reclaimed from stockpiles
- ore tonnes treated
- product quantities generated.

In many companies, stockpile management and reclaim costs are reported as mining costs, since the equipment used for the reclaim will often be part of the mining fleet and the reclaim operations are managed by the mining department staff. From the optimisation point of view, they are better seen as the first part of the treatment process, as they will typically only be performed for ore and carried out immediately prior to the material's entry to the treatment plant.

Labour costs are often included in fixed costs. Depending on the level of accuracy of the analysis, it may be appropriate to model labour numbers by fixed and variable relationships relative to various physical activities, and then apply a cost per worker to the number of workers thus derived. There might need to be different categories of workers with different costs for different classes of work.

If there is a need to account for inflation rates for various cost categories, or for the possibility of a step-change price shock – for example, for fuel or energy – it may be necessary to model each category separately in addition to the physical cost driver splits.

Although fixed costs are, as a general rule, the same for all levels of activities across the full range investigated, some are fixed across smaller ranges. These are sometimes referred to as capacity or step variable costs. An example is the labour cost associated with truck haulage. As a mine gets deeper, the truck hours needed to maintain the tonnage targets increase. When the required hours exceed the hours available for the existing number of trucks, an additional truck must be acquired, as well as extra operators and maintenance personnel to maintain it. The costs of these workers will then behave like fixed costs, incurred per time period until the haulage distance increases further and another truck and its associated labour is acquired.

It is important to note that the split between fixed and variable costs may be dependent on the overall time frame. Costs that are fixed in the short term will often be variable in the longer term. Consider, for example, the labour cost of the maintenance workforce. Once a short-term plan has been developed, such as the annual budget, the number of vehicles in the mining fleet, their work hours and their estimated maintenance requirements will be determined, and the maintenance workforce size will be set accordingly. This will set the total labour cost of the maintenance workforce, which will not change if mining and treatment activities fluctuate within the budget period. If asked what proportion of the maintenance labour cost is fixed, the maintenance manager, whose focus is potentially short term, will probably advise that it is 100 per cent or a value close to that. However, if in the longer term the mining rates were to halve or double, the maintenance requirements, and hence labour cost, would also be expected to halve or double. In the longer term – which is what is implicitly being evaluated for strategy optimisation – this particular cost is probably close to 100 per cent variable, not 100 per cent fixed.

Other cost behaviours are possible. Non-linear cost versus activity relationships are sometimes used. For example, the annual cost is derived by multiplying a specified value by the quantum of a specified activity (often ore tonnes treated), raised to a fractional power. Site administration and general overheads are sometimes estimated by such a relationship. In the author's experience, the value used as the exponent is empirical and site-specific, and depends on which costs at what level of aggregation are being modelled. Over a small range of activity, such non-linear cost behaviours may be approximated as a simple fixed plus variable relationship, based on the formula of the tangent to the cost versus activity curve at some base case level of activity. The resulting cost estimates will then be less accurate the further the level of activity is from the base case and the degree of non-linearity in the underlying relationship.

It is rare for mine cost accounts to record costs as fixed or variable (or non-linear), and even if they did, one would have to ask whether the distinction was based on short- or long-term considerations. Simply asking the person who is best able to provide information regarding the split of fixed and variable cost components may not elicit the right answer (as we just saw with our maintenance manager). Strategy optimisation teams must ensure that they obtain information that is appropriate for their task. The problem may perhaps be avoided by carefully framing the questions to be asked. As indicated, asking what the fixed–variable split is may be unintentionally ambiguous (as it does not mention time frames or activity levels). A better question might be to ask what it is that we really want to know: how will the costs actually behave with changing physical activities? What would happen to these costs if the quantity of the associated activities were to halve or double? For short-term estimates, such as annual budgets, a substantial proportion of costs will be fixed. For longer-term strategy optimisation, we should expect that the majority of costs, especially operational costs, will be variable.

This discussion of operating costs has presented some potentially complex relationships. As has been noted, the optimisation team must assess carefully how complex the overall cost model needs to be to adequately discriminate between cases evaluated. The units in which costs are expressed (per tonne, per ounce, etc) in various company reports cannot be assumed to describe how they behave. The evaluation team must examine the cost data provided and, when necessary, reallocate or reclassify, to ensure that the cost model sufficiently accounts for how costs will behave over the wide range of situations to be evaluated.

As with other aspects of the task, where there is uncertainty in the data – whether relating to the actual cost or the appropriate split of fixed and variable components – these should be recognised as scenario parameters to be flexed to ascertain their impact. Variations in assumptions will change the reported values, but do they change the strategic decisions to be made?

CAPITAL COST INPUTS

Capital costs have been discussed in some detail in Chapter 2. The accounting distinction between capital and operating costs is in many ways irrelevant for strategy optimisation; it is how costs behave that matters. Chapter 2 highlights the distinction between project capital – typically a lump sum expenditure to acquire some new capability – and sustaining capital, which may be thought of as irregular maintenance expenditure to preserve the capabilities of the existing production facilities.

Sustaining capital

Sustaining capital will typically behave similarly to variable operating costs, and should be treated in the same way as operating costs. The main reason to maintain the capital/operating distinction will be to correctly handle the tax treatment of costs – they will be either expensed immediately when incurred or depreciated over a number of years. The depreciation rules will vary from jurisdiction to jurisdiction, and discussion of these, other than to identify an issue that the evaluation team will have to deal with, is beyond the scope of this book.

Sustaining capital, therefore, will require input data that specifies costs incurred per unit of activity, similar to operating costs. The unit of activity is a physical item that causes the production facilities to wear out so that they need capital repairs or replacement. In an open pit mine, most sustaining capital will be rock-related. An underground mine will have a mix of rock and ore-related sustaining capital, and the treatment plant a mix of ore and product-related capital.

Although sustaining capital may appear to be fixed costs expressed as dollars per year, the reality is that very little will be a function of time elapsed – most will be the result of wear and tear from handing rock, ore and product through the production facilities. The exception is typically in the general and administrative overheads area, where buildings, furniture and fittings and office equipment do deteriorate with time or become obsolete.

Often it is adequate to model sustaining capital by simple fixed and variable cost relationships for each area of the production system (mine, plant, etc), but occasionally more detail is warranted. Fleet replacement is a particular class of sustaining capital that might require detailed calculations, depending on what information the project sponsors want to extract from the study. This might require calculations of, for example, operating hours for each truck per year and hence detailed rebuild and replacement schedules. For the strategy optimisation evaluation this detailed analysis can usually be replaced by lease-style calculations to model the sustaining capital costs,[12] even if vehicles are owned rather than leased.

Project capital

While project capital costs will usually not be included in direct calculations required for any of the cut-off derivation processes described earlier, they will, however, be included in an optimisation evaluation so that the value measures account for all costs. Project capital for new capabilities may need to be specified for sequential upgrades of an item of plant, or as mutually exclusive alternatives for acquiring various capabilities. In a simple analysis it may be adequate to model this capital expenditure as a single lump-sum amount, typically in the year before the capability acquired becomes available for use in the operation. In a detailed analysis, it may be appropriate to model a spending pattern over a number of time periods prior to the capability becoming available.

The timing of any such project capital expenditure for the provision of additional capability will often be a strategic decision to be optimised. The data specifications and modelling constructs must therefore be able to handle these requirements.

The analysis should also account for the impact of acquiring some new capability on future maintenance, operating and sustaining capital costs. There will be different effects for different types of project capital. For example, adding a process into the existing sequence of processes in the plant without increasing the throughput would result in additional variable maintenance, operating and sustaining capital unit costs, and perhaps fixed costs. Duplicating an existing process to increase capacity might increase fixed costs, but would probably not change the variable maintenance, operating and sustaining capital unit costs. In the mine, acquiring an additional truck to deal with the increasing depth of mining would probably not increase variable maintenance, operating and sustaining capital unit costs per tonne-kilometre hauled or per truck hour, but would increase the cost per tonne of rock mined.

In some situations, increases in capacity are suggested but the associated capital costs have not been engineered. Non-linear cost versus activity relationships are often used to derive an initial estimate. The estimated cost for a proposed capacity is derived by multiplying the engineered cost for a base case capacity by the ratio of the proposed and base case capacities raised to a power less than one. Six-tenths seems to be the default

---

12. The resulting cash flow timings may not be precise, but errors will usually be minor in the context of the overall project cash flows, in total or in any particular year. The mathematics underlying the leasing costs ensures that there is no impact on the accuracy of any NPV calculated. It may be appropriate to find the best strategies using simple lease-type cost relationships, and calculate the actual replacement schedule and cash flows separately for the small number of cases to be reported in detail.

consensus value for the exponent for this type of process, giving rise to what is sometimes called the *six-tenths rule*[13] for capital cost estimating.

Working capital

Working capital may have an impact on cash flows and value. Simplistically, working capital can be thought of as the result of the difference in timing between expenditure of costs and receipt of revenue. Operating costs will need to be expended for some time before first revenues are received. Since models typically generate physical activities, working capital is often estimated by specifying a lag (usually) for revenue cash flows between the time of production and the receipt of revenue, and a lead or lag for cost cash flows (depending on stock levels, consumption rates and payment terms for cost components). First fill – the initial supply of various consumable items to make all parts of the production process operable – may be a significant working capital item that requires separate specification.

SOCIAL ISSUES

Depending on the operation's location, social issues may have an impact on the costs and timing of the start-up of operations. Items that need to be accounted for, in terms of dollars and time, may include:

- relocation of towns, villages and social infrastructure
- provision of facilities for the local area, such as schools and hospitals
- provision of job opportunities in the region, both directly in the operation and in the supply of goods and services to the operation
- compensation to people affected by the operation
- requirements and expectations of the various levels of government for returns such as taxes, duties, employment and provision of infrastructure
- economic sustainability of the region following closure of the operation.

Many of these issues could be considered as options or scenarios for strategy optimisation. For example, it may be useful to identify the value of the operation with or without the relocation of people or infrastructure. Similarly, the time required to reach agreement on many issues can generate a loss of value by deferring the project. This sort of knowledge can assist decision-makers to assess the time or money that is worth spending on such matters. For a marginal project, these issues could make the difference between whether it proceeds or not.

The initial capital and, if relevant, ongoing costs of these social issues are just as much a cost of the operation as the direct mining and treatment costs. They should be accounted for in the evaluation in the same way as similarly behaving operational costs.

REVENUE ITEMS AND ECONOMIC PARAMETERS

Revenue items include forecasts of product prices and sales terms and will be specified by the company commissioning a study. There may be alternative sets of price projections, often referred to as base case, optimistic or upside, and pessimistic or downside. The

---

13. The six-tenths rule may also be applied to operating cost estimates, and the exponent may vary for different components of the cost. See Chapter 5 of the AusIMM's *Cost Estimation Handbook* (Lanz, Seabrook and McCarthy, 2012).

strategy optimisation team should not be constrained to use only these official forecasts. An important aspect to be derived from the study is identifying what would cause strategic decisions to change, and price is often a decision driver. The optimisation team should be free to evaluate prices that are well outside the range of official forecasts, to identify how robust optimum strategies based on the official forecasts actually are.

If the project will be a major producer in the market, it may be appropriate to have relationships between the price and amount of product generated.

Other economic parameters affecting both revenues and costs include exchange rates and inflation rates, which will also be specified by the company commissioning a study. There can potentially be a number of inflation rates, such as international (perhaps applied to product prices) and domestic (applied to cost items). There may also be different escalation rates for prices and particular categories of costs, such as labour, fuel and energy, in addition to the general inflation rates.

Discount rates are an important input for deriving value measures such as net present value (NPV). As with prices, these will be specified by the company commissioning the study. Net cash flows and discount rates can be in real or nominal terms, and before or after tax. It is important to ensure that cash flows and the discount rate applied to them correspond.[14] There can be concern about the impact of discount rate on strategy. As shown in Figure 6.9 and the associated discussion, the discount rate will impact on the value of NPV, so if purchase or sale price is the issue, getting the discount rate right may be crucial; however, the discount rate may have little impact on operating strategies. As with other uncertain parameters, the discount rate could be treated as a scenario variable and flexed to determine its impact on strategic decisions.

TAXATION

Taxation generally needs to be included in a strategy optimisation analysis. Revenues and operating costs are usually accounted for easily, while capital expenditure will be accounted for by depreciation deductions to derive a taxable profit. Depreciation rules will vary from jurisdiction to jurisdiction. In many jurisdictions, working capital is not deductible for tax purposes. At the level of accuracy of most studies, it will be satisfactory to aggregate capital expenditure into a smaller number of categories than would be used for formal tax accounting.

Although there are many intricacies in taxation law, it is rarely necessary to take that into account in a strategy optimisation evaluation. Indeed, one could suggest that if the strategic decision depends on the intricacies of the tax effects, it is an entirely different order of problem from what would normally be necessary to determine the optimum strategy for such items as cut-offs, mining and treatment rates.

---

14. It is beyond the scope of this book to discuss discount rates and discounted cash flow calculations. The reader is referred to any standard textbook on project evaluation if more information is wanted. For completeness here, it is noted that the NPV derived from nominal cash flows and a nominal discount rate should be the same as one derived from real cash flows and a real discount rate; however, the before-tax and after-tax NPVs can be different and either could be the larger. It is incorrect to apply both real and nominal discount rates to the same cash flow and call the results real NPV and nominal NPV.

Royalties are for convenience classified here as a form of taxation. Many companies record them as a cost, often expressed as a cost per tonne of ore. As described in Chapter 3, they are usually classed as either:

- *ad valorem*, expressed as a percentage of the net price after deduction of allowable costs, which are typically off-site realisation costs such as treatment and refining charges, and perhaps freight and transport from site to smelter, though some jurisdictions charge ad valorem royalties on gross value of product
- tonnage-based, expressed as a simple dollar charge per tonne of product, or sometimes per tonne of ore
- profit-based, expressed as a percentage of net profit.

In some jurisdictions, royalty specifications may be a combination of these, or the rate specifications may be more complex than a simple unit cost or single percentage. The optimisation model will need to generate royalties at an appropriate level of accuracy.

Expressing the royalty as a cost per tonne of ore may be satisfactory for financial reporting, but unless the royalty is actually levied as a charge per tonne of ore, this is not the appropriate way to model it for strategy evaluation.

END-OF-LIFE ISSUES

The major end-of-life issues that need to be accounted for are redundancy payments, mine closure and rehabilitation costs, which include ongoing commitments, and salvage values of both capital items and stocks of consumable items.

If the project has a long life, the impact of these on NPV could be negligible and they may be safely ignored, although some companies' procedures require these items to be taken into account, regardless of the life of the operation; however, if the life is relatively short, these cash flows may impact significantly on value. The discussion of Lane's limiting cut-offs in Chapter 5 has indicated how high closure costs may drive a reduction in cut-offs to extend mine life and defer incurring closure costs.

The longer the remaining life of the operation, the more uncertain the data for these cash flow items might be. Well-estimated costs for what is current best practice for many of these issues may turn out to be totally inadequate years later when accepted best practice may have altered. The impact of ongoing resource discoveries on the life of the operation will add to this uncertainty. Again, in such a situation, the values can be flexed to ascertain whether they have an influence on current decisions.

It may be necessary to derive relationships that describe how these costs may vary if the life of the operation changes. For example, salvage values can reduce if the age of items disposed of increases, and redundancy payments can vary if the workforce's average number of years of service varies with the operation's life. Ongoing rehabilitation during the operating life may impact the end-of-life rehabilitation costs if the life changes.

The difference between capital expenditure cash flows and the associated accounting depreciation charges is generally well understood. It is equally important to distinguish between accounting provisions for mine closure costs during the life of the operation and the actual cash flows at the end of the life. There may, however, be cash outlays during the operating life to meet closure bond requirements, with drawdown of the bonds helping to fund closure costs. In this case, depending on interest and discount rates, the cash flows and accounting provisions may be similar. The way cash flows are actually incurred should be identified and modelled appropriately.

BUILDING AND USING THE OPTIMISATION MODEL

Having identified the options and scenarios to be accounted for and all the data inputs and relationships, they must then be combined into a model. Depending on the optimisation techniques and software used, the model may be set to show the results of various combinations of inputs, and/or be processed in some way so that the optimum strategy can be identified. Chapter 13 discusses various optimisation methodologies that might be used. For now, it is sufficient to note that some methodologies may help produce multiple combinations of options and scenarios, so that hills of value (such as illustrated in Figures 6.6 to 6.10) can be generated to identify the optimum strategy, indicating both the best value and the response of value to changes in strategic decisions (options) and external parameters (scenarios). Other methodologies may simply identify the set of options that optimises the value measure, given the set of assumed scenario values.

Time periods

An important consideration for the optimisation model is the lengths of the time periods used. Existing mine planning and scheduling software applications can create schedules with effectively continuous time scales, theoretically allowing activity durations to be expressed to fractions of a second. Strategy optimisation models, however, typically work with sequential time periods of specified durations. Study sponsors often like to have a mix of durations in final reports. For example, it may be preferred to have the first two years reported monthly, the next three years quarterly and annual periods from the sixth year on. The question then is: is it appropriate to conduct optimisation modelling using the final reporting time periods? The answer will frequently be no.

As a first consideration, as described for mine scheduling inputs, it will be rare for a change in strategy to be made instantaneously. There will be a period at the start of the evaluation time frame that needs to have all its physical activities locked in and, in some cases, a reduced amount of locked-in activity extending into later time periods. A significant portion of the initial time frame required to be reported in short time periods (for example, monthly) may therefore be taken up by locked-in schedules.

Regardless of any initial locked-in activities, the optimisation model will need to be capable of scheduling reasonably realistically any of the large number of alternative cases that could potentially be evaluated. The scheduling rules will therefore have a number of approximations. The implication is that even if very detailed schedules are able to be generated, it is unlikely that monthly outcomes are accurate predictions of what will happen in that month some time into the future.[15] Conducting the optimisation with monthly time frames, although giving an appearance of accuracy, is not achieving it. One could argue that generating accurate, detailed short-term plans is not the task of a strategy optimisation study anyway. Depending on the optimisation methodology

---

15. Even with schedules produced by experienced engineers of detailed budgets and forecasts for one well-defined plan, the reality is that with the normal variabilities encountered in mining, they will prove to be inaccurate even a few weeks or months into the future, and are often obsolete by the start of the budget year. Schedules produced by rules designed to be suitable for a range of inputs cannot be more accurate than that.

employed, monthly time frames may result in an unacceptably large number of periods handled by the optimisation software.

Annual time frames are often ideal for reporting, but in many cases they will be too coarse for optimisation modelling. For example, there will often be scheduling rules where one activity cannot commence until another has been completed. The simplified scheduling process in the model is often such that any conduct of the predecessor activity in a time period precludes the commencement of the successor activity in that same time period. Consider the situation where the predecessor activity is completed early in a time period. If the time period is a year, then it may be almost a year before the successor activity can commence. This may be unacceptable for realistic scheduling. Monthly time periods can alleviate this problem but, as noted, may be impractical. The author has found that quarterly time frames pose a good balance between schedule reality and the number of periods modelled. A delay of up to three months or, on average, six to seven weeks (half a quarter) between predecessor and successor activities represents a reasonable activity duration overrun or unplanned delay between successive activities.

The accuracy of the derived schedule and the length of the modelled time periods will also depend on the duration of activities relative to the average potential gap between activities generated by the scheduling rules. If modelled activities take on average, say, one month, but the average enforced delay between sequential activities is six weeks, quarterly time frames will be unacceptable. The average duration of activities in the schedule need to be greater than some multiple of the average duration of activities, and that multiple will itself depend on the desired accuracy.[16]

If, on the other hand, the predecessor–successor relationships can be modelled in a way that permits portions of predecessor and successor activities to be scheduled in the same time period, some of these problems will not exist. Whatever the situation, the optimisation team has to identify the capabilities and limitations of the software, the complexity of predecessor–successor and other scheduling constraints and the level of accuracy of the study, and hence determine the best durations of sequential time periods to be used. It will often be possible to use different durations for different time periods, typically shorter periods earlier in the operation's life and longer later, when mining blocks may be larger and have less accurate data.

Shorter time periods modelled by the optimisation software (such as quarters) can be aggregated into longer periods (such as years) for reporting purposes. Where the required reporting periods are shorter than what is needed for optimisation, the ideal situation is for the optimisation team to convince the project sponsors that the shorter time periods are inappropriate and the optimisation time periods can be used for reporting. If this cannot be done, the longer periods can be simply divided into the required number of shorter periods, with all activities from optimisation model periods spread evenly across the shorter reporting periods. If this is inadequate for reporting, it may be necessary to do detailed schedules at the level of required time granularity in order to report early years' information for cases that are to be reported in this way. As noted, many of the activities in these shorter time periods may in fact be locked-in activities that will be the same for all cases reported.

---

16. Simplistically, the accuracy of the schedule duration or overall production rate will be given by the ratio of the average inter-activity delay duration to the average activity duration for activities on the critical path.

Calculating cash flows

There are many ways in which cash flows can be estimated. All will depend first on having a good model of the physical activities that are actually driving the costs. Assuming that we have a reliable model of the physical activities, how then should cash flows be modelled? This topic could form a book in its own right. The following discussion, therefore, is a high-level guide to complete the major issues facing a team conducting an optimisation study. The discussion describes what might be thought of as the most complex and most accurate cash flow model. What is done in practice will depend on the level of accuracy of the study. The effects of common simplifications in cash flow models are described following the description of the ideal model logic:

1. Obtain variable unit costs, fixed costs and revenue parameters in the currencies in which they are incurred. It may be necessary to distinguish between currencies, for example, for expatriate and local labour costs and for consumables sourced from various locations. Where possible, costs (and revenue items) that are quoted in one currency but are incurred in another, such as petroleum products and product prices, should be identified and modelled in the original rather than quoted currency.

2. Apply the appropriate cost and revenue relationships to the physical quantities scheduled to derive periodic (monthly, quarterly, annual, etc) costs and revenues in uninflated terms; that is, in today's money at today's cost levels, ignoring the future effects of general inflation and escalation, in the currencies in which they are incurred.

3. Apply factors to account for inflation and escalation, still working in the original currencies, to derive cash flows in nominal terms; that is, in money of the day, the values that would be recorded at the time the costs are incurred or revenue obtained, assuming our estimates are accurate. Typically, there is a single rate for general inflation, but different escalation rates may be appropriate for different cash flow items such as product prices, labour, power, fuel and consumables. Escalation rates will be expressed relative to the general inflation rate and may be positive or negative. General inflation and escalation may also be combined into differential inflation rates for various cash flow items. These factors may also incorporate step changes associated with price shocks if any are predicted. It may be necessary to take account of the age or date of the data used for each item. Often, product price projections will be relatively recent and expressed in today's money; however, cost data might be taken from the previous financial year's cost reports and be more than a year out of date. If relevant – for instance, if there is a high inflation rate – it may be necessary to apply inflation factors to items of data to ensure that all items are expressed in money of the same date, typically *time zero* in the model.

4. If necessary, convert all cash flow items to the currency of the country taxing the operating entity, usually the country where the operation is located. This should be done using nominal exchange rates; that is, the exchange rates between currencies that would be seen at the time in each period. Exchange rates are often quoted in real terms, but differences in inflation between countries will alter the relative buying powers of currencies in addition to any real changes that are forecast.

5. Calculate depreciation for tax purposes – and hence taxable profit and tax, and nominal net cash flow – in the currency of the taxing country.

6. Convert this to the reporting currency in nominal terms, again using nominal exchange rates with nominal cash flows. In some cases it may be necessary to do further tax calculations for the country where the company is based.

7. Deflate the nominal net cash flow after tax to real cash flows using the appropriate deflator for the reporting currency or country.

8. Discount the real after-tax cash flow using the real after-tax discount rate to obtain the NPV.

Implications of simplifying assumptions

Simplifying assumptions are often made in cash flow analyses, but they may affect the accuracy of the outcomes:

- *Ignoring general inflation* – inflation erodes the value of depreciation. Although costs are inflating, once capital is spent it is depreciated on the basis of the cost incurred at the time of acquisition. Depreciation is therefore expressed in terms of the money of the year of expenditure, but those amounts are then treated as nominal amounts when determining tax liabilities in later years, when the purchasing power of the money has declined due to inflation. Ignoring general inflation therefore overstates the value of depreciation, reduces the tax payable and therefore results in an overstated value. For comparative purposes, this might not be an issue if the options considered have similar levels of capital intensity and spending patterns; however, if the comparison is between a high-capital/low-operating cost option and a low-capital/high-operating cost one, the relativities could be distorted by this simplification.

- *Ignoring differential inflation or different escalation rates* – if cost inflation is greater than revenue inflation, margins will reduce and the mine may have to close earlier than indicated without accounting for the inflation effects. The value of the operation will be overstated. Conversely, if cost inflation is less than revenue inflation, margins will increase and the mine may have a longer life than indicated without accounting for the inflation effects. The value of the operation will be understated.

- *Ignoring taxation or conducting a pretax evaluation* – this may potentially create invalid comparisons between options. It may not be a major issue if the options considered have similar levels of capital intensity and spending patterns; however, as with general inflation, if the comparison is between a high-capital/low-operating cost option and a low-capital/high-operating cost one, there could be problems.

- *Using over-simplified cost driver relationships* – simple cost driver models are often used in mining economic evaluation models. Often, particularly in high-level studies, a single unit cost per tonne will be used. This ignores the fixed and variable split of costs and will not therefore adequately account for changing activity rates over time. A fixed cost per year plus a variable cost per tonne of ore model is also often used, but this ignores changes in ratios of physical quantities. Chapters 2 and 3 have discussed the various costs that may be associated with rock, ore and product, and earlier sections of this chapter have indicated the need for more detailed physical relationships to be identified to handle the effects of strategic decisions to be made. Simple relationships may be adequate if there is little future variation from current ratios of various physical quantities and activity rates over time: these are implicit in the current unit cost per tonne. If there are variations over time, simplifying assumptions can lead to wrong conclusions and hence suboptimal decisions.

In all these situations, there will be a trade-off between simplicity, accuracy and reliability. There is no simple answer to this issue. In part, its resolution will depend on the level of accuracy of the evaluations, as highlighted in Chapter 7. It may be that there is one level of accuracy needed in the data, assumptions and relationships to achieve a reliable result, but another more detailed level required to prove that this is so (or perhaps to silence sceptics).

WORKING WITH DIFFERENT COMMODITIES

Base and precious metals

Virtually all the discussion to this point directly applies to base and precious metal operations. Although there may be constraints within an operation that affect how quickly a new strategy is implemented, as discussed, there are no constraints imposed by the market. Products will typically be sold within the terms of existing contracts or freely on the open market. Changes of operating strategy suggested by cut-off and optimisation studies will be internal to the operation and have product effects that can easily be absorbed by the market without special considerations.

Bulk commodities

In many cases, the requirement for bulk commodity operations is to maximise the amount of material above a contracted quality specification, making the problem largely one of scheduling, stockpiling and blending. Optimisation studies for plans of such operations may not need to optimise cut-offs to maximise value in the same way as base and precious metal operations, but can require some cut-off criteria to control the properties of blends produced as the grade distribution mined varies over time.

However, if variations in resource quality translate to variations in value, Lane's methodology or strategy optimisation studies that do involve cut-offs may apply. With bulk commodities such as iron ore and coal, some form of beneficiation or washing is conducted to upgrade the ore as mined to a saleable or more profitable end product. The only difference between bulk commodities and base or precious metals in this regard is that with bulk commodities a large proportion of the ore (perhaps even 100 per cent) becomes final product, whereas with the latter, a relatively small proportion of the ore becomes product, but this does not invalidate application of the methodologies.

Bulk commodity sales contracts may provide little flexibility for product quality and perhaps volumes, though, depending on the state of the world economy, there may be some potential for open market sales of additional product. A wide-ranging cut-off and strategy investigation, however, may identify potentially better strategies for volumes and product mixes that could add value in the longer term. The cut-off concept may have to be expanded to account for the allowable grades in the final product of both the primary commodity sold and deleterious impurities. Changing cut-offs should therefore be thought of as changing not only the minimum grade of the main product but also the maximum grades of the contaminants.

For example, increasing an impurity grade limit will result in a higher impurity grade in the product; this may perhaps make it unsaleable with current contracts, but could be saleable with new contracts, albeit with a lower price. However, the increase in tonnage may more than outweigh the loss in the unit value of the product, resulting in increased revenue and profit. This new product would become the focus of marketing activities for future periods after current contracts expire. The locked-in period for activities in a strategy optimisation may therefore be longer than for a base or precious metal operation, but conceptually there may be little difference between the processes used.

Cut-off optimisation is not inapplicable for bulk commodities, as is sometimes postulated. It simply requires a different outlook and time frame from base and precious metals.

Market-limited products

Some products, such as diamonds and specialty metals used in niche applications, may be constrained by an external market limitation. In one sense, this is the case with all products, but for many, an individual mine will often be a small contributor to the global supply. Its profitability and operating strategies will not be affected by the physical market limitations and influence of its own outputs on the market.

Discussions in this chapter and Chapters 5 and 6 in particular have recognised a product-related constraint, but noted that it is usually a physical constraint within the operations and downstream product-handling facilities. Yet the possibility of a limit on the amount of product that can genuinely be sold into the global market was recognised. Those discussions therefore generally cover market-limited products.

In some instances, an operation's output may impact the price of the product. Simplistically, the discussions assume a specified product price, but there is no reason to exclude the assumption of a price varying in relation to the amount of product generated by the operation. If such a relationship is derived, it can easily be accounted for in the strategy optimisation processes described in Chapter 6 and in the iterative processes relayed for Lane's methodology in Chapter 5.

CHAPTER SUMMARY

This chapter has described the data requirements and procedures to perform a strategy optimisation in detail. It is not possible to fully optimise the mining strategy without considering the metallurgical treatment processes and vice versa. A full strategy optimisation requires an integrated evaluation of all parts of the production process across the whole life of the operation.

The corporate goals must be borne in mind throughout the optimisation process. They indicate what is required to derive value measures for those goals, and the processes and data to do that. As well as goals, it is vital to identify all the options to be evaluated; data and relationships must support the full range of values.

It is necessary to take account of the influences of uncertain parameters in an optimisation study. As with the most uncertain parameter in a mining strategy evaluation, the product price, it's the norm to estimate the most likely scenario and flex the price inputs over a range. A change in price or any uncertain input will change the value of any operating plan, but the issue is not the absolute value, but how changes might cause the optimum plan to vary. If the decision-changing situation is found to be outside the likely range for an uncertain parameter, it may be important for defining the value, but not the optimum strategy. If it lies near the most likely value or relationship, steps must be taken to reduce the uncertainty or consider the trade-offs between the benefits and costs of making decisions based on the right or wrong assumptions.

Following is a summary of the types of inputs and processes typically required.

Geology

The major activities related to geology for an optimisation study are:

- create a reliable model for the range of cut-offs to be investigated
- generate potential orebodies at each cut-off
- identify domains that require different mining methods, scheduling constraints, metallurgical treatment, etc.

If the model is unreliable for some cut-offs, rigorous analytical processes can still be conducted using the existing model. If the optimum strategy is then found to lie within the range of cut-offs where the geological model is deemed to be reliable, the identified optimum will be reliable. If, however, the optimum strategy lies outside the reliable range, identify the best value within the reliable range, the additional value potentially available, and whether that justifies the cost of extending the reliable range of the model.

Mining

The major activities related to mining for an optimisation study are:

- identify suitable mining methods
- create realistic mining shapes and aggregated mining blocks, and hence reserves for each cut-off or pit limit and mining method
- produce conceptual mine designs and schedules for selected representative cases (for model calibration)
- aim to maximise the information obtained while minimising the work to be done
- develop suitable interpolation relationships for other cases.

For open pits, the aggregated mining blocks will be bounded by pit shells corresponding to each major pushback stage and by a number of elevations, spaced at a multiple of the bench height. The optimisation software will select the blocks that are to be mined and schedule them in such a way as to generate the ultimate pit size and optimum mining sequence. It will usually be necessary to define haulage distances for each mining block to both ore and waste-dumping points. The optimisation will decide what is ore and what is waste, so both distances need to be available to the optimiser. More complex information may be required depending on dump locations, rock types and the like.

Underground, mining blocks will typically be defined by a combination of ore zones, sublevels, northings or eastings, geological structures, changes in the size and shape of the orebodies and boundaries defining geological domains within the rock mass or orebodies. The mining inventory for each cut-off must take account of both the ore loss and dilution to generate realistic stope designs and the elimination of outliers of material above cut-off that would be uneconomic to extract.

Development requirements for each underground mining block will be needed. It may be relevant to specify what proportion of each category needs to be completed before the mining block can start producing. Trucking distances may vary if different haulage and hoisting options are to be evaluated. Quantities of backfill of various types may need to be specified if mining methods being evaluated have different fill types or proportions of the orebody to be filled. Underground data requirements may also include such items as access methods or topologies.

Scheduling rules need to be specified for both open pit and underground optimisations.

Processing

The major activities related to processing for an optimisation study are:

- specify recovery relationships for the range of cut-offs and hence feed characteristics to be evaluated
- identify constraints and how to remove them for the ore and all product streams
- determine the impact of debottlenecking on recovery, product quality, operating and sustaining capital costs, etc.

Because of the wider range of feed characteristics modelled in an optimisation study, better relationships may need to be developed for the study than those used in the operation for a narrower range. Scenario analysis may be necessary over any uncertainty range in the relationships to identify whether strategic decisions will change if relationships vary. The increasing focus on geometallurgical properties of mineral resources leads to greater information being taken into account to optimise operating strategies. This potentially increases the complexity of the analysis to account for it.

Operating costs

A suitable cost model is required, typically incorporating fixed costs (related to time elapsed) that do not vary across the range of activity levels evaluated, and variable costs driven by physical activities such as development metres (underground) or waste mined (open pit), ore tonnes mined, truck tonne-kilometres or operating hours, backfill tonnes (underground), ore tonnes milled and product quantities.

Capacity or step variable costs may be relevant. These are fixed across a portion of the whole range, with step changes at threshold levels of activity.

The general principle is that variable costs need to be specified separately for any physical parameter that does not vary in direct proportion to rock, ore or product quantities. An overall average cost per tonne will include costs associated with all parameters, assuming that the ratios of all quantities are unchanging. If the physical ratios change with changing strategies, the unit cost will as well. This introduces a circularity into the analysis, resolved by relating costs to the parameters that drive them.

The way in which unit costs are reported in company documents usually does not describe how they behave. The evaluation team must examine the cost data provided and, when necessary, reallocate or reclassify, to ensure that the cost model accounts for how costs actually behave.

Capital costs

Capital costs are separated into sustaining capital, including fleet replacement and project capital. The effect of project capital on future maintenance, operating and sustaining capital costs must be identified. Potential discoveries from exploration expenditure, in or near the mine and regionally, may impact on plans for known resources and often needs to be evaluated.

The distinction between capital and operating costs is in many ways irrelevant for strategy optimisation; what matters is how costs behave. Project capital is a lump-sum expenditure to acquire some new capability. Sustaining capital may be thought of as irregular maintenance expenditure for existing production facilities. Sustaining capital behaves similarly to variable operating costs, and should be treated like operating costs. The main reason to maintain the capital/operating distinction for an optimisation study will be to correctly calculate tax.

Working capital may have an impact on cash flows and value. It may be thought of as the difference in timing between expenditure of costs and receipt of revenue – operating costs will need to be expended for some time before first revenues are received. Working capital is often estimated by specifying a lag for revenue cash flows between the time of production and the receipt of revenue, and a lead or lag for cost cash flows (depending on stock levels, consumption rates and payment terms).

Social issues

For strategy optimisation, many social issues could be considered as options or scenarios. It may be useful to identify the value of the operation with or without the expenditure or the loss of value due to deferring the project because of time required to reach agreement on social issues. This knowledge can assist decision-makers to assess the time or money worth spending on these matters.

Financial and economic inputs

These inputs include such items as product prices and sales terms, taxation and discount rates. Although corporate policies may specify values to use, an optimisation study team should not be constrained to use only official forecasts. The study should show how robust the recommended plans are. To this end, it might be appropriate to indicate how far, for example, prices could vary from the forecasts without affecting the optimum strategy.

End-of-life issues

These include issues such as redundancy payments and rehabilitation costs and commitments. If the project has a long life, the impact of these on NPV may be negligible and safely ignored, although some companies' procedures require that they be taken into account, regardless of the life of the operation. If the life is relatively short, these cash flows may be significant. Lane's limiting cut-offs indicate how high closure costs may drive a reduction in cut-offs to extend mine life and defer incurring these costs.

The longer the remaining life of the operation, the more uncertain the values for these types of cash flow items. Well-estimated costs for what is current best practice may turn out to be inadequate many years later when accepted best practice may have changed. The impact of ongoing resource discoveries on the life of the operation will also add to this uncertainty. The values can be flexed to ascertain whether they have an influence on current decisions.

Data adequacy

Generating adequate data for an optimisation study may require a significant amount of human-generated mine design, and availability of technical staff to do this work may be a practical constraint on the overall strategy optimisation process. The guiding principle for data generation is: maximum information from minimum work. Exactly how this balance is achieved in each case depends on the complexities of the mineralisation and mining methods, the engineering judgement of the technical staff and the corporate culture of the company undertaking the study. If the optimum solution found by the software is one without engineered designs, there should be sufficient input into other cases from which the data for the optimum case has been interpolated so that results can be presented with confidence.

CHAPTER 9

Specifying the Grade Descriptor

AN INTRODUCTION TO GRADE DESCRIPTORS

In this book, the term *grade descriptor* refers to any parameter that describes the value of a block and to which we may apply a cut-off to separate material into different categories. The purpose of the grade descriptor is to assign values to volumes of rock (such as those in the geological block model) with larger numbers representing greater value. Grade descriptor values should rank blocks in the same order as their true value, identifying that *this bit* is more valuable for processing than *that bit*.

The grade descriptor does not have to be an absolute true value, and often it won't be. Any measure of grade is a ranking tool only. Relativity is the important issue, but only in rank, not proportionality. As will be shown, even single metal grades will not express value ratios accurately; for example, material with a grade of, say, 2 g/t will not be worth twice the value of 1 g/t material. Grades expressed as dollar values – such as net smelter return (NSR), dollars per tonne – may also not describe the actual value gained by treating that material.

The specification of a grade descriptor is essential to a strategy optimisation study for a polymetallic or multiproduct mine to generate a single value to quantify the relative value of each block. For a simple single-metal deposit (for example, disseminated gold mineralisation), the actual grade (gold grams per tonne) is usually quoted, naturally achieving the required ranking; however, there are many grade descriptor issues that are associated with all styles of mineralisation, including single-metal deposits. Much of the discussion in this chapter will be relevant only for multiproduct deposits, but other aspects (such as grade descriptors not expressed on a per tonne basis) are relevant to all styles of mineralisation.

It is not essential that there be a linear relationship between the grade descriptor and the true value; non-linearity is often the case. Even with a simple gold deposit, if there is a non-linear relationship between recovery and grade, there is a non-linear relationship between value and grade. With a linear relationship, the relativities of the grade descriptor value and the true value may differ. For example, if a gold treatment plant operates with a constant 0.5 Au g/t tailings grade, an *in situ* grade of 2.0 g/t will have a recoverable grade of 1.5 g/t, while an *in situ* grade of 1.0 g/t will have a recoverable grade of 0.5 g/t. The ratio of *in situ* grades is 2:1, but the ratio of values (assumed for now to be represented by the recoverable grades) is 3:1.

In polymetallic deposits, common practice is either to generate an NSR dollar value for each block, or apply multipliers to the metal grades for each constituent to derive either a dollar value (typically per tonne) or a metal equivalent (typically expressed as the grade unit of the dominant metal). The typical value equivalence formula is:

$$Value = a \times Metal\ A\ grade + b \times Metal\ B\ grade + c \times Metal\ C\ grade + \ldots$$

Mathematically, this is a linear function of a number of variables, which in this specific application are the metal grades. Geometrically, the formula describes a plane in multidimensional space. In this formula, the value derived will often be a dollar value; however, if a metal equivalent grade – say, based on Metal A – is required, the formula would be:

$$Metal\ A\ equivalent\ grade = Metal\ A\ grade + b' \times Metal\ B\ grade + c' \times Metal\ C\ grade + \ldots$$

where:

b'   = b/a

c'   = c/a, etc

There are several ways in which the multipliers are commonly derived, and the calculations, advantages and disadvantages of each are described in the following subsections. In this discussion, any value used as the grade descriptor will be termed a *grade*, regardless of whether it is a single-metal grade, a metal equivalent, or a dollar value, unless the context indicates otherwise.

SELECTING THE CORRECT GRADE DESCRIPTOR

The importance of selecting the correct grade descriptor is illustrated as follows. Figure 9.1 shows the types of error that can occur if cut-offs are applied to a calculated grade that does not rank blocks in the same order as their true values. The calculated values are often derived from an equivalence formula while the true values may be NSR dollar values.

It is assumed that the underlying metal grades are correct. The variations in Figure 9.1 are in addition to any grade estimation errors. Inappropriate multipliers can undo the time and effort spent by resource geologists to reduce these errors.

The horizontal and vertical lines intersecting on the trend line through the data points represent cut-offs applied to the calculated and the true (but possibly unknown) values. Points in the upper-right and lower-left quadrants are correctly classified as ore or waste. Points in the upper-left quadrant are truly ore, but are incorrectly classed as waste on the basis of the calculated grade. They are excluded from the reserve and represent a loss of ore. Points in the lower-right quadrant are waste incorrectly classed as ore. They are incorrectly included in the reserve, and represent a degrading of the reserve with subgrade material.

FIGURE 9.1

Typical plot of true and calculated grade values.

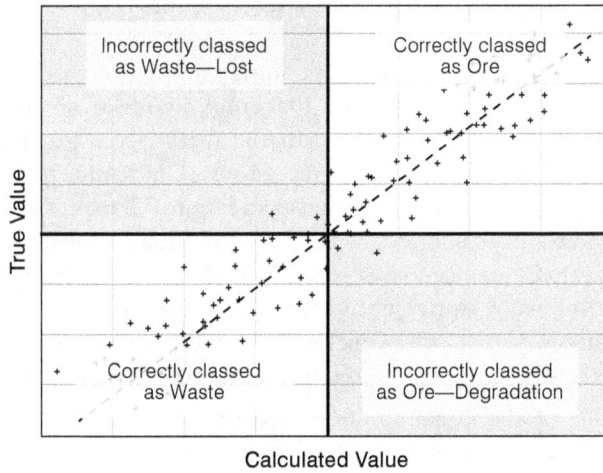

This implies that at any specified cut-off, an unsuitable grade descriptor will produce a resource or reserve with a grade lower than it could be. It's uncertain whether the tonnage will be higher or lower – this will depend on the nature of the data point distribution at the cut-offs applied to the true and calculated grades; however, for any given tonnage, the grade will be lower than it could be. Figure 9.2 illustrates this – the result of a case study from a base metals operation where the original equivalence formula (Formula 1) created a spread of true values for an equivalent value. A second formula (Formula 2) was then derived to reduce that spread.

It can be seen at low cut-offs and high tonnages that there is little degradation effect; the two grade versus tonnage lines are close or coincident. But with high cut-offs and low tonnages, major improvements result from changing the equivalence formula. For

FIGURE 9.2

Tonnage–grade variations with different equivalence formulas.

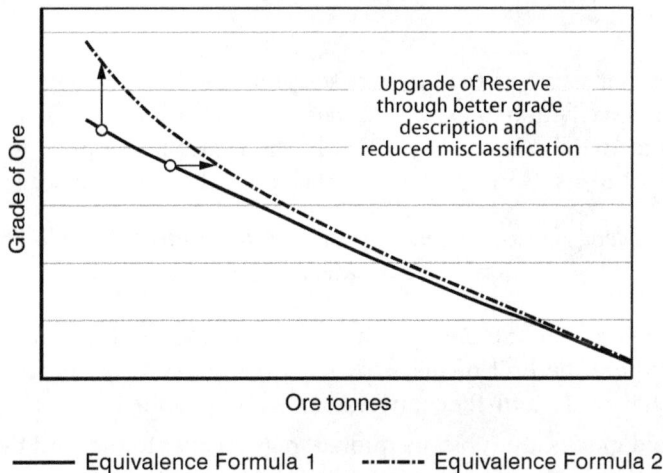

any given tonnage the grade has improved, and for any given grade the tonnage has increased.

If there was a constant ratio in all blocks between the metal grades contributing to the equivalent value, the relative weightings would be immaterial – all sets of weightings would generate equivalent values showing the same rankings of blocks of mineralisation. In the simplest case, all constituents other than one could be given zero weightings, and the metal grade of the remaining constituents would be a good grade descriptor, ranking all blocks of rock the same as their true values. The scatter plot in Figure 9.1 and differences between grade versus tonnage curves in Figure 9.2 would cease to exist. Grade contours (in two dimensions) or boundary surfaces (in three dimensions) at different cut-offs would remain nested in concentric sets regardless of the weightings applied and the resulting numerical values of equivalent values that could be generated. The numerical value given to a boundary or surface enclosing a volume of rock would change with different weightings applied, but the boundary or surface itself would not.

In reality, the ratios of the constituents' metal grades of a polymetallic deposit do vary from block to block. Figures 9.1 and 9.2 result from combining these grade ratio variations and different weightings to constituents of the mineralisation. Grade contours (in two dimensions) or boundary surfaces (in three dimensions) at different cut-offs no longer remain nested in concentric sets with different weightings applied. There is now the potential for what was ore in one case to be waste in another, so there is also the potential for sets of boundaries at different cut-offs using different weightings to cross over each other rather than remain concentric. If one of the weighting sets (or calculation methods) generates the true value, other weighting sets will generate the scatter plot and the existence of points in Figure 9.1's error quadrants. This effect is discussed in more detail shortly.

TYPES OF POLYMETALLIC GRADE DESCRIPTORS

In the following discussion, dollar-value equivalent grades are assumed, but there is generally a simple ratio between a dollar value and a metal equivalent grade. As indicated, to obtain a metal equivalent, the dollar-value multiplier of the basis metal is usually divided into the dollar-value multipliers of other ore constituents, so that these are expressed in terms of their values relative to the basis metal.

In situ value

Using an *in situ* dollar value as the grade descriptor has the advantage of simplicity. It merely applies the specified metal prices to the metal grades of any block, and its *in situ* dollar value is immediately determined. That is, the multiplier applied to each metal in the equivalence formula is its price. The typical *in situ* value equivalence formula is:

$$Value = Metal\ A\ price \times Metal\ A\ grade + Metal\ B\ price \times Metal\ B\ grade$$
$$+ Metal\ C\ price \times Metal\ C\ grade + \ldots$$

When metal prices are in another currency, the exchange rate has to be included in the calculation. Where the unit of mass for the price quoted differs from the unit for describing the metal grade, a further conversion factor is applied.

For example, gold grades are typically quoted in grams per tonne, and the gold price is quoted in US\$/oz (troy). The exchange rate from US dollars to the local currency must be applied to convert the gold price in US$/oz to a local currency price per ounce. This is then divided by the 31.10348 g/oz conversion factor to derive a local currency price per gram multiplier to apply to the gold grade in grams per tonne. Similar conversions will apply to base metals, where prices are often quoted in US cents per pound, to derive multipliers to apply to percentage grades. Care must be exercised in these cases to ascertain whether the percentage grades are recorded as percentage values with magnitudes between 0 and 100, or as decimal fractions between 0 and 1. The latter might then be displayed as percentages (which can be done with software applications such as Microsoft Excel™).

The main disadvantage of the *in situ* value is that it fails to take account of the proportion of each constituent that is recovered, and the net revenue then received for what is recovered. There is potentially a spread of true values for any given *in situ* value, as shown diagrammatically in Figure 9.1 (assuming that the horizontal axis is *in situ* value), which can in turn generate the effects shown in Figure 9.2. The magnitude of the spread will depend on the variability of the ratios of the constituent metals and of the recoveries and payabilities of each.

Recoverable value

Using a recoverable value as the grade descriptor again has the advantage of simplicity – the same processes are applied as for *in situ* value, except that metallurgical recoveries are also applied to each metal. This is a simple process if recoveries don't depend on grade: the multiplier applied to each metal in the equivalence formula is its price multiplied by its recovery. The typical recoverable value equivalence formula is:

Value = Metal A price × Metal A recovery × Metal A grade + Metal B price × Metal B recovery × Metal B grade + Metal C price × Metal C recovery × Metal C grade + …

Where recoveries are grade dependent, simple grade multipliers may be inadequate and a more complex function of grade might have to be applied. The recovery terms in the formula may need to be functions of the metal grades.

Compared with *in situ* values, recoverable values have the potential to reduce the spread of true values for any given recoverable value.[1] Yet recoverable values still fail to account for differences in payments for recovered metals. Applying the specified metal price to recovered grades implicitly assumes that the same proportion of the price is received for each metal. This is usually not the case. Sales terms for base metal concentrates, for instance, have different payable percentages for the various metals, as well as treatment charges (TCs), refining charges (RCs) or combined (TC/RCs) that represent different proportions of the various metal prices. Different grades of metal concentrates will also result in different freight costs per tonne of metal, even if the cost per tonne of concentrate is the same.

As with *in situ* values, the differences between recoverable and true values will depend on the variations in the ratios of grades of the contained metals and in the proportions of the metal prices actually received for each metal.

---

1. The author has, however, encountered situations where constituents with low recoveries have high payable proportions and vice versa. In one of these cases, the *in situ* values were better at ranking blocks similar to the true value than recoverable values. Every case must be judged on its own merits.

Payable value

Payable values take account of all the metallurgical losses, product-associated costs and payable proportions of the recovered products. For a base metals operation selling concentrates, this is represented by the NSR, less concentrate freight, transport and selling costs, which may be referred to as the NSR at the mine gate. If a mine has its own smelter and refinery, the costs that would be incurred by a concentrate seller – represented by the payable percentage and the TC/RCs, plus any penalties for deleterious elements – are incurred as downstream losses of recovery and direct operating costs, but the principle is the same for both types of operation.

Simplistically, the formula for recoverable value could be enhanced by including a payable term for each metal, but often the payability is a more complex relationship than can be accounted for in a formula.

As has been noted, common practice on mine sites is to express all site costs as a cost per tonne of ore, even though the ore tonnage has little or no bearing on many of them; several costs, such as site administration and much of the labour cost across the operation, are fixed, varying with time rather than volume of ore. Similarly, once concentrate has been separated from the ore stream, many costs are related to concentrate or metal quantities, not ore tonnes. Relating all costs to ore tonnes may be reasonable for certain reporting functions; however, it can be dangerously misleading if used for operational decision-making.

The best grade descriptor (in terms of ranking blocks of rock) will be the one that identifies the net return from the block. It takes account not only of metal prices and metallurgical recoveries, but also costs in the concentrator and downstream in the smelter, refinery, etc that will vary according to the amount of metal recovered from the block (that is, the costs dependent on metal grades), not on average costs resulting from average feed grades. High-grade blocks will incur higher costs per tonne of ore for product-related variable costs than low-grade blocks. This is not a concern for a single-metal deposit; however, for a polymetallic deposit, these variations can impact greatly on the value ranking of blocks where the ratios of constituent metal grades vary. To correctly rank blocks, it is essential that variations in costs associated with various products be accounted for. The only way to do this is by using a payable grade or value.

There are several ways to determine a payable grade. At many operations, complex recovery relationships and product sales terms are built into the block model calculations to generate individual NSR values for each block, assuming that the total mill feed has the grade and other characteristics evaluated. Where this is possible, it will be the best way to determine the payable value of each block; however, this may not be an easy process to implement because of complex flows and relationships within the metallurgical processing routes from ore to final products.

NSR-type payable grades will commonly be expressed as dollar values. However, there is no reason why a metal equivalent grade cannot be developed for an NSR-type grade. If an equivalence formula were used, the multiplier for the basis metal, which is effectively the value of one grade unit of the basis metal, would be divided into the dollar-value multipliers for all other metals, as indicated earlier in this chapter. To derive an NSR-based metal equivalent, the NSR value of a feed grade of one grade unit for the basis metal and zero grade for all other constituents could be derived and this value divided into NSR-based dollar value grades.

Inaccuracies common to recoverable and payable grades

Inaccuracies may exist even with the most accurate NSR calculations of payable grade. Although generally unrecognised, these same inaccuracies apply to all other grade descriptors, even when simple metal grades are used for a single-metal deposit. These errors can arise from the recovery–grade relationships, which are often derived at an operating mine from historical records of production performance, and are therefore based on the average grades of feedstocks composed of ore blends from many sources. There is no guarantee that the average recovery relationships can be applied to individual ore sources or rock blocks. This is, however, frequently done, and in reality there may be no alternative. As long as the potential problems are recognised they may not be accounted for, but errors can be introduced if they are not. The increasing availability of geometallurgical information may assist with reducing these errors, but at the time of writing, non-linear processes for estimating geometallurgical properties of individual blocks from sample points, and for recombining these to predict the behaviour of ore blends, are not well established.

If the recovery is a linear function of head grade, the recovered product from a blended feed, using the recovery calculated for the average grade of the blend, is equal to the sum of the product quantities recovered using recoveries calculated separately for the grades of the components of the blend. Here, applying the recovery formula to individual block grades in isolation results in a good estimate, not only of the true recoverable or payable grade of each block, but also of all blends when combined using linear tonnage-weighted calculations.

If the recoverable grade versus head grade relationship is non-linear (for example, where recovery is expressed as a polynomial function of head grade), the blend recovery cannot be estimated by linear tonnage-weighted calculations applied to the blend components. If an incremental tonnage with a particular grade is blended with a base tonnage of a different grade, the grade of the blend will obviously differ from the grades of the component tonnages. The recovery formula can be applied to the grades of the base as well as incremental and total tonnages to determine the recovered product. It's usual that the sum of the recovered products from the components is not the same as the recovered product from the blend. The additional product recovered by adding the increment to the blend is therefore not the same as the recoverable product calculated for the increment in isolation. The difference can be positive or negative, and can vary depending on the recovery relationship and the relative tonnages and grades of the base and incremental tonnages.

There is potentially an error in the estimated true value of a block of rock if it is calculated in isolation using the recovery relationship based on average feed grade. In light of the discussion in the preceding paragraph, this error may be negligible, but it may also be relatively large, and either over-value or under-value the block. Furthermore, the error is not constant, but depends on what the block is mixed with at the time it is mined.

The major implication is that if the value of a block depends on what it is blended with, it may never be possible to specify the absolute true value of any block. In these circumstances, the best that can be said is that the recovery calculated for a block in isolation will result in a value that reasonably accurately ranks the block relative to all other blocks available for mining at the same time.[2] If the cut-off is derived by an optimisation process that selects the best cut-off value to use based on the mining inventory available at different cut-offs, such a value will still be satisfactory for ranking material and prioritising it for treatment.

If the cut-off is based on a break-even calculation, there is no way of telling whether – and if so, to what extent – a non-linear recovery relationship may under or overstate the value of marginal material relative to the break-even value. Similarly, if strategy optimisation software used requires values to be assigned to mining blocks before the optimisation is conducted (rather than blend value being calculated within the optimisation process), the result may be suboptimal.

It is therefore important to ascertain to which grade the metallurgical recovery relationship is to be applied. It tends to be to average feed grade, though, due to the practicalities alluded to before, an 'average grade' relationship may need to be applied to individual blocks. The author has encountered a situation with a highly non-linear recovery versus grade relationship, where it was assumed that small ore blocks, equivalent in size to geological model blocks, would effectively be batch processed sequentially rather than blended. It was therefore appropriate to apply the recovery relationship to individual geological model blocks. The recovered metal from the ore feed over a period of time could be reliably estimated by totalling the recovered metal individually for each of the geological model blocks in each time period, rather than by applying the recovery relationships to the average feed grade.

GENERATING AN EQUIVALENCE FORMULA

Despite the increasing capability of geological modelling software to generate an NSR for each block, in some cases the metallurgical processing may be too complex for this. Therefore it is still common to derive an equivalence formula by which equivalent grades can be generated for all blocks of rock. For *in situ* and recoverable grades, the multipliers are easily derived from forecast prices and recoveries as appropriate (though, as indicated in complex plants, recovery may not be determined by a simple relationship to apply to grades). When all payabilities and costs are clearly driven by individual products only, simple extensions applying them to the recoverable calculations for each metal separately may be satisfactory for deriving payable values. But when some costs are common to two or more products – for example, treatment and freight charges for base metal concentrates with two or more payable products – more complex calculations may be necessary.

By accounting-style cost allocations

If more complex calculations are required, it is common to assume an average feed grade and derive the NSR. Incurred costs are deducted to derive net values, and hence value multipliers or weighting factors, for each product component. Common costs are distributed among the various products. This is typically done by an arbitrary process; for example, in proportion to revenues generated, or by attributing all costs to the major product and none to the other products. The latter is common for base metal concentrates containing precious metals – although the precious metals may contribute significant value, they are an insignificant component of the concentrate weight, which is what drives

---

2. This is a qualitative, experience-based judgement by the author and is not supported by any rigorous mathematical or statistical analysis.

transport costs. This raises questions about whether some elements are being unduly favoured or penalised by this process.

If the metal ratios in the orebody are constant through its volume and over a range of grades, these arguments are academic, as the constancy of the metal ratios will automatically compensate for any cost misallocation. If the ratios are not constant, as is more often the case, it is possible to misclassify blocks of rock to a severe extent if the multipliers, and hence relative weightings of ore constituents, are incorrect. Unfortunately, at operations where this process is employed, the discussion about how to allocate common costs frequently remains focused on the resulting multipliers themselves or on the processes used. There is little point in arguing about the relative merits of alternative arbitrary processes – if they are arbitrary, none has any more rational justification than any other. Such arguments are futile without an objective way of assessing whether a set of multipliers derived using one set of assumptions about common cost allocation is better or worse than another set with different assumptions.

The issue is not the values of the multipliers or the processes used, but the effect that they have on the resulting orebody. Notionally, there is a rational procedure to be followed – generate block models with block values derived using a number of multiplier sets derived with cost allocations, and see how the orebody (or family of orebodies) varies between cases. If variations with different multiplier sets are non-existent or small, the method of cost allocation to derive the multipliers is not an issue, and any set may be used if derived using reasonable processes, even if they are arbitrary.

If the differences between orebodies derived with multiplier sets are significant, it must be determined which set is best. It would then theoretically be necessary to do mine designs, schedules and financial evaluations for alternative orebodies defined by the multiplier sets, to ascertain which gives the best result. This is a long and involved process, and cannot be done every time one of the many parameters that drive the value of an orebody changes; however, it is the logical sequence to be followed for a rigorous answer using this accounting methodology to allocate common costs.

By multiple linear regression

There is another way. As noted, the typical value equivalence formula, which is a linear function of the metal grades describing a value plane in multidimensional space, is:

$$Value = a \times Metal\ A\ grade + b \times Metal\ B\ grade + c \times Metal\ C\ grade + \ldots$$

In the same way that we require two points to define a line in two dimensions, and three points to define a plane in three dimensions, so we require n points to define a plane in n dimensions. If we were to investigate the relationship between two variables, we would obtain many more than two points to perform a statistical regression and define the straight-line relationship (the line of best fit) between variables. If we have many more than n points in an n-dimensional problem, we can perform a multiple linear regression to define the plane of best fit. In a two-dimensional analysis, a simple linear regression will generate the best fit values for the y axis intercept and the coefficient of the independent variable x. In the same way, a multiple linear regression will calculate the best fit values of the dependent variable axis intercept and the coefficients of the independent variables in n dimensions. If we have a number of sets of ore grades and their corresponding dollar values, a multiple linear regression can be performed, and the best fit values of the multipliers for the constituent grades will be determined. There is no need to make arbitrary allocations of costs. The maths automatically derives the best relationship between value and metal grades.

Mines usually have a model for deriving the final net revenue from the mine head grade, accounting for a large number of metallurgical performance and product sales terms parameters, which have an effect on the ultimate value received by the mine for a parcel of ore treated. If the processes are too complex to derive an NSR-type grade value for each geological model block, a data set of grades for a representative sample of blocks can be extracted from the block model. Assuming that each sample represents the grades and other characteristics of the mill feed, the physical, economic and financial model can calculate the net revenue generated in each case. Each combination of input grades and the resulting value will be the coordinates of a point in multidimensional space. With a number of data points greater than the number of variables, and covering the full range of individual metal grades and ratios of grades (to avoid problems of collinearity), the equation of the multidimensional plane can then be determined by multiple linear regression. Software such as Microsoft Excel™ provides a simple way of doing this.

If all the points lie close to a multidimensional plane, the linear equivalence formula model is as accurate as can be. The multipliers derived by regression will be the theoretically correct multipliers, in that the value predicted by the simple equivalence formula model will be the same as that resulting from the more complex economic and financial model. The coefficient of correlation (R-squared) calculated by the regression process will help to identify whether this is so. The closer R-squared is to 1, the better the linear equivalence formula fits the data points. This will be the case where the net payable grades are linear functions of head grade.

As well as generating linear multipliers to apply to each metal grade to derive an equivalent value, the multiple linear regression process will generate a constant term. It is unusual for equivalent grade formulas to include a constant term, but there is no reason why they should not. There are two options regarding the constant term. Since it will be applied to all blocks, it will not affect the ranking of blocks relative to each other. If simple ranking of blocks is the purpose of the grade descriptor, the constant term may be ignored, and only the multipliers for each metal applied to the metal grades. If it is desired that a dollar value grade descriptor should represent the true value as accurately as possible, the constant term should be included in the equivalence formula. This will need to be the case if, for example, sets of multipliers need to be derived for different rock types and different constants are derived from the regressions. Alternatively, the regression can be conducted in such a way as to force the constant term to be zero. Multipliers for metal grades will differ from those obtained with a constant term, and the correlation will not be as good. Consideration of the resulting change in the correlation coefficient or of values derived for actual grades, using a process similar to that illustrated in Figures 6.14 and 7.1, will indicate whether eliminating the constant term is appropriate.

Where the net payable grades are not linear functions of head grade, or for some other reason the value versus grades relationship is non-linear, one could potentially identify a more complex but accurate mathematical formula relating value and the metal grades. Alternatively, if the main purpose of the equivalent value is to define blocks above and below cut-off, a linear formula chosen to be accurate in the region of the cut-off could be a good solution.[3]

DOLLAR VALUES VERSUS METAL EQUIVALENTS

Many polymetallic mines use a dollar value as a grade descriptor. Common practice at such operations is to calculate the cost per tonne and then use that unit cost as the cut-off. This has the appearance of both rigour and simplicity, and is effectively applying a break-even cut-off.[4] As noted in the preceding subsection, there is no guarantee that the dollar value assigned to a block accurately represents its true value. The application of break-even cut-offs to dollar values of material can potentially lead to under or over-stated cut-offs.

When price predictions change (often when new forecasts are issued annually as part of the company's planning cycle), typical practice is to derive and apply new multipliers to block grades to produce new dollar values, and then re-create ore boundaries by applying the existing cut-off value to the new block dollar values. If break-even cut-offs are being used, this process is logical, but again there is no guarantee that the new dollar values accurately represent the true value, nor that the errors, if any, are consistent between the old and new values.

However, there are problems with this approach when optimum rather than break-even cut-offs are used. Optimum cut-offs are intended to achieve some particular optimised goal, such as maximising NPV, whereas break-even cut-offs merely ensure that every tonne pays for itself. As indicated in previous chapters, NPV-maximising cut-offs are typically higher than break-even cut-offs. Conventional wisdom regarding changes in cut-offs when prices change is based on the behaviour of break-evens; changes in cut-offs are expected to be inversely proportional to changes in prices. Yet the outcomes of optimisation studies show that, for a single metal deposit using the metal grade as the grade descriptor, the NPV-maximising optimum cut-off will change by a significantly smaller proportion than the inverse proportion of the price change. Figure 9.3 illustrates the problems that can arise when using dollar values.

The figure shows value versus cut-off curves for two price scenarios, the higher prices being 20 per cent higher than the lower prices. There are two cut-off axes for dollar values derived from each price set. Since all prices are assumed to increase by 20 per cent, the values on the high-price axis are all 20 per cent higher than the corresponding values on the low-price axis. For example, $120/t on the high-price axis corresponds with $100/t on the low-price axis. The curves indicate that with low prices, the optimum cut-off is $100/t and, if the grade descriptor values are unchanged but the metal prices increase by 20 per cent, the optimum cut-off drops to $93/t, a seven per cent reduction.

---

3. The regression would be conducted using only the grades of sample blocks, for which the true values lie within a small range above and below the cut-off value.

4. There is an operation with a number of polymetallic orebodies that are mined by separate mines feeding a single concentrator. Staff in a centralised section are responsible for deriving NSR-type grade descriptors for the blocks in the block models of all orebodies. These are derived to represent the net value of the ore at the point of delivery to surface at each mine. Junior technical staff are tasked with determining the cost per tonne at their mines. These costs are then applied as break-even cut-offs to the NSR-type payable value grades. By virtue of these procedures, the company effectively allocates all responsibility for defining what is and is not its ore first to the market, which sets prices and hence ore values; and second, to its junior technical staff, who then determine what will be classified as ore. The company in question is not alone in doing this.

FIGURE 9.3

Value versus cut-off curves with different prices and dollar value grade descriptors.

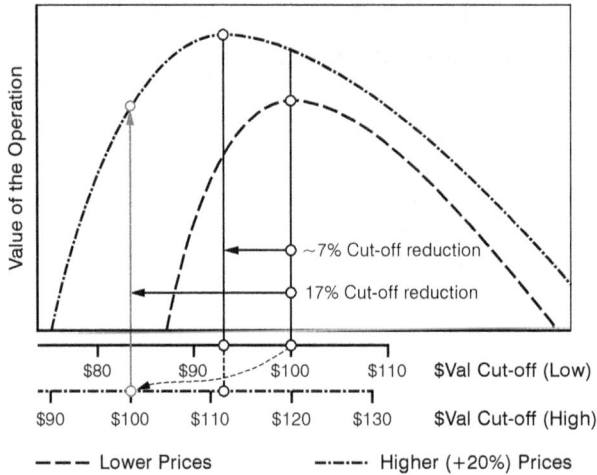

With operations using a dollar value grade descriptor, the block model dollar values will often be adjusted – in this example, upwards by 20 per cent – and the existing optimum cut-off will then be applied to the new dollar values. It can be seen that $100/t on the higher-price cut-off axis is equivalent to $83/t on the lower-price axis. The net effect of this, as with break-evens, is to reduce the effective cut-off by 17 per cent, which will be suboptimal since the optimum cut-off should reduce by only seven per cent. The optimum cut-off with the higher prices and increased dollar values in the block model should actually be $112/t, an apparent increase in cut-off of 12 per cent. There is potentially a significant loss of value by applying the old $100/t cut-off to the new higher-price dollar values.

Some readers might be wondering about the validity of a cut-off using old dollar values when there are new price forecasts with new dollar values. The initial response is that the old dollar values are simply a set of values ascribed to each block of rock to indicate those that are more valuable than others. As indicated earlier in the chapter, there is no requirement for the relationship between true value and grade descriptor value to be linear, merely that the grade descriptor values rank blocks of rock in the same way as their true values. So if old dollar values rank blocks of rock as accurately as the new true values, it is unnecessary to change the grade descriptor.[5] To illustrate the matter further, consider the same situation with a single-metal resource.

Suppose a gold operation has been working with a gold price of $1000/oz, and that the metallurgical recovery is 93.3 per cent for all gold grades and the unit cost is $90/t. Assuming a conversion factor of 31.1 g/oz, the break-even grade is therefore 3.00 g/t. The optimum cut-off has been found to be 4.00 g/t. Although the deposit is a single-metal deposit, there is no reason why we cannot determine dollar values for the mineralised material. With the information given, there is a very simple conversion: $30/t per 1 g/t of gold grade. The dollar value break-even is therefore $90/t (the unit cost, as it should be) and the optimum cut-off is $120/t.

---

5. When to recalculate metal equivalents or dollar values is discussed later in this chapter.

Consider that the gold price is forecast to increase by 20 per cent to $1200/oz and all other physical and financial parameters remain the same. The break-even gold grade will reduce to 2.50 g/t, a 16.7 per cent reduction, as expected with a 20 per cent price increase. Suppose that with this price increase, the optimum cut-off is found to reduce by seven per cent, to 3.72 g/t. What happens with the dollar values? With the new price, the dollar value increases by 20 per cent to $36/t per 1 g/t of gold grade. The dollar value break-even with the new price is therefore still $90/t, as it should be – the costs have not changed. The optimum cut-off with the old dollar value relationship will reduce to $111.60/t (that is, 3.72 g/t × $30/t per g/t), a seven per cent reduction, the same as the gold grade cut-off reduction. However, the optimum cut-off with the new dollar value relationship will increase to $133.92/t (that is, 3.72 g/t × $36/t per g/t), an 11.6 per cent increase, the same as in the preceding example.

There are several options for dealing with this, but first this apparent paradox needs to be understood. Figure 9.3 and the discussion following it illustrate the main principles, but the underlying considerations need to be explicit. The key issue is that even though the old and new prices are used to derive grades expressed in dollars per tonne, they are two totally unique grade descriptors. One is Year X $/t, while the other is Year Y $/t. To understand the distinction, consider a copper–nickel (Cu-Ni) deposit.[6] Both base metal grades are typically expressed as percentages. That does not mean that we could derive a cut-off to apply to the copper grade one year, then apply the same numerical value as the cut-off to nickel grades the following year. Cu per cent is a different grade descriptor from Ni per cent, and we would quite rightly never consider applying the same cut-off value to both simply because they are both expressed as percentages. Why then should we apply the same cut-off value to both Year X $/t and Year Y $/t just because they are expressed as dollars per tonne? Clearly it is irrational and simply wrong as a general principle, but it is often done. It will, however, be appropriate in the special case where the dollar value is a true value and the cut-off is intentionally a break-even cut-off and the costs have not changed.

Understanding the source of the paradox, we can now consider how to deal with it. The first possibility is to completely redo the optimisation study with new price forecasts and new dollar values in the block model, but this is usually not a trivial process. Price forecasts will typically be updated at least once a year by most companies, perhaps more often by some, but as noted in Chapter 7, redoing a strategy optimisation is not usually done every year, and certainly not more frequently. Even if the full optimisation is not redone, the forecast price change can be accounted for in several ways. All options require that the latest optimisation be simply rerun using the new price forecasts, so that the optimum cut-off with the current grade descriptor and new prices can be identified, as illustrated in Figure 9.3. This will identify the proportional change in the optimum cut-off using the current grade descriptor. The options are then:

- leave the dollar values in the block model unchanged, and reduce the cut-off applied by the appropriate percentage (seven per cent in the examples)
- increase the dollar values in the block model to account for the price change and reduce the cut-off applied by the appropriate percentage; then re-increase the cut-off by the price change percentage (in the examples, 20 per cent) to generate the net effect

---

6. Or any other polymetallic combination where the net values per grade unit of the various constituents are significantly different.

of the price change and the required reduction in the optimum cut-off due to the price change (11.6 per cent in the examples)[7]

- convert dollar values (both in the block model and as the cut-off) to metal equivalent grades, leave the block model metal equivalent grade descriptor values unchanged and reduce the cut-off applied by the appropriate percentage (seven per cent in the example).

Benefits of metal equivalent grades

Equivalent grades have the potential to be more stable grade descriptors over time than dollar values, and their use can potentially avoid some of the pitfalls of dollar values described in this section. Although the prices may change, equivalent grades will only change significantly if the proportional changes in prices of the various grade constituents and the ratios of grades are different, and if two or more constituents of the ore contribute notably to the equivalent value. For a single-metal ore, the grade does not change when the price changes, though the cut-off applied may. So, for a multiproduct orebody, although there will often be times when the grade descriptor should be changed when price forecasts change, the *prima facie* argument is that the grade descriptor values need not automatically change merely because the prices have. The need for change should be demonstrated, not assumed as a matter of course. This is true even if the grade descriptor is expressed in dollars per tonne; once a grade descriptor value has been calculated, it does not need to be changed unless the relative rankings of blocks change because of the price changes.

It is desirable to have a stable, unchanging grade descriptor. Technical staff are able to become familiar with the grades of the deposit without having to adjust to grade values that change on a regular basis. The use of a metal equivalent grade also serves to break inappropriate links between dollar value grades and unit costs, which lead to unthinkingly applying break-even grades – it is too simple to equate costs and values that are expressed in dollars per tonne.

Metal equivalents are recommended in preference to dollar values. Yet, since the two will usually be related by a simple factor (the multiplier of the major metal in a dollar value equivalent formula, or the dollar value of material with a grade of one grade unit of the basis metal and zero grade for all other constituents contributing to the total dollar value) there should be no problems with either so long as potential problems with dollar values are understood and addressed. It should be noted that the multiplier of the major metal in a dollar value equivalent formula will change as price forecasts change; so, although metal equivalents may remain stable, dollar values may change significantly from time to time. Again, this is a primary reason for using metal equivalents and changing them as infrequently as possible.

There are views that metal equivalents are in some way technically inferior to dollar values. Given the simple relationship between dollar values and metal equivalents, this would seem to be an ill-informed argument. There are at least four types of dollar values: *in situ*, recoverable, payable values derived by an equivalence formula and payable values derived by a full NSR-style calculation. The same applies to metal equivalents. The author therefore suggests that the view that metal equivalents are inferior to dollar values is based

---

7. Note that the mathematical procedure is not additive (-7 per cent +20 per cent = +13 per cent) but multipli-cative (93 per cent × 120 per cent = 111.6 per cent); that is, an 11.6 per cent increase, or 12 per cent rounded to the nearest percentage.

on an assumption that, for example, dollar values can be accurate NSR values, while metal equivalents are – in the minds of its detractors – by definition less accurate *in situ* values or similar. This is incorrect; metal equivalents can be derived by exactly the same processes as dollar values and are not necessarily less accurate or technically inferior.

RECALCULATING EQUIVALENT GRADES

When should equivalent values – whether metal equivalents or dollar values – be recalculated? If a break-even cut-off is intentionally being used with a dollar value grade descriptor, it will be appropriate to recalculate dollar values whenever the metal price forecasts change. Otherwise, the author recommends that recalculations should only be done when there is a significant change in the nature of the set of grade boundaries at different cut-offs.

Figure 9.4 shows two sets of grade boundaries generated for several cut-offs for two price sets. In this figure, the nature of the grade distributions of various metals and the changes in prices are such that the boundaries for both price sets are of the same family – they are nested or concentric and do not cross over each other. The same boundaries could be used for both price sets, the only change being the value attributed to each boundary. For circumstances like this, it is unnecessary to generate new grade descriptor equivalent values. The old grade descriptor values can be used with the new prices, though a new cut-off value might need to be specified.[8] If the grade descriptor is a dollar value, suitable cut-offs can be applied to that number in the same way as to any other grade descriptor. That the unit of the grade descriptor is dollars per tonne does not imply that it should be used any differently from other grade descriptors.

Figure 9.5, on the other hand, shows a situation where the nature of the grade distributions of various metals and the changes in prices are such that boundaries for the two price sets are very different – they are not nested or concentric, but rather cross over each other. For circumstances such as this, it is necessary to generate new grade descriptor equivalent values.

FIGURE 9.4

Grade boundaries with different prices and equivalent grade cut-offs – insignificant changes.

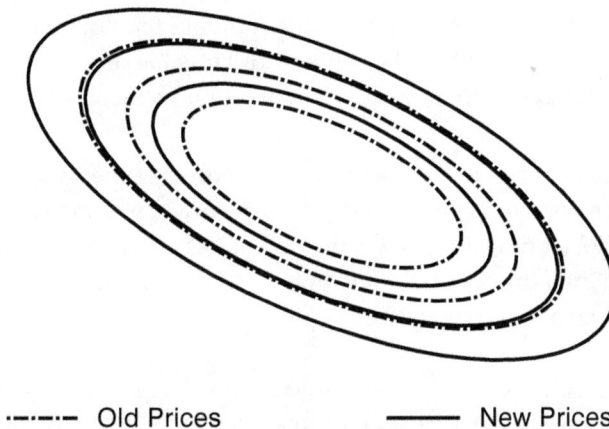

·——·——· Old Prices   ———— New Prices

---

8. A new cut-off value would not be required if, for example, the optimum cut-off were a balancing cut-off.

FIGURE 9.5

Grade boundaries with different prices and equivalent grade cut-offs – significant changes.

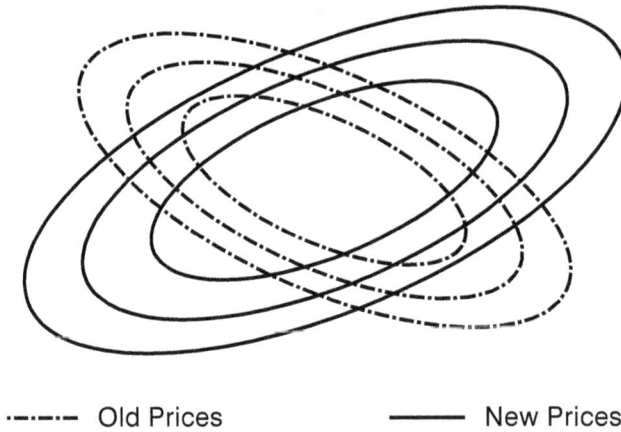

·—·—·— Old Prices ———— New Prices

Figure 9.4 and Figure 9.5 may be seen as the two extremes of a continuum of possible orebody boundary changes resulting from price changes. If predicted price changes generate large changes in the nature of the orebodies, as illustrated in Figure 9.5, it may be necessary to attribute dollar values or metal equivalent grades to each block in the geological model for two or more price sets, so that the optimisation software can take proper account of the changing relativities of values in parts of the deposit over time. In many cases, the impact of price changes on the orebody shapes will be more subtle, perhaps with small variations in boundaries where there are variations in the ratios of grades of metal constituents. Each case will be different and it will be up to the technical staff to determine how much change in the nature of the orebodies would warrant a change of grade descriptors.

Did the cut-off increase or decrease?

When prices increase and dollar value grade descriptors are being used, the dollar values are often recalculated and, of course, they increase. Often costs will have increased since the last cut-off specification, so a break-even cut-off, expressed in dollars per tonne to match the grade descriptor, also increases. After applying the new break-even cut-off to the new dollar values, it is found that the average dollar value of the ore above cut-off has increased as expected, but the ore tonnage is greater and its metal grades are lower than before the change.

This is often the cause of some bewilderment. The normal expectation is that when the cut-off is increased, the ore tonnage will reduce and its grade will increase, but in this case the opposite is true. What has happened is that the prices and dollar values have risen by a greater proportion than the cut-off value applied, so the cut-off has effectively reduced relative to the increase in dollar values.[9]

9. By way of analogy, consider a person marooned on a desert island. They wake up one morning and discover that the sea level is lower than it was the previous day and the island is bigger. With no outside reference, they cannot say whether the land has risen or the sea level has fallen. In the dollar values situation described, the land has risen (by the rate of the price rise) and the sea level has also risen (by the cost increase), but not as fast as the land has risen. The exposed dry land is larger. The tonnage above cut-off is larger and therefore also lower grade.

The issue, however, is the same as the change in optimum cut-offs using dollar values. The dollar values derived with different price sets are two unique grade descriptors, so the relativity of the cut-offs applied are irrelevant. The cut-off has not increased; rather, a new cut-off has been applied to blocks whose values are defined by a new grade descriptor.

If it is convenient to say whether the cut-off has increased or decreased when the grade descriptor has changed, the aspect that will define the cut-off change is the effect on the orebodies, not the numerical values of cut-offs applied to distinct grade descriptors. If the tonnage increases and average metal grades reduce, the cut-off has effectively been lowered. Conversely, if the tonnage reduces and average metal grades increase, the cut-off has been raised.

BEYOND GRADE OR VALUE PER TONNE

Although grades, metal equivalents and dollar values are typically expressed as per tonne values, the best ranking parameter for blocks of rock will be value per unit of constraint. If the aim is to generate cash, selecting blocks that produce the most cash when filling the constraining process will be a good strategy. For example, if two blocks of rock have the same grade and other metallurgical performance parameters, but one mills at half the rate of the other, it will therefore generate half the number of dollars of revenue as the other for each hour of milling consumed. Using value per unit of constraint grade descriptors will enable sequencing strategies and cut-offs to specify the best-value material to be targeted earlier in the operation's life, delivering the maximum value achievable, limited by the constraint.

If the real constraints are genuinely tonnage-related, a grade or value per tonne is still a good grade descriptor; however, constraints expressed as tonnages are often surrogates for something else. For example, the capacity of a treatment plant is usually expressed in terms of tonnes per hour or per year after accounting for breakdowns, maintenance and other delays. But the real constraint in a treatment plant is the available treatment hours or the milling power available. If there is only one ore type, or if all ore types have the same treatment rate (or power consumption), the ore tonnage capacity is a satisfactory surrogate for available hours, and a value per tonne – subject to the various concerns discussed earlier in this chapter – will accurately rank blocks of rock relative to each other.

If different material types have different throughput rates or recoveries, net dollars per hour may be the best grade descriptor to accurately identify that *this bit* is more valuable than *that bit*. As an example, there was an operation that had several rock types, all of which had different hardnesses, and hence distinct milling rates. The rock types also had different mineralogies, and hence different recoveries and products. The grade descriptor developed was net revenue per mill hour.[10] King (1999) refers to these types of grade descriptors as cash flow grades.

Other situations may be cited. For example, in a refractory gold plant with a roaster, the constraint may be the rate at which sulfur can be burnt. In this case, the gold/sulfur ratio can be a useful grade descriptor. In this situation, there may actually be two constraints: a

---

10. This was making use of the best available geometallurgical data before the term *geometallurgy* existed, or at least was in common use. The current focus on geometallurgy can only serve to advance the information available, and hence, improve grade descriptors and the making of informed strategic decisions.

milling constraint, common to most operations, and the roaster sulfur constraint affecting the treatment of sulfide concentrates. The first of these may be seen as an ore stream constraint and the second, a product stream constraint from a Lane-style or strategy optimisation point of view. When it comes to specifying an appropriate grade descriptor, both may be relevant. Low-grade or, at least, low-sulfide ore will typically be affected by the milling constraint, whereas high-grade or high-sulfur ore may be effectively constrained by the roaster.

Where there are multiple constraints such as this, it may be fitting to generate multiple *value per unit of constraint* measures. Cut-offs applied to each may be appropriate. In many cases the cut-offs will specify minimum grades to be classified as ore, but in others, a second (subsidiary) cut-off may represent a maximum value to be classified as ore.

In complex cases, linear programming or other similar algorithms, rather than cut-offs applied to grade descriptors, may be needed to identify material that delivers maximum value when it is necessary to account for such items as:

- multiple constraints
- blending requirements
- specified feed or product qualities.

This will be addressed in more detail in Chapter 13, which discusses various strategy optimisation methods.

HANDLING VALUE-REDUCING CONSTITUENTS

There are a number of constituents of the ore feed that may reduce its value. The author suggests that the four classes of such materials are those that:

1. result in additional costs in the metallurgical treatment process, downstream tailings handling and environmental management activities
2. reduce the recovery of the revenue-generating constituents
3. reduce the quality of the product generated and sold
4. result in applying penalties that reduce the net revenue received for the product sold.

The initial knee-jerk approach to all these situations is that their negative impacts on value should be represented in the values generated for the grade descriptor; however, this is not always applied in practice, and not necessarily true. The appropriate treatment will depend on how each value-reducing effect applies.

Penalty or loss-causing elements in the product – the last of the four categories – will often only generate the penalty if the grade of the penalty element is above a specified threshold value in the product sold. If payable values are derived for individual blocks by an NSR calculation, the effects of penalty elements are often included in the NSR dollar value. This is typically done by applying the sales terms to the product that would be generated if the total mill feed had the grade and other characteristics, including the effects of penalty elements and any threshold grades, for penalty payments; however, it is rare (in the author's experience) to find negative multipliers for penalty elements in linear equivalence formulas to derive polymetallic grade descriptor values. Rationally, if it is appropriate to include penalty effects in one calculation method, it should also be included in an alternative, albeit simpler, method.

For both calculation methodologies, if there is a penalty threshold, technical staff should consider whether to include penalty elements in the derivation. It is the average grade of a batch of the product, which in turn will usually depend on the grade of the ore blend, that will influence whether the penalty is incurred or not, not the grade of any individual block. Therefore, the value of an individual block should not be downgraded because it has a high-penalty element-grade if the average blended grade will usually be below the threshold. But if typical ore blends will generate penalties, penalty calculations should be applied to all blocks. Rather than setting the grade below which no penalty will apply, the threshold level should be used to create a penalty credit or negative penalty for low-penalty element-grade blocks. Their low grades will result in a reduction in the average grade of the penalty element, thus reducing the penalty that would be payable. Their individual block values should therefore be increased to reflect that benefit, just as the values of high-penalty element-grade blocks should be reduced.

If there is no threshold grade for generating losses – typically for the first three types of loss listed – all grade values for loss-causing constituents will generate losses. These loss-causing effects should, if material, be included in the grade descriptor values. If the value is derived by an equivalence formula, these constituents will be included in the formula with a negative multiplier.

As an example, the author is familiar with a number of polymetallic sulfide base metals operations where the metallurgical recovery of the base metals is negatively correlated with sulfur grade. In one case, a complex multistage NSR calculation was applied to take account of the sulfur grades (and other non-revenue-generating elements) in deriving dollar values for each block. In another case, the sulfur grade was included in an equivalence formula with a negative multiplier, derived by multiple linear regression.

COSTS VARYING WITH LOCATION IN UNDERGROUND MINES

The discussion of break-even cut-offs (Chapter 3) indicated that because unit costs may vary with location, cut-offs may also do so. Indeed, it may be that every stope has its own cut-off to account for the unique set of costs incurred to mine it (but subject to the warnings of the dangers of this in Chapter 7).

To simplify the cut-off optimisation, it may be more suitable to use a dollar value grade descriptor with variable mining costs deducted in addition to the usual deductions. This would effectively make all blocks of rock with the same dollar value equally valuable and equally desirable to add to the ore stream, regardless of their location in the mine. This is of course the intention of using any grade descriptor – to be able to state simply that this is worth more than that by comparing their grade values only, without having to do any further complex calculations.

It may therefore be appropriate to use a dollar value grade descriptor even for a single-metal deposit, where metal grade would be the more usual grade descriptor.

If including mining-variable costs in a dollar value grade descriptor, the simplest way may be to deduct all such costs. In reality, since the implied intent of the grade descriptor in this situation is to remove the effects of location from the grade descriptor value, it is only necessary to deduct location-dependent costs. If the mining method and associated activities are the same throughout the orebody, the location-dependent costs may only be haulage costs from each location to some common point in the ore flow to the mill. If mining methods vary throughout the mine, differential costs associated with different methods may also need to be accounted for. However, as indicated in the discussion of individual stope break-even grades late in Chapter 7, cemented fill costs should be viewed as an average cost for the method and not as costs for individual stopes depending on whether they are cemented or uncemented.

For cut-off derivation purposes in an open pit operation where the schedule has already been decided, location-dependent costs are generally irrelevant. Open pit variable mining costs are committed by virtue of the decision to mine to a particular pit limit and are typically associated with rock. They do not vary depending on whether that rock is classified as ore or waste, so are not relevant to the ore–waste decision; that is, the cut-off, decision. Underground, the material under consideration is ore, and it is appropriate to account for differential ore-related costs when applying cut-offs or, as discussed here, calculating dollar values. However, location-related costs for mining rock in an open pit will be relevant for determining the optimum mining schedule, and as discussed in earlier chapters, sequences, mining and treatment rates and cut-offs should ideally be optimised together.

CHAPTER SUMMARY

The ultimate purpose of the grade descriptor is to assign numerical values to volumes of rock so that larger numbers identify the more valuable material and smaller numbers the less valuable material. Ideally, the values used for the grade descriptor will rank blocks in the same order as their true value, to identify what is more valuable. For a simple single-metal deposit, the actual grade naturally achieves this ranking purpose. It is not essential that there be a linear relationship between the grade descriptor and the true value, and non-linearity is often the case.

In polymetallic deposits, common practice is either to generate an NSR dollar value for each block, or to apply multipliers to the metal grades for each of the valuable constituents to derive either a dollar value (per tonne) or a metal equivalent (the grade unit of the dominant metal). The typical value equivalence formula is:

$$Value = a \times Metal\ A\ grade + b \times Metal\ B\ grade + c \times Metal\ C\ grade + \dots$$

If a metal equivalent grade – say, based on Metal A – is required, the formula would be:

$$Metal\ A\ Equivalent\ Grade = Metal\ A\ grade + b' \times Metal\ B\ grade + c' \times Metal\ C\ grade + \dots$$

where:
b' = b/a
c' = c/a, etc

There are several ways in which equivalent grades are derived. The most common are:
- *In situ* value, which accounts only for grade and price.
- Recoverable value, which accounts for grade, price and metallurgical recovery.
- Payable value, which accounts for grade, price and metallurgical recovery, and also the net amount received by the seller after accounting for downstream recoveries, treatment costs, freight charges and the like. These may be generated by either full NSR-style calculations or by an equivalence formula.

Accurate derivation of the equivalent grades is essential to ensuring that the grade descriptor values rank material in the same way as their true values. If this is not done, misclassification of ore and waste can occur – material that should be classed as ore is rejected as waste (representing ore loss), and material that should be classed as waste is incorrectly classified as ore (representing degradation). The effects of these are cumulative. The best grade descriptor will be the one that generates the maximum grade for a given tonnage, or the maximum tonnage for a given ore grade.

Common practice is to derive multipliers for equivalence formulas through accounting-style calculations using the average head grade. This may be satisfactory if all costs and recoveries can be directly attributed to individual products; however, if some costs are common to two or more final products (such as a bulk concentrate with two base metal components, or a base metal concentrate also containing precious metals), accounting-style distribution of common costs may not generate the best set of multipliers. In this situation, it may be better to derive the values for a number of potential ore grades by extracting a sample of blocks from the block model, finding the net values of each grade combination and then using multiple linear regression to calculate statistically rigorous multipliers. These are then applied to the constituent grades in the block model to generate the best estimates of true value.

Depending on the mathematical nature of the recovery relationships – and if recoveries apply to the blended grade or to the grades of individual small-tonnage components of the ore feed – it may be that the net recovery of any block is dependent on what it is blended with. It might therefore never be possible to state the true grade or value of a block of rock in isolation.

Although both dollar values and metal equivalents are equally valid grade descriptors and either could be used with appropriate care, the use of metal equivalents where possible is recommended. By analogy, a single-metal grade does not change when the metal price changes. The cut-off grade might change, but as seen in earlier chapters this is not necessarily the case; even if it does, it may not be in inverse proportion to the price change as would be the case for a simple break-even cut-off grade. With a polymetallic deposit, there is also no reason why the grade descriptor should change simply because prices change. If the grade distributions of the valuable constituents and their price changes are such that the general shapes of the orebody boundaries defined by different cut-offs remain generally of the same family (concentric or nested), there is no need to change the grade descriptor. The cut-off may change, as with a single-metal deposit, but it might not. If, on the other hand, the shapes of the families of orebodies vary significantly with price changes, it may be necessary to change the grade descriptor.

There are two main reasons to use metal equivalents instead of dollar values. First, from a practical point of view, their use potentially reduces the frequency of changing the grade descriptor values, and all technical and operating staff have a better opportunity to develop an understanding of the grade distribution within the deposit. Second, it breaks the nexus between a grade quoted as dollars per tonne and costs quoted as dollars per tonne, with the almost inevitable equating of the two in a de facto break-even style of cut-off definition. Even if dollar value cut-offs have previously been optimised, the tendency is to update the grade descriptor dollar values with new price predictions, but apply the same dollar value cut-off as before. This is wrong in principle for two reasons. First, the optimum cut-off has been effectively varied in a break-even style of change, and we know that the optimum cut-off will often not behave in that way. Second, the dollar values derived from two price sets, though both expressed as dollars per tonne, are for different dollars – they are as different as grades for two base metals, though these might both be expressed as percentage grades. The same numerical value would not normally be applied as a cut-off to two grade measures just because they happened to have ostensibly similar, though actually different, units.

Although grades, metal equivalents and dollar values are expressed as per tonne values, in general, the best ranking parameter for blocks of rock will be *value per unit of constraint*. If the aim of the operation is to generate cash, selecting blocks that produce the most cash when filling the constraining process would usually be a good strategy. For example, if two blocks of rock have the same grade and other metallurgical performance parameters, but one is harder and mills at half the rate of the other, it will generate half the number of dollars of revenue as the other for each hour of milling consumed. In these circumstances, the capacity of the plant is best expressed as available hours. Nameplate tonnage capacities effectively assume one particular nature of the ore feed. Using *value per unit of constraint* grade descriptors will enable sequencing strategies and cut-offs to specify the best-value material to deliver the maximum value achieved, limited by the constraint.

It may also be appropriate to include penalty or loss-causing elements in the grade descriptor value calculation. This should be considered carefully. If there is a penalty threshold in the blended product and the average penalty grade will be below the penalty level, an individual block of rock should not be penalised if it has a high penalty element grade. In this case, penalty elements should not be included in the grade descriptor. If the average grade of the penalty element in the product sold is likely to be above the threshold, all grades of the penalty element should be accounted for, relative to the penalty threshold. Low-grade blocks will reduce the grade of the penalty element in the final product and hence the penalty, and such blocks should be credited with their value-adding potential. Penalty elements may be accounted for in full NSR value calculations, and can also be applied with negative multipliers in equivalence formulas.

CHAPTER 10

Goals and Value Measures

MEASURING THE ATTAINMENT OF THE CORPORATE GOALS

What measure do we optimise? The simple answer must be: whatever the corporate goal is. Yet that apparently simple response hides a number of difficulties. It effectively assumes that we know two points: what the corporate goal is, and what measure/s of value to use to determine whether it's been achieved in the past, as well as guiding decisions to achieve it in the future.

Chapters 3 to 5 have identified how different cut-off derivation processes have goals that are either explicit or implicit in the processes:

- *break-even analysis* has an implicit goal of ensuring that every tonne classed as ore pays for itself, whatever that might be taken to mean
- *Mortimer's Definition* has explicit goals to ensure that every tonne classed as ore pays for itself, and that the average grade of ore treated delivers a specified minimum profit per tonne.
- *Lane's methodology* (in both its simple and complex forms) has the explicit goal of maximising net present value (NPV).

All these cut-off derivation methods assume that all other strategic decisions have been made. Chapter 6 has introduced strategy optimisation as a process to identify the strategy that best delivers the overall corporate goal. This accounts not only for the cut-off policy but also for all drivers of value that the company can make decisions about, such as mining and treatment rates, timing of expansions and so on. Chapter 6 also briefly covered the concept of making trade-offs between conflicting goals to select a preferred overall outcome, while not necessarily optimising for any one goal.

This chapter discusses aspects of quantifiable goals that the company might nominate, and how they might interact with each other. Chapter 12 will discuss a goal that is not so easy to quantify: finding the balance between maximising the upside gain and minimising the loss potential, depending on whether the predictions of future conditions in the optimisation evaluations on which strategic decisions are based eventuate.

Note that valuation of exploration properties is outside the scope of this book. As indicated elsewhere, most discussions deal with sufficient levels of certainty to derive mining and treatment plans acceptable for the level of study undertaken. It is also implied that the resource is known well enough to compare alternatives to identify the best strategy. One exception is the flexing of additional ore sources that might subsequently be proved by exploration activities; however, this is not done to value those low-confidence resources, nor to identify an optimum strategy for exploiting them. Rather it is to identify whether their eventual realisation would impact on the mining and treatment plans being developed for higher confidence resources.

VALUE, GOALS AND CONSTRAINTS

Corporate goals are often expressed in phrases such as *maximising the value of the shareholders' investment*. Industry analysts can be heard to say in public forums that cash is king. How do these sentiments translate into actions when it comes to project evaluation and determining strategy at operating or planned mine sites?

Value creation in different types of companies

Ultimately, the real value of an operation (not of an exploration property) depends on the stream of free cash generated over the course of its life. The owners of the operation cannot spend tonnes, ounces or years of mine life; only the cash that the mine generates. The actual net cash flows over the life of an operation are only known with certainty after the operation has closed; but a measure based on the best estimates of those cash flows will be a good gauge of its value before or while it is producing. NPV is arguably the best single number surrogate for quantifying a series of cash flows, but there are caveats to that statement. Discounted cash flow (DCF) methods, which also account for the time value of money, are therefore used in nearly all developed economies. NPV and internal rate of return (IRR) are the two most common value measures considered.

In most developed economies, directors of companies are legally required to act in the best interests of shareholders. This implies conducting mining operations to maximise the cash generated for shareholders over the life of the operation, accounting for the time value of money. Maximising returns to shareholders involves companies being good corporate citizens and maintaining the social licence to operate. This requires spending some of the business income on benefits to the community, protecting the environment, and so on, so as to maintain that status and licence. Shareholders receive nothing if adverse public opinion forces a mine to close. How much goes towards this social licence to operate to maximise value for shareholders is beyond the scope of this book; however, if cash flows from the operations are maximised, it becomes easier to both maintain the social licence to operate and to maximise returns to shareholders.

Shareholders receive returns in two ways: dividends and capital gains resulting from share price increases. Many smaller companies never pay dividends, so all the return to shareholders comes from capital gains. As many companies are managed to improve the share price, the operating strategy must focus on which behaviours will be rewarded by the market. It is more like a chess game – a sequence of steps that build value – with each step having its own capital and strategic constraints aimed at improving measures rewarded by the market, namely in share price increases rather than by maximising cash returns from the operations.

There are, perhaps, three types of companies operating in the resources industry. Some are purely involved with exploration, and both valuation experts and the market, via the share price, attribute value to these. At the other extreme are companies that only produce from one or more operations, but virtually all resources companies have some amount of exploration activity to extend the mine life or increase production. Exploration companies generally can only obtain fund exploration activities by raising capital from the market. Operating companies generate revenues from their operations, and net cash flows can be used, simplistically, either to pay dividends, invest in other production activities (such as new operations or plant expansions) or spend on exploration. For this second type of company, the primary aim is to maximise the value of the operations, with the increase in resources through exploration being one means to that end, but by no means the only or major means.

The third type of company lies between these two points, and may have at least one operating mine and one exploration program funded by the net cash flow from the operation. It may be thought of as an exploration company that does not have to revert to its shareholders on a regular basis to obtain more funds to continue its program.

For any company with (or planning) an operating mine, the company's shareholders are best served by the cash generated by the operation being maximised. Whether it is to be returned to shareholders as dividends, reinvested in further value-adding projects or injected into an exploration program, more cash is preferable to less cash.

The goal of maximising NPV should therefore be the main focus of all planning activities, particularly for strategy optimisation studies. This is relevant for any operation, regardless of the type of company overseeing the operation. Most companies, however, have multiple, often conflicting corporate goals. As will be discussed, many of these other goals are negatively correlated with cash generation. Seeking to maximise them, therefore, will result in a reduction in cash generated. We will return to this issue throughout this chapter.

Goals and constraints

Measures of value other than NPV are frequently evaluated and enter into the decision-making process. These may include undiscounted cash flow and factors such as ore tonnes and contained metal, mine life, unit operating costs and various return-on-investment measures, of both DCF and accounting types.[1] Minimising initial capital outlay and generating sufficient cash at the right times to meet debt servicing and other commitments will frequently be major concerns. Although maximising NPV is often the stated goal, in practice it is often subordinated when it may adversely affect other measures, particularly if these other measures are perceived to drive share prices. By implication, achieving satisfactory outcomes for these measures must also be seen as important de facto corporate goals.

---

1. Figure 6.9 and the associated discussion in Chapter 6 indicate how different discounted and undiscounted cash flow measures, unit costs and production volumes for a range of cut-offs might be traded off to deliver a good outcome for several of these value measures without necessarily optimising the strategy for any one of them.

It is necessary to understand whether these others are genuine goals or simply constraints. For example, do we wish to simultaneously maximise NPV, mine life and metal in reserve while minimising capital expenditure? Or do we wish to maximise NPV subject to life, reserves and capex being within specified acceptable bounds? The distinction is important and will lead to a different philosophical approach to the optimisation problem. As we have seen, it is very unlikely that one strategy will simultaneously optimise all the value measures described. If all are goals, qualitative value judgements will have to be made regarding the trade-offs required to arrive at the strategy best satisfying disparate and conflicting goals – even though it may not actually be optimal for any of them. If, on the other hand, there is one goal but multiple constraints that act as filters on the cases evaluated, only those that are acceptable in that they satisfy the constraints are assessed to determine which achieves the goal.

One might then ask, if the goal is to maximise returns for shareholders by maximising the NPV of future operations, why should additional constraints be placed upon strategies that potentially make the value-maximising strategy unacceptable? Or why should other conflicting goals cause a shift away from the strategy that delivers the primary goal?

One possible answer is that these may represent real practical or social-licence-to-operate conditions. For example, the mine will not be able to start if the required capital is more than the company can raise, or if the life of the mine is less than the government or local social requirements will deem acceptable.

Of more concern, however, is that other goals used may be easy-to-measure parameters that are thought to be correlated with less easily quantified, but more important, higher-level goals. It is therefore critical that companies identify what their overriding goal is and focus on it. If the focus is turned to more easily quantified measures, that correlation must be clearly demonstrated, not assumed on the basis of simplistic arguments.

In recent times, reported comments from industry analysts and senior company executives at mining financial conferences and in other presentations suggest there is now industry recognition that the earlier focus on the amount of valuable product in reported reserves and maximising production did not lead to increased cash generation for shareholders, despite metal prices being at all-time highs. With prices having since fallen, many companies are struggling to maintain output and stay profitable.

It is not suggested that company officials have deliberately embarked upon strategies that will reduce shareholder value. Rather, these strategies that target measures other than maximising NPV have been implemented in the mistaken belief that the measures can be used as surrogates for maximising cash generation. This also indicates that these same beliefs have been held by industry analysts who in a very real way drive the prices of company shares up and down. It would take a brave CEO to commit to a plan to maximise cash flow that opinion-forming analysts perceive to be value-destroying, and which could, as a result of the analysts' comments, drive the company's share price down.[2, 3]

A number of measures that are frequently used for supporting decisions, and which appear to be correlated with maximising cash returns for shareholders, in fact are not. Worse, in many cases, is that they are negatively correlated – improving the measure that is being focused on actually reduces cash returns for shareholders, as indicated earlier.

Although NPV is the primary cash-flow-related value measure in common use, option value and real options valuation are more frequently mentioned, at least by thought leaders in the industry. In the author's experience, there is a lack of understanding of what these measures can do, which may be due to their limited usage. They are major topics in their own rights, but brief overviews will be given and some relevant issues for cut-off and strategy optimisation discussed.

The preceding discussion is directly applicable in societies where maximising value for shareholders is the legally required aim of a mining company. Companies will pay the legal minimum in taxes and royalties and the least amount possible to maintain a social licence to operate, with all other returns from the operation vesting in shareholders; however, it may be that in other societies and countries, national policies for employment and utilising resources and the like drive investment decisions (and hence strategy optimisation goals) to maximise the resource while achieving the minimum acceptable return for investors to encourage investment to occur. Simplistically, companies will pay the bare minimum to investors, with all other returns going to the community in the form of taxes and royalties to governments, payments for goods and services, wages, infrastructure and community facilities. Even in this case, benefits to the community will tend to be maximised by maximising the cash generated by the operation.

Maximising NPV for the operation, therefore, is assumed to be the goal for strategy optimisation in this book, regardless of the economic system within which the operation is working.[4]

Preceding discussions have focused mainly on corporate measures of value, but there is no reason why a company evaluating a project should not also investigate returns to government and other stakeholders. If all parties obtain maximum value with the same strategy, there should be a harmonious relationship between the company and stakeholders – at least in regard to the best operating policy. On the other hand, if the company and another stakeholder, such as the government, gain maximum benefits

---

2. It is the author's perception that, for a short time in the industry press, many companies were reporting that their earlier strategies focusing on growth were clearly wrong and that the focus would now be on cash generation. This very quickly changed to a period of cost cutting, followed by a sudden realisation that cost cutting for the sake of cost cutting can be counter-productive, and the new focus became productivity improvement. At the time of writing, productivity is being reported as the biggest challenge facing the industry.

3. In the author's experience, the expected negative market response to reporting a reduced reserve that results from increasing the cut-off from a break-even to an NPV-maximising optimum is often deemed to be more important than the reduced cash generated by remaining at the break-even cut-off. This makes the change in strategy almost impossible.

4. This discussion does not purport to be a rigorous consideration of the economics of mining in different economic systems. In particular, it ignores the situation where national requirements for generating export income and foreign exchange from the mining industry outweigh the concerns of the operating entity and local community.

from different operating strategies, such an analysis will at least indicate the range within which negotiations need to take place.[5] Alternatively, external stakeholder requirements can be viewed as constraints within which the company's NPV is to be maximised.

To conclude, it is worth reviewing the discussion about the goals of break-even and Mortimer-style cut-offs – to ensure that every tonne pays for itself and, for Mortimer's Definition, that the average grade of material generates a specified minimum profit. Perhaps these goals are better seen as constraints. If so, these cut-off derivation methods then have an implicit goal of maximising the ore tonnage or contained metal, subject to the condition that every tonne pays for itself and, for Mortimer, that the average grade of material generates a specified minimum profit. In a similar way, when the method of cut-off derivation is specified by law, it also acts as a constraint on the corporate strategic planning. Value is to be maximised, subject to the constraint that the cut-off is a certain value or lies within a particular range.

DISCOUNTED CASH FLOW MEASURES OF VALUE

Net present value

Before we consider maximising NPV, we should understand some of the characteristics of that value measure and how it is applied generally in investment decision-making.

If a project returns an NPV of zero, by definition it has repaid the capital invested and delivered a rate of return equal to the discount rate used. The weighted average cost of capital (WACC)[6] is often used as the discount rate. If a project returns an NPV using the WACC (NPV at WACC) of zero, all the investors (debt and equity providers) have received their repayment and their required rates of return. In the absence of risk, on a stand-alone basis, a project with an NPV at WACC of zero ought to be acceptable.

In practice, however, this is rarely the case; a positive NPV is usually required for a mining project to cover downside risk. The question is then: what should the acceptable NPV be? How can we account for cash flow risks not being what are predicated by a simple model? If we do not undertake any quantitative evaluations to address this issue, any decision can only be subjective. The best way to quantify this risk is to conduct some form of stochastic simulation of inputs and outcomes, so that a probability distribution of NPVs is gained from the evaluation model. This is discussed more in Chapter 11, but for now we can say that this quantifies the probability of the NPV being greater or less than any particular value. We can then answer two questions:

1. What is the probability that the NPV is less than zero? That is, what is the probability that the investors are not satisfied, in that they fail to receive the return of their investments and required rates of return thereon?

---

5. The author was involved in a study for an operation with three major shareholders, three levels of government receiving taxes and other contributions and several categories of land owners compensated for the operation's impact on them. Virtually all stakeholders were best off within a narrow range of operating strategy options, with little rational reason for conflict.

6. The WACC is composed of three rates of return: the return on debt (closely related to the average interest rate payable on all forms of debt), the risk-free rate of return (the return on long-term government bonds) and the extra return required to justify the risk of investing in the company rather than risk-free investments. Readers are referred to standard texts on discounted cash flow evaluation for more detailed discussion of the WACC.

2. What is the expected value (NPV) of our project? That is, how much additional value do the shareholders expect, over and above their required rate of return?

In many companies, the issue of risk is dealt with simply by adding a premium[7] to the discount rate to account for the perceived risk of dealing with certain commodities or in certain countries. As we shall see, the NPV typically reduces with an increasing discount rate. The NPV using the WACC plus a risk premium (NPV at WACC + RP) will usually be less than the NPV at WACC. This requires projects to be more robust – or less risky – to generate a zero or positive NPV at WACC + RP.

In light of the discussion of goals and constraints, we could use a constrained goal of:

- maximising the NPV at WACC, subject to the condition that the probability that the NPV at WACC is less than zero does not exceed some specified level of confidence, or
- maximising the NPV at WACC, subject to the condition that the NPV at WACC + RP is greater than zero.[8]

It is important to note that, as indicated in Chapter 7, most feasibility studies are precisely what the name says and nothing more. They merely seek to demonstrate to the satisfaction of various stakeholders that a certain option for project development is technically and financially acceptable, and it is therefore feasible for the project to proceed as defined by the options in the study. If a feasibility study fails to demonstrate this, project sponsors will usually seek other ways of developing the project to make it economic; however, if the project is apparently healthy and robust, there will typically not be an attempt to find a set of options that provides a significantly better outcome. It is nevertheless common to hear that a project being developed after a favourable feasibility study is being optimised, but this simply involves finding better or cheaper ways of implementing the strategy. It rarely, if ever, takes the form of finding a better strategy once all the time and effort has gone into justifying the strategy evaluated in the study.

The implication is that most mine plans, if they have not been generated by a strategy optimisation study focused on NPV maximisation, will be based on a strategy that has (at some time, but not necessarily recently) been demonstrated to generate an acceptable NPV. Although maximising NPV may be the stated goal, the strategy selected will often not have been rigorously demonstrated to do that. The same can be said of most of the other measures used by the company – acceptable results will have been demonstrated, but the best possible outcomes are not necessarily being pursued by the strategy adopted.

In concluding this discussion of NPV, there is often resistance to its use on the basis that, by discounting the effects of cash flows in the future, it drives decisions towards short-term gain at the expense of the longer term. To that end, some companies also consider total net undiscounted cash flow, which may be thought of as NPV with a discount rate of zero.

NPV is also criticised as being an inappropriate value measure for long-life projects for the same reason. Its use tends to lead to decisions that potentially shorten the mine life, to the detriment of future generations. One possible way of addressing this problem

---

7. To the best of the author's knowledge there is no theoretical justification for doing this, nor the numerical values used as risk premiums. The practice is apparently legitimised by long and widespread usage only.

8. The author suggests that the second goal, not uncommon in practice, is implicitly trying to achieve the more rigorous specification in the first, but without any quantitative justification for why a particular value for the risk premium will correspond to the specified maximum probability that the NPV at WACC is less than zero.

is to derive NPVs of the remaining life for alternative strategic options every five or ten years into the future. It then only needs to be recognised that maximising NPV now and maximising it 20 or 30 years hence are different goals. In the same way that trade-offs can be made between different value measures now, they can also be made for values measured at different points in time (as best we can estimate them). There will be trade-offs between an acceptable reduction in value for today's shareholders and a perceived increase in value for future generations.[9]

It should be noted that nothing in this section, or indeed this book, suggests that a company's share price at any time should be the NPV determined for the known resource divided by the number of shares. Even if this figure is known to the market or someone external to the company, the market will apply various premiums and discounts to account for perceptions of exploration success[10], price and cost movements and capabilities of the management team, for example. It may, however, be appropriate to conclude that if the management team has determined the strategy that maximises cash generation potential for the known resource as represented by the NPV, it will continue to do so for future discoveries. Simplistically put, intentionally maximising NPV now should indicate that NPV will be maximised in the future too.

Internal rate of return

The IRR is, by definition, the discount rate that results in an NPV of zero. When a project returns an NPV of zero, the implication is that it has repaid the investment and delivered a rate of return equal to the discount rate. The definition therefore indicates that this is the rate of return generated internally by the project without reference to external factors such as the sources and costs of capital invested or the returns required by investors.

For a project to be classed as satisfactory, its IRR must be equal to or exceed a minimum value, typically referred to as the *hurdle rate*. As a principle, the hurdle rate used to identify an acceptable IRR should also be the discount rate used to obtain the NPV. For a single project on a stand-alone basis, a simple acceptable-versus-not-acceptable decision can be made by comparing the IRR with the hurdle rate or the NPV with zero. Subject to certain caveats (discussed later), the NPV and IRR should lead to the same acceptable-versus-not-acceptable decision for a project on a stand-alone basis. This does not, however, imply that the strategy that maximises NPV will also maximise IRR; they are two distinct goals. However, if the capital investment is similar in cases evaluated, there will usually be a close correlation between NPV and IRR, but if there are significant differences between capital requirements, NPV and IRR may not be correlated.

Some companies specify a hurdle rate higher than the discount rate for NPV. In these situations, it is common to find that the discount rate used is the WACC, while the hurdle

---

9.   The author feels obliged to ask why current shareholders should forgo value now so that value is available for shareholders in 20 or 50 years' time. Operations in any long-established mining field are often profitable in areas that were uneconomic 30 or even 20 years ago, due to technological and economic changes, and the rate of technological improvement is not slowing. To try to predict what we or our successors will be doing 20 or 30 years hence seems somewhat presumptuous. Identifying the value to sustainability is an issue that has yet to be properly addressed by the industry as a whole, and this book won't attempt to do so.

10. As indicated earlier, valuation of exploration assets is not within the scope of this book, as it is a specialised field in its own right; however, it may be appropriate to ascertain whether today's optimum strategy is influenced by whether various exploration targets subsequently measure up to expectations.

rate is the WACC plus the risk premium. The stated rationale is that the higher hurdle rate is applied to filter out risky projects, and the relativities of alternative projects that pass that test are made on the basis of their true value – NPV at WACC. From a strategy optimisation point of view, IRR maximisation is not the goal. Rather, it is used as a filtering constraint. The goal becomes: maximise NPV at WACC, subject to the condition that IRR is greater than or equal to WACC + RP.

Though often a key measure used by many companies, there are several problems associated with IRR, causing most commentators to class it as inferior to NPV. Some of these issues needing consideration by project teams and decision-makers when seeking to optimise cut-offs and other decisions are described below.

If there is no cash outlay (that is, no negative cash flow), there is no IRR – it is infinite. This will occur if comparing options for a project already operating with different patterns of positive net cash flows, for example. IRR values can only be determined for cash flow patterns that have at least one period with positive and one period with negative cash flows.

Mathematically, there may be as many discount rates creating an NPV of zero as there are changes in the sign of the cash flow. For a simple project with an initial negative cash flow followed by all positive cash flows, there is only one change of sign (from negative to positive), and therefore one possible IRR. If NPV were plotted as a function of discount rate, it would be positive at low discount rates, reduce with increased discount rates, pass through zero NPV at the IRR, and then become progressively negative with further increases in discount rate. This is illustrated in Figure 10.1.

In these circumstances – cash outlay followed by cash inflow – comparing the IRR with a specified hurdle rate will correctly identify whether an option is acceptable when considered in isolation; but there is no guarantee that alternative options will be correctly ranked using IRR.[11] A company is, of course, free to specify 'maximising IRR'

11. Consider a 100 per cent return on a $10 investment – $10 – and a ten per cent return on a $1000 investment – $100. A lower rate of return on a large investment may generate more dollars than a larger rate of return on a

FIGURE 10.1
Net present value versus discount rate for one change of sign in cash flow pattern.

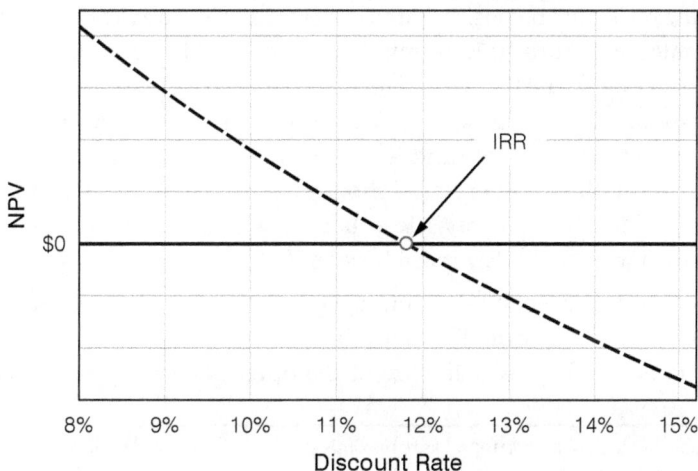

as a corporate goal if it wishes, but it must recognise that as a general rule, the strategy that maximises IRR may not be the one that maximises NPV.

If the sign of the cash flows changes twice, there may be two IRRs. This cash flow pattern is not uncommon: negative cash flows for the capital outlay to develop the project, positive cash flows during the life of the operation and finally negative cash flows for mine closure and post-closure rehabilitation. For a realistic mining project, if there are two IRRs, the NPV will typically be negative at both low and high discount rates and positive for an intermediate range, as illustrated in Figure 10.2.

FIGURE 10.2

Net present value versus discount rate for two changes of sign in cash flow pattern.

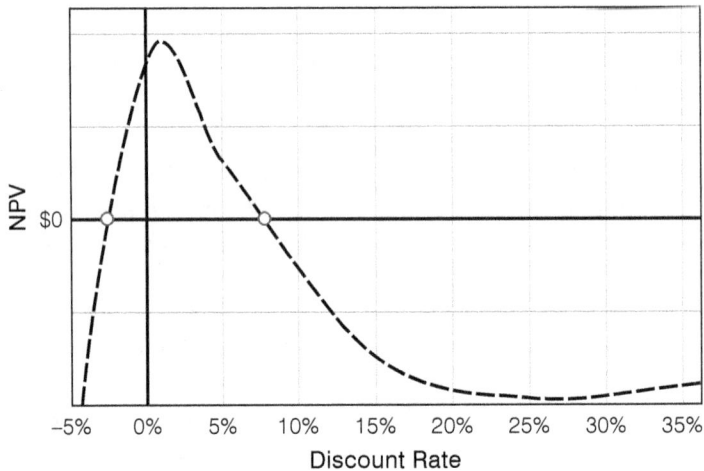

When there are two IRRs, as a general rule, neither of them can be safely compared with the hurdle rate to decide whether the project is acceptable on a stand-alone basis. In practice, though, the issue is of less concern and the higher of the two rates can be used as the IRR – but always with caution.

Generally, the overall undiscounted net cash flow (the NPV at zero discount rate) is positive, and the lower of the two IRRs is negative; in the realistic range of positive discount rates, the NPV may be positive at lower rates and negative at higher rates, similar to what is illustrated in Figure 10.1. Yet the NPV might not be continually reducing with increasing discount rate across the full range of non-negative discount rates. Nevertheless, in this case, it would probably be safe to use the higher of the two IRRs as a valid IRR for comparison with the hurdle rate and the IRRs of other options. There is, however, no guarantee that this will be so. Every case should be evaluated to identify how the NPV behaves with changing discount rates to establish whether two IRRs occur and, if so, whether one of them can be safely reported as an IRR.

The situation becomes even more complex when there are more than two changes of sign in the cash flow pattern. This may occur for a single project if there are, for example, variations in prices over the life of the operation or capital injections during

---

smaller investment, but a rational investor seeks to maximise what can be spent. A high rate of return is intangible, the result of a mathematical calculation. It cannot be spent to create any benefit for its owner.

the producing phase. Multiple changes of sign are also likely when comparing the incremental cash flows generated by expansion options to identify additional value from each injection of capital. If patterns of cash flows are such that there are multiple IRRs, it's usually impossible to select any to compare with the required rate of return without further investigation. The exception might be where by chance there is only one positive IRR and the NPV-versus-discount-rate curve is similar to Figure 10.1 for all positive discount rates.

Finally, the mathematical formula for calculating NPV for any specified set of cash flows over time is a complex polynomial function of the discount rate. Finding the roots of this formula – the values of the discount rate that generate an NPV of zero – is rarely achievable by analytical methods, and typically done by a converging iterative numerical process. Processes used in software packages do not indicate whether there are multiple IRRs, so there is no guarantee that any IRR found is valid. If there are multiple IRRs, the value finally found for the IRR reported will depend on the initial value assumed for the discount rate for the first calculation.[12] There is also the possibility that, even when there is a valid IRR, it might not be found. This could be due to an unfortunate choice of starting value[13] for the IRR for the first iteration of the calculations, or the iterative process reaching an internal limit on the number of iterations to be performed before it has converged to an acceptable outcome.[14]

In summary, IRR will sometimes be correlated with NPV, but as a general rule, the strategy that maximises IRR won't also maximise NPV. As shareholders can only spend dollars, not percentages, maximising NPV is usually preferable to maximising IRR.

Present value ratio / capital efficiency

The present value ratio (PVR) is defined as the ratio of NPV to the initial capital outlay. If the capital outlay is spread over several years, it may be more appropriate to define it as the ratio of NPV to the present value (PV) of the initial negative cash flows. For the special but common case where the initial cash flow is only at time zero, this more complex definition reduces to the initial definition, since the PV of an outlay at time zero is numerically equal to that outlay.

While NPV and IRR are universally recognised acronyms, the PVR is less well known and goes by a variety of names within companies that use it. Capital efficiency (KE) and capital efficiency ratio (CER) are other terms the author has encountered.

---

12. For the curve in Figure 10.2, for example, starting with a discount rate less than that generating the maximum NPV will converge to the lower NPV = 0 rate of return. Starting with a rate between those delivering the maximum and minimum NPVs will converge to the higher NPV = 0 rate of return. Starting with a rate greater than that generating the minimum NPV will not find an IRR, or will return an error value.

13. This would be the case if the gradient of the NPV versus discount rate is equal or close to zero at the initial discount rate used.

14. If encountering situations where a valid IRR should be found but sometimes is not, it can be useful to calculate the NPV at several discount rates, effectively creating the coordinates for plotting an NPV-versus-discount-rate curve similar to Figure 10.1 or Figure 10.2. This will indicate the shape of the NPV-versus-discount-rate curve and hence whether there should be a valid IRR. If so, a simple linear interpolation between the highest two adjacent discount rates generating positive and negative NPVs will give an initial estimate of the discount rate generating an NPV of zero. With this used as the initial value for the iterative IRR calculation process, such as the Microsoft Excel™ IRR() function or Goal Seek process, a valid IRR will almost certainly be found.

The PVR is primarily used as a project ranking tool to allocate scarce capital among competing projects. If there are limits on the capital that can be spent, each potential project must be assessed with others so that scarce capital can be allocated to projects that generate the greatest NPV from the total capital invested (assuming that maximising NPV is the corporate goal).

To achieve this, ranking options by PVR and selecting those with the highest PVRs to consume the available capital will produce the highest possible NPV from that capital. Ranking by IRR is often used and may result in the same decisions, but is no guarantee of maximising NPV, while ranking by PVR will.

The author has not encountered any instances of a strategy optimisation seeking to maximise a project's PVR, but suggests it would suffer from similar issues to those described for maximising IRR.

It is nevertheless important to consider the PVRs of incremental capital outlays for each project to maximise a company's total investments in its projects. If there is only one project for the company to invest in, the strategy that maximises value for that project will be the optimum strategy. Various points on the hill of value (for example, for different production rate targets for the HoV in Figure 6.5) will have different capital requirements and NPVs. The information used to generate the HoV can be used to calculate the incremental capital expenditures required to increase the production rate and incremental NPVs. This will then provide an estimate of the PVR of each incremental capital investment in the project. Typically, these will reduce as the gradient of the HoV flattens towards the top. For a company with a number of projects and limited capital, the maximum NPV for the available capital might be achieved by investing in a number of projects at a smaller size than their individual optimums, rather than in one or a few projects at their stand-alone NPV-maximising sizes.

ACCOUNTING MEASURES

A variety of financial accounting measures may be specified as corporate goals.

Profit measures

Profit is determined by standard accounting practices and methods of calculation. Acronyms that might be encountered include:

- NOPAT – net operating profit after tax
- EBIT – earnings before interest and tax
- EBITDA – earnings before interest, tax, depreciation and amortisation.

Rates of return

Accounting rates of return are typically calculated as profit (which may be any of the profit measures noted) divided by the book value of either total assets or owners' equity – the latter being net assets, which are total assets less liabilities. Some of the acronyms defining rate-of-return measures include:

- for return on total assets
  - ROA – return on assets
  - ROCE/ROFE – return on capital/funds employed
- for return on owners' equity

- ROE – return on equity
- RONA – return on net assets.

Problems with accounting measures of value

Accounting measures suffer from a number of common problems, of which the most important follow.

Accounting measures are generated on an annual basis and may vary from year to year. When comparing strategic options, it's a case of deciding which pattern of results is better than another pattern. One could consider total profits and average returns over several years for an average, but apart from long-term totals, there is no analytical way to combine a series of accounting measures over time into a single measure to reliably compare and rank. Therefore, there is no way to distinguish between two options having the same total profit, but with one making positive profits over its whole life and another with losses in early years but higher profits in later years. Accounting measures could be produced for successive time frames, but optimising short-, medium- and long-term outcomes are different goals; a way of trading off benefits now and later would be needed. It was to overcome these sorts of problems that discounted cash flow measures were developed. They provide that trade-off rigorously, mathematically and economically, to account for the time value of money.

Even if these problems can be overcome, the results will still be impacted by other issues. For example, accounting measures can be affected adversely by past history, in particular by the carrying value of assets in a company's books and the way that these are handled by depreciation and amortisation. It is a general principle that past spending is a sunk cost that should not affect future decision-making. The company's task is to maximise the value generated going forward, and if this is done, the adverse effects of past decisions – if any –will be mitigated to the maximum possible extent.

The main ways in which past decisions may legitimately affect future decisions, and hence need to be accounted for in forward-looking evaluations, are:

- *The book value of assets for tax purposes.* This will be depreciated for determining taxable profits in the future, and future decisions will potentially affect mine life, hence the depreciation deductions and the quantum and timing of future tax payments.
- *Past actions or decisions resulting in future cash flow items.* These include rehabilitation and closure costs as well as costs sometimes described as legacy costs, possibly resulting from past agreements with workers or government authorities. If such costs vary with future decisions, they need to be accounted for. If they are unaffected by any future decision, for strategy optimisation purposes they may be ignored. They will be the same regardless of the decisions and will therefore not drive them. The values of all options will differ by the same amount if such cash flows are included or excluded, so finding the strategy generating the maximum value does not depend on their inclusion in the analysis; however, these types of costs might need to be taken into account for ensuring profitability using, for example, the processes for applying Mortimer's Definition in Chapter 4.

Carrying past decisions forward, which is implicit in standard accounting measures, may distort the understanding of what adds most value in the future. Past accounting policies can result in remaining book values that incur losses, despite positive cash flows and NPVs. Similar outcomes may result from stockpile valuation policies and methods.[15]

Perfectly legitimate variations in accounting treatment may affect whether assets are depreciated quickly or slowly. Rapid depreciation will result in lower profits in earlier years and higher profits in later years. It will reduce the value of assets faster, thereby lowering the divisor applied in rate-of-return calculations in earlier years relative to slower depreciation. The level of dividend payments will also affect the total company assets, which can impact on accounting rate-of-return measures. Consequentially, these essentially arbitrary decisions[16] can affect the reported values of these measures.

If comparing projects within subsidiary companies or business units operating for different periods of time – and therefore having different levels of asset book values relative to production capacities – it may be impossible to rank projects across the whole company using accounting measures of profitability and rates of return.

In addition, it will usually be difficult, if not impossible, to relate accounting rates of return to the company's required rate of return as expressed by its WACC. Some companies try to overcome this by calculating a value measure by deducting from an accounting profit measure a *return on capital* amount. This is calculated by multiplying the book value of the assets by the WACC. It purportedly represents the return required by providers of capital as a cost to the operation. Any resulting value is then viewed as an economic value created for the benefit of shareholders. This type of measure tends to be more a short-term measure of management performance and use of assets, intended to drive managers to make best use of existing assets, rather than spend more capital, thereby increasing the assets to achieve similar outcomes. Some companies have used this as a project valuing and ranking tool in the past, but being derived from an accounting profit measure, it suffers from all the same problems as accounting measures.

These measures will generally be produced using the company's financial accounting and public reporting accounts, where there may be some latitude in the way matters are handled, so long as there is consistency. In most jurisdictions, taxation law allows for different treatment to attract mining investments; for example, accelerated depreciation. Accounting measures of value are not generally calculated using the tax accounts.

Finally, if care is not taken, focusing on annual accounting measures can drive an even shorter-term outlook than is sometimes identified as a concern with NPV, as has been noted. Share prices are often affected by annual or quarterly profit announcements, so company directors taking short-term actions to keep share prices up may actually reduce the long-term potential of their operations and, hence, long-term shareholder value.

In summary, accounting measures are not demonstrably correlated with long-term cash generation as measured by NPV. Future results can be affected by past practices. The accounting measures are usually for annual periods, which makes the problem

---

15. The author was once associated with an improvement project at an operating mine where, although all *do something* options generated positive cash flows, they also produced negative financial accounting outcomes. The impact on publicly reported profits was deemed more important than the cash generation, despite poor corporate cash flows being an issue, so the *do nothing* option was selected instead.

16. The author is not suggesting that accounting practices are arbitrary – there are strict rules and accounting standards that must be abided by; however, within those rules and standards there is still considerable latitude for companies or accountants to apply equally legitimate practices that, within the context of this discussion, can be viewed as arbitrarily impacting on the value measures derived.

either considering short-term measures only or, if measures over a long time are used, qualitatively deciding which pattern of measures is better than another.

COST MEASURES

Operating cost measures

Minimising cash operating unit costs is an important aim for a company. It is a worthy aim because it increases net cash flow and profitability, and helps to ensure viability at low points of the metal price cycle; however, there are two ways of looking at this. One is the process of attempting to reduce the unit cost of acquiring inputs and reducing consumption rates. This is a productivity focus that should be ongoing in any operation. It was discussed in Chapter 7 in relation to budgeting for the mine planning process, where it was identified there are limited ways to minimise costs such as reducing:

- The unit costs of inputs, which is good as long as the cost reduction is not accompanied by a reduction in quality and hence an increase in both usage and total cost.
- Consumption rates, as long as usage has not already been reduced to the fully efficient consumption rate that cannot be improved upon. Any further attempt at reduction, for example, by restricting supply, can only result in reducing the associated activity, which will be counter-productive.
- The physical activities scheduled. If the plans from long term to short term have been developed to achieve the corporate goal, any reduction in activities must result in a deviation from the plan that delivers this goal.

So there are both good and bad cost reduction programs. Cutting out fat is good; cutting out muscle is not.

The other way of looking at minimising unit costs is as an outcome of a strategy optimisation, with this being the goal of the study. In the same way we can find the NPV from several strategies and select the one that delivers the maximum NPV, we can find the unit costs from various strategies and select what delivers the minimum unit cost. The implication is that the underlying productivities and cost drivers are the same in each case, and that any differences are fully accounted for in the analysis. Variations in unit cost are then the outcome of the interplays between physical activities and costs in the different plans, not the result of cost reduction measures; however, the degree of success by improvements in productivity could certainly be flexed as a scenario parameter to ascertain whether changes in the underlying cost driver relationships would influence the strategic decisions or not.

Any strategy optimisation model that has determined any of the value measures already discussed should permit calculating any unit cost measures desired. Some minor additions may be needed to split costs out, though, so that they can be slotted into the relevant definitions of cost measures. It is also necessary to select the unit costs that are to be the indicators of relative values: per tonne of ore treated? Per unit of product?

These unit cost measures will be reported for each period, so will suffer from the same sorts of problems as annual accounting measures of value discussed. They can also be calculated for a number of years. Being ratios of dollars spent to physical quantities, the correct treatment is to sum both the costs and the physical quantities for the time frame in question and then divide the total dollars by the total physical quantity, not to take the arithmetic average of the unit costs of the years in question.

Perhaps better still, if a single value was needed for comparison purposes to be able to say that one strategy is better than another (like NPV, the single number derived to represent the value of a series of cash flows over time), one could use a discounted average unit cost. Instead of simply totalling up the dollars and physical quantities, these are discounted and then summed separately to obtain present values of both costs and physical quantities. The measure resulting from dividing the discounted costs by discounted quantities would represent an average unit cost that takes account of the time value of money and the fluctuations in both physical quantities and cost efficiencies over the life of the operation.

The concept of discounted tonnes or present value tonnes is unusual for many, but consider the following situation. Suppose the cost per tonne were constant over the life of the operation but the tonnage varied annually. The total dollars would vary from year to year, being the products of the unit cost and the annual tonnages. If we were to determine the present value of these costs, the dollars for each year would be multiplied by a discount factor, and the results added up to obtain the present value of the costs. Each term in the sum would be the product of the year's tonnes and the discount factor, both different each year, and the unit cost, the same for each year. Mathematically, therefore, we can remove the constant unit cost from each term, leaving the product of tonnage and discount factor – that is, discounted tonnes – for each year. When totalled up, present value tonnes are obtained. This figure is then multiplied by the constant unit cost to derive the present value of the costs.

Reversing this process, in order to derive a unit cost measure for the life of the project that takes account of the time value of money, it is necessary to discount both the total costs and the tonnes. For the example just described, we would expect that the discounted average unit cost would have the same value as the constant unit cost that occurs every year. If we divided the total discounted present value of the costs by the undiscounted total tonnage, we would not get the expected average value. To correctly match costs and physical quantities accounting for the time value of money, both dollars and physicals must be moved to the same point in time using the same mathematical principles. The second part of Table 3.1 in Chapter 3 illustrates how a present value for net capital costs at time zero is moved to where the tonnes are mined by an equivalent annual calculation, from which a time value of money unit cost is derived. Exactly the same unit cost would have been obtained if the tonnes had been discounted to time zero and the total discounted tonnes divided into the present value of the capital at time zero.

From the point-of-view of cut-off and strategy optimisation, there are several issues related to cost minimisation. If using break-even cut-offs, a unit cost reduction allows the cut-off to be lowered and more resource to be defined as ore. As we have seen, simply lowering the cut-off because the unit cost has changed is not necessarily a good strategy. This will lower the head grade, so the average margin per tonne of ore will reduce. The nature of the grade versus cut-off relationships will determine whether the reduction in costs is greater or less than the reduction in revenue. If the operation is constrained by the amount of ore it can treat, the annual revenue must reduce, and again the annual profit may increase or decrease, depending on the grade versus cut-off relationships. The mine life will potentially increase, but this may not necessarily result in additional value, whether measured by discounted or undiscounted cash measures. As was seen in Figure 6.10 and the accompanying discussion in Chapter 6, reducing unit costs and cut-offs may reduce revenue and hence value by an even greater amount.

It is good if cost reductions are achieved through waste reduction and productivity improvement programs, but the effect is notionally to increase margins, similarly to an increase in prices. Both of these should, in principle, lead to a new optimisation study so that the effects can be assessed across the range of options available, not just the current plan. As with prices, as discussed in Chapters 5 and 6, a change in costs may not lead to any change in cut-off. For example, the effective cut-off may be a Lane-style balancing cut-off, which is unaffected by cost and price changes. Or the effective cut-off may be a Lane-style ore-limited cut-off, in which case the effect will be to increase the value of the operation and hence the opportunity cost. This in turn will lead to any change in cut-off being less than the proportional change suggested by a simple break-even calculation, as illustrated in Figure 9.3 and the related discussion in Chapter 9.

Games can be played by operating staff in the face of cost reduction demands or incentive schemes with a short-term focus. Transferring operating costs to capital is intellectually dishonest and, if the company's accountants do not transfer it back in the tax books, may result in capital depreciation charges over time. This will defer tax deductions that could have been expensed immediately, thereby destroying value in the long run –all just to see operating costs come down. Furthermore, if break-even cut-offs are used and capital is excluded from the calculation, the cut-off will be artificially depressed and not cover the costs associated with generating ore and product, driving value further down the low cut-off slope of the HoV. For cut-off and strategy optimisation purposes, the critical issue is how costs behave, not how they are classified for accounting reports.

Spending capital and doing difficult things in the operation to set it up for improved performance later may be referred to as short-term pain for long-term gain. Cost-cutting programs, however, are often the opposite: short-term gain for long-term pain. Who has not seen instances of deferring exploration drilling, access development into new mining blocks underground or waste stripping in open pits so that cut-offs have to be lowered in existing areas to fill the mill with low-grade material. Removing cement from fill underground is another recurrence that improves short-term costs but results in either sterilisation of reserves or increased dilution and difficulty in recovering pillars, with accompanying increased costs and ore loss in the longer term. Reducing technical staff in the planning team will also save costs, but if they have been performing, the impact of reducing the planning effort may not become evident for months or years. Eventually the lack of long-term planning will impact the operation, and it may take years to return to a value-maximising plan. The technical team costs are easily seen in cost reports. The benefits of this team are are less easy to quantify; they are in creating better plans and avoiding costly mistakes.

Cost reduction and minimisation are emotive topics in our industry, and any technical person with more than a few years' experience will know of cost reductions that have destroyed significantly more value in the longer term than they initially created. While not decrying attempts to minimise costs, there is too great a focus on this, particularly for short-term management incentive schemes (see Chapter 7 on mine planning processes). By all means, assess ways of reducing costs, but let it be within the context of a top-down long-term to short-term planning process, so that short-term changes that look good can be evaluated across all time frames to see their ultimate effect on the project's value.

To conclude on operating cost minimisation measures, we return to the bigger picture of strategy optimisation. Figure 6.9 in Chapter 6, taken from a real case study, illustrates how the unit cost resulting from different cut-offs behaves as cut-off varies. This analysis assumes that the same cost structures apply for all cut-offs or are fully accounted for in the analysis. The cut-off that delivers the minimum unit cost is somewhat higher than that which maximises NPV, which in turn is higher than an industry standard break-even. This type of behaviour has been seen in a number of studies for various operations. The sample size is too small to draw a universal conclusion, but it suggests that there may be some correlation between NPV maximisation and unit cost minimisation; however, that is the result of finding the strategy that minimises unit costs for a given cost structure, which is just as complex a problem as finding the strategy that maximises NPV. It is not the same thing as a cost-cutting program to change the underlying cost structure. That may be good, but if taken to extremes can be counter-productive.

Capital cost measures

A number of capital cost measures are used by different companies. The most common are the total capital, initial capital, ongoing capital and maximum cash outlay.

The initial and ongoing capital measures may correspond with project and sustaining capital as defined in Chapter 2. In other situations, initial capital is what is spent before the project is deemed to be operational, and ongoing is any capital spent after that, which may be a mix of project and sustaining capital. While the definitions in this book are what should be applied within the strategy optimisation and cut-off derivation processes, it is entirely up to the company concerned as to how its value measures are defined. The strategy evaluation project team must therefore be careful to ensure both the evaluations and reported results make use of appropriate definitions, and to highlight definitions used in the optimisation process that are different from the company's normal usage.

As with operating costs, minimising capital costs has two aspects: reducing the capital spent for any set of options, and identifying the capital costs associated with various strategies, based on similar assumptions about how capital costs behave, and hence the strategy that generates the lowest capital outlay.

Great care should be taken when trying to minimise capital expenditure. All things being equal, reducing the capital will increase both the NPV and IRR. But all things will not stay equal forever. If a project has been well-engineered, a relatively small reduction in capital may be possible through smart detailed design or astute procurement. Substantial reductions in capital can only be achieved by building an alternative, typically with less capacity or reduced capability. Earlier chapters have stressed the importance of developing relationships between the parameters going into a project evaluation and the strategy optimisation model. The relationship between capital spent and system capability is an important one. NPV and IRR will not increase indefinitely with reducing capex; there must be a reduction in these value measures at some point, and the value reduction will typically be more than the cost reduction.[17]

The maximum cash outlay, or the maximum (in absolute terms) negative cumulative cash flow, will indicate the total funds that must be provided by investors. While minimising this will be important, the value as defined by NPV will override it because the assumptions surrounding the calculation of NPV imply that whatever capital is needed is available at the cost of capital implied by the discount rate. However, there

---

17. It is dangerous to base decisions on how much capital to spend at the point at which a simplistic IRR versus capital flex curve crosses the hurdle rate. The author has seen a situation that suggests a major project failure may have occurred from a capital spending decision resulting from such a process.

may be a practical reality that the company is only able to raise a specified amount of capital. While maximising NPV might therefore be possible, it may be constrained in practice by the requirement to not spend more than a certain amount of capital. Minimising the capital outlay is therefore likely to be applied as a constraint to limit which project options can be included in the strategy optimisation based on another measure, such as maximising NPV.[18]

PHYSICAL AND TIME-RELATED MEASURES

Physical and time-related measures are frequently based on conventional wisdom that can be unhelpful. They are typically measures that, all things being equal, would improve the economics of the operation. In reality, all things are not equal, and the assumption that these measures are positively correlated with economic value are false. These measures tend to be used emotionally rather than rationally, particularly when economic value maximisation drives down these measures; however, many of these measures are used by industry analysts. Since they are not correlated with value, focusing on them appears to have the perverse effect of causing value-adding strategies to be criticised and value-destroying strategies to be applauded. Strategies that should cause a company's share price to rise drive it down, and vice versa.[19]

Payback measures

The payback period, often simply referred to as payback, is a measure that indicates the time taken for the cumulative undiscounted cash flow from the project to equal zero. In other words, it is the time taken to repay all the capital invested.

Simple examples that illustrate the process usually have a single capital investment at time zero, followed by a series of cash inflows. Defining the *time zero* of the payback period is therefore trivial – it is the project's time zero. The author is unaware of any commonly accepted definition of payback for a project where the initial capital expenditure is spread over a number of years and continues to be spent even after revenue-generating activities commence; however, given the nature of the measure and what it is attempting to do, a logical time zero for the payback might be at the time of maximum net cash outlay or the time when the revenue stream starts – it is the revenue stream that is doing the paying back. It is possible that early cash inflows are smaller than the initial operating costs, and significant capital expenditure continues after revenue generation starts. The maximum cash outlay may therefore occur some time after the start of revenue generation.

Payback with an extended capital spending period could also be measured from the project's time zero, notionally when capital expenditure starts. However it is defined, a

---

18. One of the games that is sometimes played is to move time zero forward to a point where a significant amount of capital has already been sunk, after which the NPV, IRR and remaining capital to be spent are acceptable. This can be done by classifying the initial capital as exploration needed to obtain resource information to produce a reliable reserve. It may well be true that more information is needed to justify investing in the project, but if additional costs in excess of what is required to obtain the information are incurred, care should be taken to avoid being intellectually dishonest in making the exploration and project development decisions.

19. Much of this criticism is based on anecdotal evidence or selective examples. Nevertheless, it is still a company's prerogative to optimise these measures. This will, in fact, be a rational response if those whose pronouncements drive the company's share price choose to focus on measures that are not correlated with economic value rather than measures that are.

clear explanation of the definition for (or from) decision-makers and consistent use in the evaluations and reporting of results is essential.

On its own, the biggest drawback of the payback period is that it doesn't account for cash flows after that time. The operation could close immediately after payback is reached, or continue on with marginally positive cash flows, or have healthy net cash flows for many years. All would be reported with the same payback. It is therefore not correlated with maximising cash flow.

Payback is unlikely to be used on its own as a value measure, instead used in conjunction with other measures to form a broader view about the project's behaviour than can be obtained from any single measure. It is likely to be used as a constraint or filter (as described in the discussion of IRR for hurdle rates) and like such hurdle rates is seen as a risk-reducing measure – the lower the payback, the less risky the project. The maximum payback period is therefore likely to constrain the project options deemed worth including in strategy optimisation based on another measure, such as NPV.

Although the payback period was originally devised as an undiscounted cash flow measure, discounted payback is also used. It is defined as the time taken for the project's cumulative discounted cash flow – that is, the cumulative NPV – to equal zero. It represents the time taken to repay the capital invested and deliver to providers of that capital their required rates of return. If the project is located in a country with significant sovereign risk, the discounted payback can be an important risk assessment tool; for example, what is the likelihood of the project being expropriated before the minimum satisfactory return – an NPV of zero – has been obtained? Other comments relating to undiscounted payback also apply to discounted payback.

Potentially a useful measure for risk minimisation, payback measures nonetheless do not usually relate to any corporate goals. They may also be inconsistent with IRR targets. It can be shown mathematically that, for a hypothetical project consisting of a single cash outlay at time zero, followed by an infinite life with equal net cash inflows each year, the (undiscounted) payback period is the reciprocal of the IRR expressed as a decimal fraction, and vice versa. A payback period of three years for such an investment therefore delivers an IRR of 33.3 per cent. There is a potential conflict between a desired payback of, say, three years and a hurdle rate of 12 per cent. The latter implies an acceptable payback period of 8.3 years.

However, few mining investments are like the hypothetical just described. The common practice of targeting high-grade areas in the orebody early in the mine's life may result in above-average cash flows in early years, which may then deliver a short payback; longer-term patterns of grades and cash flows, on the other hand, may result in an IRR that is significantly less than the reciprocal of the payback period, but still greater than the specified hurdle rate. The 'Payback = 1/IRR' relationship should not been seen as an absolute for a mining investment, but it is a useful rule of thumb. A departure from it in specified IRR and payback targets should raise concerns to be investigated.

Mine life

Mine life is another measure based on conventional wisdom that can prove unhelpful. Maximising mine life is often seen as an important goal for an operation; however, long life does not necessarily improve a mine's economics. If an increase in mine life comes from successful exploration programs so that the life is extended at the same production rate and head grades as before, this will be good, though, changing the production rate or cut-off could be a better option. Varying the mine life for an unchanged resource by changing, for example, cut-off or the mining rate is not necessarily ideal.

One might ask the question: which of the following scenarios would be preferable? Receiving $1 000 000 a year for seven years (a total of $7 000 000), or $500 000 a year for ten years (a total of $5 000 000)? The longer life generates less net cash. When the author has asked this question in public forums (at conferences and the like), a surprisingly large number of people prefer the second option. Perhaps it is an indication that long life is perceived by many to be positively correlated with value creation, despite often demonstrated to generate both less cash and a lower NPV than a shorter life.

Figure 6.9 is reproduced here as Figure 10.3 to illustrate this situation. As indicated, the figure is from an underground gold mine study and is typical of the outcomes of many strategy optimisation studies.

FIGURE 10.3
Various value measures versus cut-off grade.

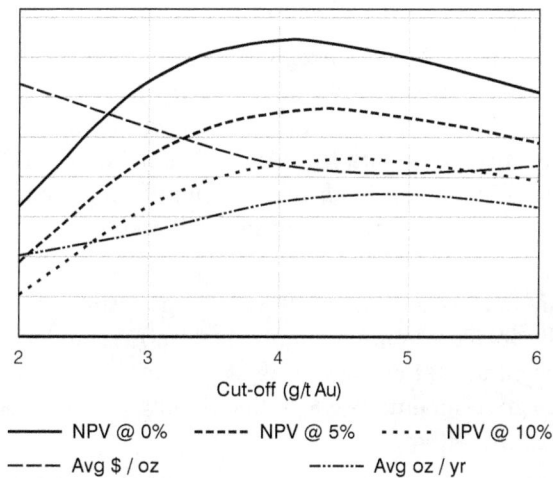

In Figure 10.3 the NPV vertical axis starts at zero. The intervals on the vertical axis represent equal increments, which in the following discussion are referred to as value units. Consider the 'total cash generated line', the top convex-up line for NPV at zero per cent discount rate. At a cut-off of approximately 2.5 g/t, the cash generated is five value units. Similarly, at a cut-off of approximately 3.4 g/t, the cash generated is seven value units on the vertical value axis. The higher cut-off generates 40 per cent more net cash flow than the lower cut-off – the ratio of total cash flows generated at the two cut-offs is 7:5, as in the question posed. Mine life is not shown on this chart, but it is reasonable to expect that the life with a reduced reserve tonnage at a higher cut-off will be less than a higher tonnage at a lower cut-off. The author can report that the ratio of mine lives at the two cut-offs was of the order of 7:10, as in the hypothetical question posed. The situation described is not a hypothetical situation – it is real and it is common.

As can also be seen from this figure, additional free cash can be obtained by increasing the cut-off to 4.1 g/t. This would potentially reduce the life even further but would also increase the average annual net cash flow. It is also worth noting that the company that commissioned this study had a break-even cut-off of 2.5 g/t. Clearly, with net undiscounted cash flow peaking at a cut-off of 4.1 g/t, mining and treating material with a grade less than this is consuming more cash than the revenue generated by it. Those lower grades do not break even with the total costs of operating the mine. The optimum break-even cut-off (representing the cut-off that maximises total net cash flow) clearly accounts for costs that were excluded in the company's official break-even cut-off calculations. As noted in Chapter 3, there are no common formal definitions of various break-even grades, so it cannot be said that the company is doing anything untoward. The reserve resulting from a 2.5 g/t cut-off still generates a positive NPV, which would qualify as being economic, so those tonnes and grades could be quite properly reported as the reserve and still adhere to the international codes for public reporting.

Note that there is an implicit assumption in these discussions that the alternative mine lives resulting from different strategy options are for the same resource. As illustrated in Chapter 6 and in Figure 6.8, there will be an optimum cut-off and an optimum production rate to maximise economic value. This will automatically determine the mine life associated with that optimum strategy. Any attempt to increase the mine life for the underlying resource must involve reducing cut-off or production rate. Consider for simplicity the two-dimensional NPV and life versus cut-off curves at a particular production rate. Increasing value is correlated with increasing life when the cut-off is reducing and is higher than the optimum cut-off. When the life is increasing by reducing the cut-off and the cut-off is less than the optimum, the NPV is decreasing and is therefore negatively correlated with life. Since virtually all operations are operating with cut-offs that are at most optimal, and frequently less than optimal, they are operating in the region where an increase in life will result in a reduction in NPV for the same resource.

Maximising mine life may be an important goal for governments in some parts of the world, as it potentially increases employment, taxation revenue and the like; however, depending on the way that various levels of governments receive their returns, the rationale may apply to governments as well – that is, long life is not positively correlated with value for governments either.

In Chapter 6, reducing the cut-off below the optimum to extend the mine life was the decision of management, as illustrated by the HoV in Figure 6.8. In this case, there was a qualitative reason for doing so, based on a management perception of issues for the social licence to operate that were less easily quantified. This was an informed decision, making use of the quantitative information provided by the study and management's understanding of the wider issues. It is a real-world example of a decision to maximise NPV, subject to the mine life not being less than a specified number of years in order to maintain the social licence to operate. Nothing discussed in this section should be interpreted as criticism of that type of decision. What is being criticised is the unsupported assumption that life and economic value are positively correlated and increasing mine life is automatically a better option. That will often not be the case.

One of the reasons advanced for preferring a longer mine life is the extra time to find more value-adding resource. For now, the reader is referred back to the precondition of the 'which would you prefer' question. 'All other things being equal' includes equal probability of finding more life-extending resources, regardless of the strategy selected, and being able to bring it into production in a timely way. The extra life, therefore, does not in this analysis provide more opportunity to find extra resources. This issue is discussed further in Chapter 14.

Reserve quantities

Closely related to mine life and sharing similar features is the maximisation of the reserve, described by either the ore tonnage or, more commonly, the contained product such as ounces of gold. Like mine life, this is also based on conventional wisdom.

As with mine life, it's a positive outcome if additional resources come from successful exploration. If they come simply from reducing the cut-off for the current resource, that is not necessarily good. Consider again Figure 10.3. The horizontal axis is cut-off, which increases from left to right. As cut-off increases, the metal in the reserve reduces. Although the relationship between metal in reserve and cut-off for a given resource may not be linear, there is nevertheless perfect negative rank correlation between the two; the higher the cut-off, the lower the amount of contained metal in the reserve and vice versa.

The horizontal axis also represents metal in reserve, increasing from right to left. Starting from the right-hand edge of the chart, value increases with more ounces, but only down to the cut-off that delivers the maximum NPV. Further increasing the ounces by reducing the cut-off results in reducing value, which is then negatively correlated with reserve. It is rare to find a mine or deposit with a cut-off higher than the one that maximises NPV. The usual case, therefore, will be for any increase in reserve tonnage or contained metal derived from a given resource to result in a reduced NPV.

If maximising the ounces in the reported reserve is truly an important goal[20], the optimum strategy is obvious – set the cut-off equal to zero. In other words, mine all the material with any hint of mineralisation, or perhaps all the mineralised lithological units. This is the purely rational response to that goal, and although it may produce a smile, it is not necessarily unreasonable. Depending on the grade distribution of the deposit, it is quite possible for a plan that mines all the mineralised lithological unit to still have an average grade high enough to generate positive net cash flow and NPV. The reserve thus defined again could legitimately be reported publicly as the Ore Reserve.

The response of many to the suggestion of a zero cut-off is that many of the reserve ounces would be losing money. The author fully agrees, but also points to the fact that, as illustrated in Figure 10.3, so too are many of the ounces included in what would be a perfectly acceptable reserve at a 2.5 g/t break-even cut-off – in this example, all the ounces in the material with grades between 2.5 and 4.1 g/t. The common experience of optimisation practitioners and mine planning personnel is that the pressure is always on to reduce the cut-off and maximise the reported ounces, often accompanied by a fear that a reduction in reported ounces will lead to a fall in the share price.

Again, more ounces in reserve are good if they come from exploration success. Maximising ounces is clearly not economically sensible for a given resource; rather there is an optimum reserve size that maximises returns. So long as the market continues to value companies by their reported reserves[21], the rational management response will be

---

20. There are a number of gold mining companies that, at least until recently, publicly stated this to be a major corporate goal.

21. In an online discussion post for the Money Mining group on LinkedIn in September 2011, Gerald Whittle likened judging a mining company by its reserves, rather than its cash-generating capability, to judging a restaurant by what it has in its refrigerator rather than by what it puts on the table for patrons. We could extend the analogy – a lot of what we have in our refrigerators might look good on the outside, but when inspected carefully, it turns out to be rotten and only fit for discarding. If we are foolish enough to eat it, we can end up with food poisoning. In light of the discussion, the analogy to reserves and a company's financial health is, I hope, obvious!

to maximise the reserves tonnes and contained product, even though it destroys cash. The company's ultimate aim is to make money. The conventional wisdom, reasonable enough at first glance, is that more ounces will lead to more money, but as illustrated, not all ounces are the same, and many of them lose money rather than make it.

Stability of production, equipment use and earnings

Stability can be considered from two viewpoints: internally by the company, focusing on physical measures in the short-term; and externally by investors and analysts, focusing on financial outcomes, though key underlying physicals may be considered. The latter is usually only possible when the company reports results publicly to the market.

Internally, companies tend to look for stability in feed grades to their mills and in fully utilising their capital assets. Feed grade stability is not unreasonable, as metallurgical performance – typically measured by recovery and product quality – will tend to be better if the nature of the ore feed is steady and the processes can be tuned to suit. Of course, the mine is working with a natural material with inherent variability, so some fluctuation is expected. If the mine can be operated in such a way that changes occur slowly and are well forecast, they can usually be managed with little, if any, loss of performance. It is rapid short-term or unexpected large changes that result in losses.

At the level of analysis conducted for cut-off determination and strategy optimisation, time periods used are such that the short-term loss-causing variability is smoothed out. If there are significant changes in feed properties from period to period, it would be implicitly assumed that the changes would occur over sufficient time for there to be no adverse effects on recovery and product quality and that the recovery relationships applied to the average grades are consistent with the effects of shorter-term variability experienced with current or predicted mining and treatment practices.

It is possible to evaluate, for example, the expenditure of working capital by exposing more ore sources to improve blending in order to reduce variability. Or more can be spent on in-mine exploration to improve knowledge of the resource and hence the accuracy of predicted grades and other performance-affecting parameters. There will always be a point where the extra cost is not justified by improved revenues.

Regarding the full utilisation of capital assets, geological variability and the dynamic nature of mining operations mean that it is impossible to have all assets working at maximum capacity at all times. Each part of the process may be the bottleneck some of the time but will be underutilised at other times. In principle, fully utilised assets are a good thing – underutilised assets imply that too much capital has been spent to acquire unnecessary capacity. Focusing on the full utilisation of assets may be a useful tactic in a steady-state operation, but mining operations are rarely that – variation is the norm.

A strategy optimisation will often suggest variable outcomes for physical mining and treatment parameters, and the variability may be substantial. This will occur for such items as waste stripping for an open pit. There will often be significant resistance to this – steady operation over a period of time is usually preferred. Assuming that the study has accounted for all costs, including those for the equipment required to meet the peak levels of activity, any attempt to smooth the fluctuations will impose a capacity constraint resulting in reduced value.

Why is this so? Simplistically, the mining industry tends to be very cost-focused. Figure 6.10 and Chapter 6, for example, indicate how trying to minimise cost can result in reduced value. A similar rationale applies here. It may reduce capital costs to cater for average rather than peak requirements, assuming that fluctuations can be levelled over a period of time; however, because the optimisation indicates that fluctuating output generates more value, having extra capacity when needed can create more benefit from receiving revenue early than the costs of providing that capacity. Anecdotal evidence suggests that the market also prefers stable production and earnings instead of fluctuating, despite a company's primary asset – its resource – having inherent variability.

If stable physical and financial outcomes is a desirable goal, the team developing an optimisation model must report suitable quantitative measures. Standard statistical measures of trends and variability might be appropriate. If stability measures are meant to constrain the model, limits might need to be placed on, for example, how much the results of one period can vary from the previous period. Optimising with and without such constraints may indicate how much value is lost (or not) by applying constraints that do not necessarily add value, despite common perceptions.

To the best of the author's knowledge, there are no studies indicating that value is added from stable operations at the granularity of time periods used for long-term strategic planning or public reporting. Rather, there is evidence of the opposite. Imposed stability is not correlated with maximising NPV, and will often be associated with reducing it.

SATISFYING VARIOUS STAKEHOLDERS

The owners of the company – its shareholders – would be expected to prefer maximum cash generation, perhaps as represented by the NPV, which recognises that cash now is better than cash later. Maximising NPV should maximise the potential for dividend payments and share price increase. As we have seen, many believe that there is a correlation between value and reserves. Yet this belief is ill-founded, even though it seems to influence share price, and decisions that will maximise cash generation potential can perversely drive the share price down, and vice versa.

Company employees would be expected to favour a long project life, for security of employment; however, this should be balanced by the need for that long life to be secure, which in turn requires that the company can maintain a safe level of profitability. As will be seen in Chapter 12, strategies that maximise mine life and reserves can significantly impact on cash flow riskiness. Employees therefore will ultimately be best served by a balance between the company's life and profitability.

Residents in the local area and governments would initially seem to be best served by a long project life over profitability, but the issues described for employees are also relevant. In terms of direct receipts, government taxes on company income is correlated with cash generation. This would also push government aims back towards cash generation. Royalties, however, would be associated with production, which would tend to favour optimising a reserves measure. Even there, timing of receipts by government will be an issue. The receiver of royalties may also be better served by receiving less total cash over a shorter time frame.

Where a single mining operation is the mainstay of the local economy in an area, there is a definite desire to make the mine life as long as possible. Whether this is rational or not will depend on how the mine's contribution is determined. If it is effectively a fixed annual contribution, an extended mine life would be preferable for the local community;

however, if the company's contributions are profit-related,[22] targeting a long mine life may be counter-productive.

Although there is an apparent tension between cash flow and life and reserves measures, the author suggests that, viewed rationally, all stakeholders will tend to be better served by strategies less focused on life and reserves measures than has been the case until now.

CHAPTER SUMMARY

Ultimately, the real value of an operation depends on the stream of free cash generated over the course of its life. The owners of the operation cannot spend tonnes, ounces or years of mine life – only the cash it generates. NPV is arguably the best single number surrogate for quantifying a series of cash flows. DCF methods, which also account for the time value of money, are therefore used almost universally in developed economies.

Maximising NPV involves being a good corporate citizen and maintaining what is frequently referred to as the social licence to operate. This will require spending some of the income from the business for the benefit of external stakeholders so as to maintain that status and licence. If the operation's cash flows are maximised, it becomes easier to maintain the social licence to operate and maximise returns to shareholders.

Companies frequently have a number of corporate goals. Although maximising NPV is often stated to be the key goal, it is often subordinated when achieving it is perceived to adversely affect other measures. By implication, achieving satisfactory outcomes for these measures must be seen as important corporate goals or as constraints on achieving the primary goals. Typical corporate goals and constraints can be classified under several headings:

- discounted cash flow measures, such as net present value, internal rate of return and present value ratio
- accounting measures, such as various profit measures and rates of return
- cost measures associated with both operating costs and capital costs
- physical and time-related measures, such as payback (undiscounted or discounted), mine life, reserve quantities (of ore and products) and stability of production, equipment use and earnings.

NPV is perhaps the most commonly used optimisation target. Accounting-style measures produce sequences of values, and it can be difficult to determine which pattern of outcomes is better than another. They and other measures may be useful for identifying satisfactory and unsatisfactory outcomes and can be used as constraints to exclude strategies when optimising key value measures to meet other corporate or external requirements.

---

22. Recent political events notwithstanding, an example of how mining returns to the community can potentially be managed for long-term benefit, without necessarily operating for a long time, is the PNG Sustainable Development Fund Ltd, the majority shareholder in the company operating the Ok Tedi mine in Papua New Guinea (PNG). It was established principally to apply funds coming from the mine to the development of PNG and in particular the Western Province. When the Ok Tedi Mining operation ends, its charter is to ensure that ongoing and lasting benefits remain with the people of the Western Province and PNG. These benefits are intended to flow for at least 40 years after the mine's closure. Maximum benefit for the fund and the people will come from maximised dividends received during the mine's operating life, which would be expected to be correlated with maximum cash generation by the operation. As we have seen, this is not correlated with long life.

The standard calculation of NPV uses the weighted average cost of capital (WACC) as the discount rate. Some companies add a premium to the WACC to account for the risks of dealing with particular commodities or places.

The IRR is the discount rate that results in an NPV of zero for a project, and is often reported along with NPV. It could be an optimisation target, but would be more likely used as a constraint to eliminate unsatisfactory cases from an NPV optimisation. Small projects may generate high rates of return but, because of their size, relatively few dollars of return. A high IRR is not necessarily correlated with value generation. Depending on the nature of the pattern of cash flows, a project may have multiple IRRs, none of which can be reliably compared with the required hurdle rate. Because IRR is not in general correlated with NPV, it should not be used as a surrogate for NPV.

Project evaluation, strategy optimisation and associated decision-making should be forward-looking. Accounting measures incorporate the book effects of historical actions and may give a significantly different view of alternative options compared with cash or DCF measures. Accounting measures are not correlated with cash generation measures.

Minimising costs is a common goal in the mining industry. To the extent that it can be done by reducing wastage and improving efficiencies and productivities, this is good; however, optimisation studies will show that cost reductions can result in reduced resource or production capacities, resulting in reductions in revenue that are greater than the reduction in costs. The outcome is that cost reductions that are intended to increase profitability actually reduce it. Low costs are not necessarily correlated with value generation.

Payback measures indicate how quickly a project will repay its investment, but, in the absence of other information, say nothing about how profitable it might be after that time. They are therefore used to filter out unacceptable projects rather than as an optimisation target.

Mine life and reserve quantity measures are often a primary focus for senior executives and industry analysts. The assumption is that, in the absence of detailed production plans, they are positively correlated with value. To the extent that additional ore, product and mine life come from new discoveries, this is true; however, changes in mine life and reserves will often come from changing the cut-offs of existing resources. Reserves and mine life are negatively correlated with cut-off – increasing the cut-off reduces the reserve and the life, and vice versa. NPV and most other profit- or cash-related value measures grow with increasing cut-off up to a point, reach a maximum value, then reduce as cut-off is increased further. These measures also rise then fall with increasing life, reserve tonnes or contained product. Virtually all mines are operating in the area where the cut-off is lower than the optimum cut-off, where value reduces if the reserve or the life of a given resource is increased.

In general, therefore, value is negatively correlated with reserves and mine life. The author suggests that many of the recent profitability problems of the industry are a direct consequence of the failure of senior company decision-makers and industry analysts to appreciate this fact.

Stability of physical production quantities and earnings is seen as desirable; however, optimisation analyses may demonstrate that variability generates better returns in the long run. The objection is that production activities require over-capitalisation – much of the capability, purchased at high cost, will be idle for most of the time. High levels of utilisation are thought to correlate with value maximisation. High utilisation implies efficiency (usually a good thing), but this focus on costs ignores the possibility that the revenue generated by having spare capacity for the occasions when it is needed – even if it sits idle at other times – more than pays for the costs incurred to provide that spare capacity.

It is important for companies to specify goals and value measures that will contribute to value creation. It is also important to identify measures that are thought to be correlated with value but in fact are not. Using these as optimisation targets would lead to value destruction.

CHAPTER 11

Valuation Methods

DERIVING VALUE MEASURES FOR OPTIMISATION

In recent years, a number of techniques have been considered for deriving value measures that could be used for strategy optimisation. In some cases, techniques have been available for years or even decades. *Recent* here refers more to when techniques became a significant part of ongoing discussion in the technical press and at mining industry conferences than to when they were first developed. This chapter's aim is to provide an overview of several techniques to introduce them to readers who are unfamiliar with them; to compare them with standard industry practice as the author perceives it to be; and to indicate how they may impact on cut-off derivation and strategy optimisation.

The apparent lack of support for some of these more advanced valuation methods may be because there is a lack of clarity as to what they are and what they should achieve. Highlighting this uncertainty may assist in furthering the discussion about them in industry forums. Several terms have not been used previously in this book and so may be new to some readers. Interested persons are advised to seek out specialised short courses or the internet for more detail on these topics.

To assist with distinguishing the various methods, it is useful to consider how:

- physical parameters are derived
- associated costs and revenues are derived
- risk is accounted for
- management flexibility is accounted for
- the measure of value is derived
- it may be used for strategy optimisation.

With regard to risk, it should be noted that there are two main types: systematic or market risk, to which everyone operating in the market is exposed, and project risk – risks specific to the individual project/s under consideration.

Strategy optimisation is the overall topic of this book. Chapter 10 considered various measures of value that could become the target for a strategy optimisation study. The various valuation methodologies described are intended to produce a single value, or in some cases a probability distribution of value, for some of these value measures. They are not themselves optimisation methods. Potentially, the value measures discussed in this chapter could become the target for an optimisation study. There are, however, certain implications for some of these methods when they are used to derive the optimisation measure of value.

The methods addressed are as follows:

- traditional valuation
- Monte Carlo simulation
- conditional simulation
- dynamic discounted cash flow (DCF)
- option value
- real options valuation.

TRADITIONAL VALUATION

Traditional valuation is essentially what has been discussed in Chapter 10. The key characteristics of traditional valuation techniques are:

- *Deriving physical parameters* – detailed engineering design work is done for one or a small number of alternative production strategies, such as different mining and ore treatment rates and cut-offs. Metallurgical recovery and product quality relationships are specified to determine product volumes. There are often insufficient cases to allow reasonable interpolation between cases, except where one parameter, such as cut-off or production rate, has been varied with other parameters held constant. It should be noted that two data points can at best describe a straight line relationship, and three a parabolic relationship. Since curvilinear relationships of value versus decisions are not likely to be simple low-order polynomial relationships, more than three data points will usually be needed to identify reliable relationships. The author prefers to have a minimum of five data points for a first-pass indication of how value might vary in response to changes in decisions for any one parameter.

- *Deriving associated costs and revenues* – unit costs are specified for each major physical activity. These are typically constant over time (in real or uninflated terms) unless there is either a change in the operation that would result in a change in the cost structure or a predicted change in the unit cost, such as an increase in the cost of power or fuel at some time in the future. Occasionally, an ongoing cumulative reduction in real costs of the order of two to three per cent per year is modelled to account for technology improvements. Metal prices and, where relevant, sales terms are specified for all products. Price forecasts will often show some variation over the first few years, particularly if current prices are believed to be off-trend, but are then typically projected at constant price levels into the future[1]. Inflation rates may also be specified for costs and revenues.

- *Accounting for market risk* – this is done by applying a discount rate, typically the weighted average cost of capital (WACC), to the net cash flow to derive a net present value (NPV). The WACC takes account of the risk-free rate of return on capital, the cost of debt and the return required by shareholders, which is dependent on the risk associated with the company's returns. The last of these is based on the way the share price responds to changes in the overall market. Smaller relative movements indicate lower risk and hence a lower required rate of return; larger relative movements indicate higher risk and a higher required rate of return.

- *Accounting for project risk* – premiums may be added to the WACC to account for the perceived risks of dealing with particular commodities or operating in specific countries. As noted in Chapter 10, there are no standard ways of determining the appropriate risk premiums to add. A number of key inputs that do not affect the underlying mining and treatment schedules will typically be flexed. As indicated in Chapter 6, these will be for items that the company has no control over; that is, scenario parameters such as prices and costs, and perhaps ore grade and metallurgical recovery. This indicates how the value will change in response to changes in external forces for each prespecified plan. Sensitivity analyses consider the impact of changes in the input parameters to be flexed by looking at variations of, say, ten and 20 per cent above and below base case values, each input varied on its own with others held at their base case values. There is typically little or no science applied to identify the likelihood of such movements in each parameter flexed and therefore which might have the greatest probability of influencing value. The mining and treatment plans are not able to vary in response to any of these changes.

- *Accounting for management flexibility* – there is little or no accounting for management flexibility in simple traditional valuations, apart from considering which of the few cases developed for different options delivers best value. Typical sensitivity analyses described may indicate when changes in one parameter cause one of the alternative plans to become more valuable than another. Sensitivity analyses may predict large negative values if, for example, prices are flexed to a low level.

- *Deriving the measure of value* – Chapter 10 describes a variety of value measures that might be specified. NPV is perhaps the most common value measure, and it is derived by adding the discounted net cash flows for each period, with discounting performed using the WACC or WACC plus a risk premium.

- *Semantic issues* – because an NPV is usually generated for one or a small number of physical options, it appears to have led its critics to describe this traditional valuation as NPV or DCF, or similar terms. NPV or DCF is then defined to be evaluating one physical case and one set of financial inputs to derive one single number – the NPV – for the value used for decision-making. NPV or DCF in reality specifically refers to a very small component of the overall process of traditional valuation. It is simply discounting each period's cash flows at the WACC and adding these to get an NPV. There are many other things that go into the evaluation that are required for other valuation methodologies. Evaluating a large number of alternatives, if time and

---

1. Constant predicted prices are usually expressed in real terms; however, the author has encountered some companies where the constant prices are in nominal terms, implying that real prices are assumed to be decreasing at the predicted inflation rate. Constant prices should not, therefore, be assumed to be in real terms.

resources permit, can be done to assess how value might be affected by variations in input parameters.

- *Application for strategy optimisation* – all the discussion of strategy optimisation so far has essentially assumed a traditional valuation style for each option considered. Optimisation methodologies are described in more detail in Chapter 13, but for now it is worth noting that for each scenario considered, a traditional valuation will typically be conducted; for example, to generate a data point on the surface of a hill of value (HoV) as illustrated in Chapter 6. Yet, as has been previously noted there, it is usually impractical to create each data point separately by the traditional approaches described. The evaluation model must become more flexible, able to handle changing decisions as well as changing external parameters. Having generated scheduled physical quantities by automated processes rather than engineered designs, the remainder of the evaluation process is similar to the traditional processes described. The same will apply for any other method; each of the cases evaluated internally by the optimiser will effectively be dealt with in the manner described here. The main differences between the optimisation methods is how different cases are selected or generated for evaluation, not the valuation processes applied. NPV – discounting net cash flows at the WACC – is typically the key value measure. It is perhaps this aspect of deriving the number to be reported as the value that generates most controversy.

The author is not defending the use of traditional valuation here. Indeed, the distinction between what is described in Chapter 6 as typical strategy selection and strategy optimisation is a criticism of typical strategy selection, which is similar to what is described here as traditional valuation. In the context of this section, the traditional methodology is presented as typical industry practice so that it can be compared with other methods to be discussed. The intention is to alert readers to possible semantic differences between what actually constitutes traditional practices and what could be seen as significantly more restrictive (and pejorative) definitions by some of its critics.

MONTE CARLO SIMULATION

Monte Carlo simulation was touched on when discussing NPV as a value measure in Chapter 10. Stochastic simulation may be applied in more advanced, though still traditional, studies. Probability distributions are used as data inputs for key parameters instead of single or flexed values. These will be for such inputs as costs and prices, but may also be applied to physical parameters such as metallurgical recovery, mining and plant capacities and availabilities, construction periods and exploration success. A number of commercial software packages provide stochastic modelling capability. In the simplest form, they are used as add-ins for standard spreadsheet applications.[2] Developer toolkits also exist so that the underlying technology can be used by software developers to enhance the capabilities of specialist applications.

The simulation software runs the model many, perhaps tens of thousands of times using different input values (extracted by rigorous sampling processes from the probability distributions) to generate multiple output values and hence a probability distribution of values, in particular the value measure such as NPV. Percentiles of the distribution of NPV, such as the 10th, 50th and 90th, can then be identified. Depending on the way these

---

2.   @Risk™ by Palisade Corporation and Crystal Ball™ by Oracle Corporation are two of the more common packages.

percentiles have been defined, they will represent either the NPVs that have ten per cent, 50 per cent or 90 per cent probability of being exceeded or, conversely, with a 90 per cent, 50 per cent or ten per cent probability that the NPV will be less than the percentile NPV values. The probability of making an economic loss – having an NPV less than zero – can also be established.

The simulation process is applied to any one particular case that is set up and evaluated. For strategy optimisation purposes, the process could be repeated for each case evaluated. Decision-makers might be interested to know whether the optimum strategy varies, for example, for different percentile values of NPV.

The probability distributions for uncertain value-driving inputs must be reliable. 'Garbage in, garbage out' may be a very real danger in this sort of analysis. However, as has been already noted, the lack of reliable information is not in itself a reason to not conduct such an analysis. Uncertain inputs can be flexed to see whether variations in the inputs affect decisions. If flexing the nature of the probability distributions of the key inputs across a realistic likely range fails to change the resulting distribution of NPVs to the extent that decisions would be different, then that uncertainty does not affect the decision and can be ignored for the problem at hand. Conversely, if the decision would be different, additional work must be undertaken to resolve the uncertainty.[3]

If we are dealing with a polymetallic deposit, the simulated price forecasts should also account for any correlations between price movements for different metals, adding further complexity to their modelling. Similarly, there may be correlations between the movements of both prices and some categories of costs due to overall changes in the economy, for instance. Auto-correlations between successive values for some parameters may also be important. Failure to account for these correlations can impact the reliability of the outcomes from a simulation model.

Although potentially providing more information for decision-makers, simulations also increase the decision-making complexity. Consider the situation with two options. Simplistically, using traditional evaluation methods and NPV-type value measures, the option with the larger NPV would be selected. In reality, decision-makers would qualitatively take account of issues such as perceived risk, so that an option with a lower value but perceived lower risk might be selected in preference to an option with a higher value but perceived higher risk. Monte Carlo simulation quantifies some of that risk – the option with the lower expected NPV may have a low NPV variance and hence a low probability of generating an NPV less than zero, while the option with the higher expected NPV may have a higher variance and higher probability of generating an NPV less than zero, with both probabilities now quantified. The simulation software would also permit generating the probability distribution of the difference in value between the two projects, so that the probability of one option achieving better outcomes than the other can also be quantified. Decision-makers now have better information on which to base their choices, but that may make it more difficult.[4]

---

3. This is discussed in more detail in Chapter 12.

4. Despite Monte Carlo simulation add-ins for spreadsheet applications being available for many years, the author has encountered little use of it, or demand by senior decision-makers. If not explicit, there still seems to be an underlying attitude of 'Stop bothering me with all these numbers. Just tell me the answer!' The reality is that there is rarely one simple answer to a complex problem; there are a lot of 'it depends' factors involved.

For strategy optimisation, simulation introduces additional value measures that might need to be traded off against each other. For example, we could potentially optimise or generate HoVs for the 10th, 50th and 90th percentiles and the probability of an NPV less than zero. The last might be used as a filter to eliminate options resulting in a loss-making probability above a specified acceptable level. If the strategies that optimise different percentiles of NPV are distinct, risk–reward trade-offs may need to be considered.

OREBODY SIMULATION USING CONDITIONAL SIMULATION

There are many uncertainties in any evaluation that could be accounted for, but there will usually only be a small number that have a significant impact on the results' variability. The simulation software mentioned will typically have facilities to identify the key input variables that contribute to output variability in a model. One of the major uncertainties will be the underlying geological resource. Conditional simulation (CS) makes rigorous consideration of this variability possible. Standard geostatistical techniques will generate the best estimates of grades throughout a block model, but in doing so will smooth the spatial variability. Conditional simulation restores the inherent variability within the resource, but it can only do this by generating a number of realisations of the deposit that have the same probability of being true.

At the simplest level, this information could be used to derive probability distributions of the tonnages and grades that could be generated at any particular cut-off. They could be used as input distributions in a standard stochastic simulation model, for example, in a spreadsheet. At a more fundamental level, designs could be derived for each CS realisation. Since the actual grade distribution will always be unknown but the CS realisations are all equiprobable possibilities, the logical process is that each realisation should be coupled with each of the designs for the set of realisations. This means that, for instance, if there are 100 realisations and therefore 100 designs, there are 10 000 combinations of designs and realisations. There can then also be stochastic simulation of a number of other inputs, as described for Monte Carlo simulation.

Using CS, the outputs would be represented by probability distributions of value associated with each design, taking account of the underlying uncertainties, both geologically and in other inputs. The decision-maker would then need to select the design that delivered the best outcome, but defining what that means is far from trivial. It could be defined by the maximum expected value, the maximum minimum value, the highest probability of exceeding a specified value, the lowest probability of an NPV less than zero and so on. For strategy optimisation purposes using CS, the mine design becomes yet another option that can be selected, another axis on a hill of value. The process is not trivial but it is conceivable, and as indicated in previous discussions, if a process can be described it can be modelled, in theory at least. We are perhaps only limited by existing computer power and optimisation algorithms – the description of the problem to be solved and the underlying rationale is straightforward, but the practical implementation may not be. The number of realisations used could be reduced to make the problem computationally tractable, but this will reduce the accuracy of the probabilities generated by the analysis.

DYNAMIC DCF – MODELLING FUTURE MANAGEMENT DECISIONS

All the discussions up to this point have incorporated fixed mining and processing plans. The values are generated assuming that the plan will be implemented to completion, which is typically the depletion of the identified ore reserve. Critics of simple so-called DCF analysis correctly point to the fact that, if a particular plan generates a loss, it will not be implemented in the way it is stated. It is an unreal plan that will never be implemented, so why attribute value to it? Managers will respond to the situation confronting them – the plan will be changed to minimise the loss or convert it to a profit.

Standard strategy optimisation processes, however they are implemented, will have the potential to generate a number of alternative plans using relationships built into the data or software itself. Each of the plans will usually be, hypothetically, implemented to completion, as with a single plan for simple strategy selection. The difference with strategic optimisation is that any plan that is finally identified as the optimum will have a positive value and be recommended for implementation. Even if the maximum value is negative, that in itself is still useful information. As indicated earlier, the HoV may be thought of as a multidimensional plot of the sensitivity of value to decisions. For this purpose, it is perfectly valid to show that if one were to make a particular decision the outcome would be a substantial loss. It is not wrong in these circumstances to show the effect of making certain decisions and carrying them out to their logical conclusions in the operation, even if a large loss were to be incurred.

The other criticism of the typical fixed or prespecified mining plan is that it fails to allow for the value that can be created by incorporating the effects of management flexibility. This at the simplest level includes the ability to choose not to invest or defer investment in a plan that does not create value with current assumptions. More broadly, it includes the ability to decide whether, and if so when, to do such things as:

- increase or reduce production capacities
- close, mothball and re-open the operation
- mine the next cut-back in an open pit
- develop to another ore lens or sublevel block underground
- change cut-offs, and so on.

Many of these types of decisions can easily be built into an evaluation model. For example, it is unnecessary to report only the results of depleting the identified ore reserve. If prices or costs are forecast to change, that plan may become uneconomic. It will often be trivial to build in a simple management option – permanent closure – by constructing the cash flow model in such a way that the mine is closed at the point where ongoing operation generates losses (which, contingent upon the input assumptions, might mean never opening). Depending on forecast conditions, if the best option is to close the operation immediately, values reported will only be positive or zero (or greater than negative resulting from closure costs). Similarly, it is not necessary to assume the final pit limits or depth of an underground mine. Stages of development can be modelled and each successive stage implemented only if evaluations show that it is worthwhile.

If decision relationships can be described, they can be modelled. This type of analysis is sometimes alluded to as *dynamic DCF*. The values are standard DCF measures, such as NPV derived from the total net cash flow discounted at the WACC, but the plan becomes flexible depending on assumed conditions, which may vary from case to case.

It is obviously a simple step (logically or theoretically, at least) to combine dynamic DCF and Monte Carlo simulation. The outcome of this will be that modelling the decisions to not proceed with loss-making options will result in a change to the probability distributions of outcomes. Simplistically, all options with NPVs less than zero (or less than the negative NPV associated with the costs of immediate closure) will be eliminated – the low-value tail of the distribution from an inflexible analysis will be truncated. High-value options will be retained. The overall effect will be to both reduce the range of probable outcomes as well as the probability of making a loss, which will result in an increase in expected NPV.

As soon as we start building in strategic management decisions, such as the timing and quantum of capacity changes, sizes of pits, cut-offs, ore lenses included and so on, we must identify exactly what we are trying to achieve with our evaluations. If we want to identify a single value or range of values to guide a decision regarding the price of buying or selling an operation, it is appropriate to include as many of these flexible decisions in the model as practicable; however, it is of little use in deciding what the operating strategies should be to optimise value. To do that, we must evaluate plans that do not allow those decisions to be changed automatically to generate a value/s. We want to know what the values are if we were to make certain decisions, so that we know which decisions generate the maximum values. From an evaluation point of view, the criticism that conventional plans lack management flexibility is ill-founded. It is exactly that inflexibility we wish to model so that we can assess which plan will optimise value.

The real situation is actually between these two extremes, since not all project decisions are committed to at the outset. What we really want to do is identify the values generated by each set of options, taking account of whatever future flexibility each may then either eliminate or facilitate. We must establish the time frames for which plans will become locked in by the decisions facing us now. All other decisions further into the future should then remain flexible within the model. We are only looking to optimise today's decisions, but that will include accounting for additional value to be obtained from future decisions still possible after implementing today's decisions.

OPTION VALUE

The term *option value* is usually understood to refer to what has become known as the Black–Scholes option pricing model, or B-S model (Black and Scholes, 1973). This is a partial differential equation designed to calculate the price of a class of financial derivatives known as *European put and call options*. Working on the assumption that all investments (including those in mining operations) can be modelled as combinations of various financial derivatives, the B-S model has been applied to mining operations as well. In general, the assumptions underlying most of the inputs into the B-S formula do not apply to mining operations and significant assumptions about the features of such operations have to be made to make them fit the form of the B-S model.

Most commentators seem to classify the B-S-style option value as inappropriate for valuing mining operations, but it is still referred to from time to time. The author has never had occasion to apply it and is not aware of any mining companies who do so.

REAL OPTIONS VALUATION

Real options valuation (ROV), or simply real options (RO), is a term applied to a number of techniques. The name was originally used to refer to the assessment of *real* as opposed to *financial* assets using financial valuation theory. There is a growing body of industry publications regarding the application of ROV in the mining world; however, there seems to be no generally accepted definition of what it is or isn't. Some writers appear to use the term for any valuation method that is not a simple DCF evaluation that generates one NPV value for one fixed operating plan. More commonly, the terminology is used to refer to valuation methods applying multiple discount rates to various cash flow components, rather than a single discount rate to the overall net cash flow. Samis *et al* (2012) provide a concise but comprehensive description of applying ROV in the mining industry.

The main objection to the standard DCF method, advanced by proponents of ROV, is that the single discount rate does not take account of the risks that may be associated with different projects or options. It must be recognised that when using traditional DCF methods with the WACC (or a derivative) as the discount rate, there is an assumption that the projects being evaluated do not change the overall riskiness of the company or its capital structure, so that the required rate of return on shareholders' equity stays the same and the discount rate is not dependent on the project being evaluated. This will only be true if projects are small enough to not have an impact on the company's level of risk, regardless of how risky each project might be, or, if projects do have an impact on the company's overall risk, that their levels of risk are similar to the company's level of risk, and will not change over time; however, this will not always be true. The use of a single WACC-style discount rate also assumes that the project's risk profile remains the same over time, which will also typically not be the case.

The RO approach is to recognise that the simple DCF WACC-style discount rate has two main components, one being the time value of money cost of a risk-free cash flow. The other, based on the returns required by investors, is associated with the riskiness of the investment. An underlying principle in deriving the WACC is that the way a company's share price moves is indicative of the riskiness attributed to the company by the market. As a result, there is a return related to that risk that the market requires in addition to the risk-free rate. The overall project cash flow is made up of a number of components such as revenue, operating costs and capital costs, each of which may have subcomponents and its own inherent risks relative to the market. Simplistically, to apply this approach the evaluation team must:

- identify the riskiness of each cash flow stream
- discount each stream at a rate appropriate to its individual risk profile
- total these up to obtain a net discounted cash flow adjusted for the risks associated with each stream
- discount the combined risk-adjusted cash flow by the time value of money risk-free rate
- add up the risk-adjusted and discounted values for each period to generate a single value to be reported.

The resulting value measure has sometimes been called the adjusted NPV (ANPV).

It should again be noted that these are systematic or market risks – the risks of being in the system – and not project-specific risks. By assessing the market risks of each item, the overall cash flow and value will automatically account for each individual project's riskiness by considering each of its components.

For example, metal prices will typically exhibit greater risk than costs, and the resulting revenue stream will therefore be discounted at a higher rate than cost cash flows. Consider two projects, the same in all aspects except that one has a significantly higher ore grade. Both RO and DCF valuation will show greater value for the higher-grade project. Let us assume that both prices and costs are predicted to remain constant into the future, in which case the margins, assumed to be positive, will also remain constant. DCF discounting will result in the discounted margin reducing over time but always remaining positive. By discounting the revenue stream at a higher rate than costs, the revenue will reduce faster than the costs, and may even become less, causing the margin to become negative. For the sake of the argument, let us assume that this happens for the lower-grade project but that the margin remains positive for the higher-grade project. Relatively, the lower grade project will have a significantly lower value than the higher-grade with RO than with DCF. This is a logical outcome of its being a more risky project because of the lower grades and margins.

There are two ways to deal with the negative margins that may result from applying discount rates to revenues and costs. One is to continue the valuation to the end of the planned mine life, accepting the reduction in value that comes from the negative discounted margins in later years. The other approach is to terminate the valuation at the time that the discounted margin becomes negative. The rationale here is that the project is too risky to attribute any value to operations beyond that time.

There are other ways of dealing with the risk associated with individual cash flows. For example, with a number of metals, the predicted prices can be replaced in the analysis by corresponding recognised forward prices. Since these can be locked in by a company forward-selling its production, they are certain and therefore risk-free. Consequently, there is no need to discount revenues derived from forward prices for risk; they can be added directly to other cash flows after discounting for risk, after which the risk-free rate can be applied to the total. Note that this is a way of modelling revenues for valuation; it does not imply that the company will in practice forward-sell any of its production.

There are other techniques commonly applied by ROV practitioners that can be equally well applied whether using a single discount rate applied to the net cash flow or multiple rates applied to various cash flow components. These include:

- *Stochastic modelling of inputs* – this has been described generally in the section on Monte Carlo simulation in relation to DCF valuations. ROV practitioners have contributed significantly to the practice by highlighting sophisticated stochastic models of price movements that may be used as an alternative to the two price modelling mechanisms described previously. The two most common are known as reverting and non-reverting price models.
  - *Non-reverting prices* have no long-term trend. They tend to fluctuate randomly until there is a price shock that shifts the mean value. Probability distributions based on past price behaviour are invoked to model the times between price shocks, the changes in mean value and the distribution of periodic values around the current mean. The band within which prices will vary increases symmetrically above and below the initial or current price; however, the rate of increase of the range between

maximum and minimum prices reduces with time into the future. Precious metal prices are typically taken to be non-reverting.

- *Reverting prices* have a long-term trend. The parameters describing a reverting price model are the long-term trend relationship, the rate at which the price will return from its current value back to the trend (typically assumed to be an exponentially declining rate defined by the time required to converge by half the difference between actual and trend prices) and random variations. The band within which prices will vary will typically have a constant range above and below the trend in the longer term. In the short term, it will exhibit a rapid expansion from the current price to the long-term band width. If the current price is above a declining trend line, the midpoint of the band will fall over time to the long-term trend line. If below, the midpoint will rise towards and then fall to converge with the long-term declining trend. Base metal prices are typically taken to be reverting.

Any one sequence of prices over time is referred to as a *price path*. Figure 11.1 illustrates a simulated price path for copper with reversion to a long-term trend, which is assumed to be a constant real price of US$2.00/lb in the illustration.

The exponential return to the long-term trend is illustrated by the three dashed lines from points on the simulated price path at different times in the future. The dotted lines in each case identify the ten to 90 per cent confidence band, which can be seen in the long-term to be between approximately $1.15 and $3.00. Although only three reverting price curves are shown, each time period's price is estimated from the previous period's modelled price, the rate of reversion to the trend and a random variation. Models would be recalculated for a large number of simulated price paths to generate a probability distribution of outcomes. As noted, decisions could be included in the model to take appropriate action, such as closing or mothballing the mine during times of low forecast prices.

FIGURE 11.1

Example of simulated copper price over time with reversion to the long-term trend (after Samis *et al*, 2012).

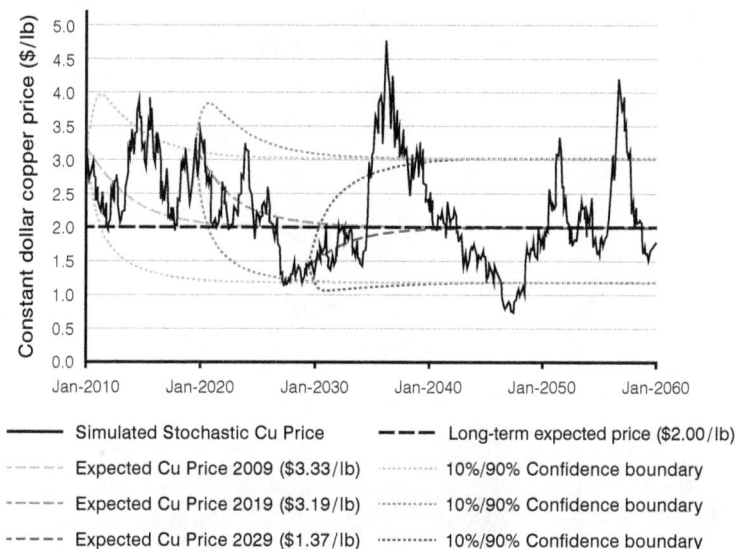

- —— Simulated Stochastic Cu Price
- –– – Long-term expected price ($2.00/lb)
- ------ Expected Cu Price 2009 ($3.33/lb)
- ·········· 10%/90% Confidence boundary
- – – – Expected Cu Price 2019 ($3.19/lb)
- ·········· 10%/90% Confidence boundary
- – – – – Expected Cu Price 2029 ($1.37/lb)
- ·········· 10%/90% Confidence boundary

- *Binomial price trees or lattices* – these might be considered to be simplified, more easily applied predictions of possible price paths than the stochastic models. Starting with the current price, it is assumed that in the next time period the price will either go up or down. There are probabilities associated with each movement, and the same occurs from each price in subsequent periods. A key feature of this method is that over any two time periods, the price resulting from an upward movement followed by a downward movement will be the same as that resulting from a downward then upward movement. The binomial tree always reconverges between the upper and lower bounds. This means that there will be the same number of possible prices in a time period as the number of the period (with the first period, using the current price only, taken to be period number one). The upward and downward percentage changes from period to period and the probabilities of each direction of change can be set so that the average or expected value of the tree mimics any desired price trend. Probabilities of upward and downward movements from each node must, of course, add up to 100 per cent.

Figure 11.2 illustrates a binomial lattice for copper price, staring with an initial price of US$3.00/lb.

The lattice shows prices increasing or decreasing by ten per cent each year. Although the central price is declining over time, it should be noted that the prices along the uppermost path are increasing faster than the prices along the lowest path are reducing. If equal probabilities were applied to price increases and reductions in each step, the expected value of lattice at any time would remain at $3.00.

Modelled cash flows would be calculated at each price in each year. If flexible operating policies are modelled, decisions would be made at each node depending on the value obtained for that period's activities at that price, plus the expected probability-weighted value of all future branches from that node. The process works backwards from the final to earlier periods. If, for example, a node on the lowest price path generates a negative value, its value could be set to zero or some appropriate negative value representing mine closure costs. This loss-minimising value would then propagate back through the lattice to the start of the model. By eliminating

FIGURE 11.2
Example of a binomial lattice for copper price.

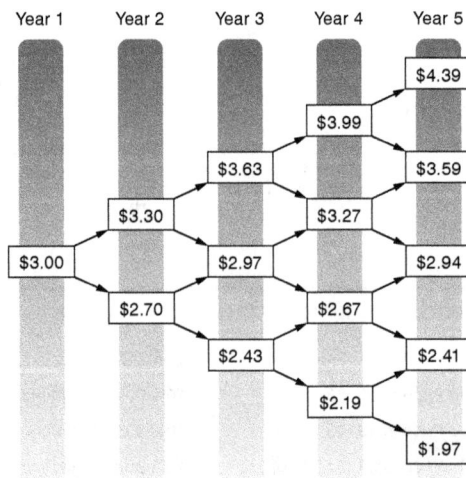

loss-making situations that would not be tolerated in practice along low-price paths, the resulting expected value would be higher than it would have been if it had been assumed that the mine must continue to operate in a loss-making state.

Although usually applied to prices, these concepts can be used for any uncertain parameter in the analysis. Similar models could be applied to inputs such as fuel and electricity, the principal costs of which are driven by the product prices of other extractive industry sectors; what is revenue for one operation, such as a coalmine, will be a cost (for example, electrical power) for a metal mine.

SOME CONCLUSIONS ON VALUATION METHODS

The price modelling constructs described are applied before any discounting is done, whether to the combined cash flow or to individual components. They may be equally well-suited to more complex applications of both DCF and ROV (unless one defines ROV to be anything that is not a simple DCF evaluation generating one NPV for one fixed mining and treatment plan).

All the other features of more complex evaluations described under Monte Carlo simulation, conditional simulation and dynamic DCF (preceding the application of a single discount rate to generate a DCF-style NPV or distribution of NPVs) can therefore be applied in analyses using different discount rates for various cash flow streams.

As previously indicated, it appears that there is no clear consensus on what ROV is or isn't, which may contribute to some of the industry confusion as to what it can do. Some ROV proponents take what could be called a maximalist approach – anything that is not the simplest form of DCF analysis is ROV. The other extreme might be called the minimalist approach – that the only distinction is the method of applying discount rates, and flowing from that, the use of forward prices in one way of modelling revenue. In this view, virtually all other modelling features often attributed to ROV can be seen to be equally valid in a thorough analysis using both ways of discounting. These features include:

- use of Monte Carlo simulation for prices, costs, capacities, construction durations, production rate ramp-ups and exploration success
- use of conditional simulation to quantify and deal with the geological risk of uncertainties in the resource model
- attribution of value to management flexibility, and flowing from that
- a need to assess what the evaluation is to be used for, and therefore
  - which management decisions, particularly in the longer term, can be modelled in response to changing conditions to attribute value to management flexibility
  - what must be modelled separately to identify optimum strategies, and hence the best near-term decisions, to aid management decision-making in the short term.

The discussion has shown that there are many ways to generate measures of value, any of which could be associated with a corporate goal and therefore a potential target of a strategy optimisation study. In principle, this is not an issue. Although it was put forward in Chapter 10 that the NPV is the key measure of value and maximising it should be the aim of a strategy optimisation study, we have now seen that there are different ways of deriving a cash-related value. There may be various probability percentiles of NPV, and ROV applies similar practices to derive dollar-denominated values that its proponents present as superior to NPV. A company may have many goals and the strategy optimisation process must be able to address all of them. It must also manage any trade-offs when optimising for different goals suggests different strategies. The value measures resulting from the more advanced techniques add to the mix.

There are a number of practical issues when it comes to implementing some of these advanced techniques.

When using stochastic simulation, we must use enough iterations for a reliable distribution of outcomes. The number will depend on the complexity of the relationships within the model and the parameters being simulated. It may be necessary to experiment by progressively increasing the number of iterations to ascertain at what point the output distributions stabilise. Further investigations can then be made using the number of iterations identified.

Another key issue when building automated decision-making (that is, management flexibility) into an evaluation model is the modelling of what will be the basis for making those decisions. To illustrate, let us consider only the variation of the metal price.

In principle, we must model two price aspects into the future. One will be the prices we predict now that will be received in each time period in our model. This price path will be applied to the estimates of production to derive our modelled revenue stream, which will contribute to the case's value. If we are modelling managerial flexibility, we must also model the prices that will be used to make those decisions. These will not necessarily be the prices on the price path that we predict will be received.

If we have a simple model with a single price forecast into the future, it is implicitly assumed that the prices in future years that are modelled now will also be the prices predicted in future years. It may not be right, but within that framework the one price forecast can be used as the prices to be received and as the prices that will be predicted at the times that various strategic decisions will be made in the future.

If, however, we are predicting prices by stochastic simulation, depending on what metals we are dealing with we will be creating a number of possible price paths into the future. As noted, there must be enough of these for a stable distribution of outcomes to be generated, and each of them will be applied to the production estimates to derive the modelled revenue streams. Each time when there is a decision to be made, there will be as many price paths from that point forward as there are iterations in the simulation.

With this method of predicting prices, it is inappropriate to make use of each of those price paths going forward as the basis of our decision-making, as was done for the simple single price forecast. Rather, for each price that is forecast for that time period, we must model future prices in the way that a decision-maker would, given that the price is then what it is predicted to be in the price path under consideration. We must then simulate another set of stochastic price paths going forward from that point, on which we are to model the decision-making. All other things being equal, we'd need to produce as many simulations of predicted price paths for making each decision as there are simulated price paths for revenue calculations. If there are n iterations needed and d decisions to be modelled, the total number of simulated price paths becomes n raised to the power of d. The size of the problem to be solved can quickly become large.

The example has considered only the simulation of prices. A similar rationale may apply to other parameters to be simulated. As indicated, increasing the number of parameters simulated will potentially increase the number of simulation iterations required to obtain stable outcomes. In practice, there will have to be a trade-off between the accuracy of the analysis and the time taken to obtain the desired results.

CHAPTER SUMMARY

There are a number of ways in which operating plans can be developed for evaluating a single set of options or multiple options to facilitate strategy optimisation. These are summarised below, but first we will consider the final measure of value. There are two ways in which discounting is applied to calculate the final NPV-type value measure:

1. Discounting the net cash flow by the WACC, with perhaps the addition of a risk premium. The WACC accounts for both the time value of money risk-free rate of return and the systematic or market risk associated with the company. This is typically referred to as DCF valuation, and the value measure obtained is the NPV.

2. Discounting each cash flow component at an appropriate rate for its own individual riskiness, adding these to obtain a risk-adjusted net cash flow, then discounting this at the risk-free rate to account for the time value of money. The value measure is sometimes called the ANPV, and the process constitutes part of what is generally referred to as ROV.

There seems to be no clear and generally accepted definition of what ROV is and, perhaps more importantly, what it is not. Some writers appear to use the term for any valuation method that is not a simple DCF evaluation deriving one NPV value for one fixed operating plan. For other writers, the main distinguishing feature of ROV is the method of discounting – multiple discount rates applied to the individual components of the cash flow.

There are a number of ways by which the operating plan, and hence the physical quantities and associated costs and revenues, can be derived. These generally do not depend on the way that discounting is applied to obtain the final value measure. Techniques or modelling principles that might be applied, separately or in combination, for both DCF and ROV evaluations include:

- A single, predetermined physical operating plan with fixed prices and unit costs. For ROV evaluations of inflexible plans, risk-free forward prices, which are fixed in the same way as a simple price projection in a DCF evaluation, are sometimes used in place of predicted prices.

- A number of alternative plans that represent a small subset of the combinations of options that might be selected for strategic decisions.

- Flexible modelling of schedules and the like so that the outcomes of any specified set of options can be generated.

- Monte Carlo simulation for such things as prices, costs, capacities, construction durations, production rate ramp-ups and exploration success. Price modelling may be simple (using random variability around a simple trend relationship), or more sophisticated (using reverting and non-reverting price models), which are typically applied to base and precious metals respectively.

- Conditional simulation (CS) to quantify and deal with the geological risk of uncertainties in the resource model. This may be used to generate reliable probability distributions of ore tonnes and grades for a specified mine design and consequent plan. At the other extreme, designs can be done for each CS realisation, and each design evaluated for each realisation of the geology. The distributions of values obtained for each design can be compared to select the best design. Selecting the best design will then become part of the strategy optimisation process.

- Attributing value to management flexibility, which involves constructing the evaluation model so that decisions can be made to: increase or reduce production capacities; close, mothball and re-open the operation; mine the next cut-back in an open pit; develop to another ore lens or sublevel underground; change cut-offs; and so on. If combined with stochastic simulations, loss-making options are eliminated, thereby truncating the negative value tail of the distribution of values, reducing the range and increasing the expected value of the measure.

If management flexibility is being modelled, there is a need to assess what the evaluation is to be used for, and therefore

- Which management decisions, particularly in the longer term, can be modelled to respond to changing conditions to give value to management flexibility. If we want to identify a single or range of value/s to guide a decision regarding the price for which we might want to buy or sell an operation, it is appropriate to include as many of these flexible management decisions as possible.
- What must be modelled separately to identify the outcomes of different strategies and hence the best near-term decisions in the short term. To optimise strategic decisions, we must evaluate plans that do not allow those decisions to be flexed. We want to know what the value is if we were to make certain decisions, so that we can identify which options create the maximum values. Including the downstream management flexibility in the evaluation will then account for future options that may have been created, retained or precluded by combinations of short-term decisions.

Each of these methods will generate different NPV or ROV-type value measures, or probability distributions of values. A company may have multiple goals and the strategy optimisation process must be able to address all of them. It must also manage the trade-offs needed when optimising for different goals. The value measures resulting from the more advanced techniques described in this chapter simply add to the mix. The probabilistic evaluations, various percentiles of value measures and the probability of the value being less than a specified value also increase the number of effective measures that might need to be considered together.

CHAPTER 12

Strategy Optimisation and Risk Management

ACCOUNTING FOR THE IMPACT OF UNCERTAIN PARAMETERS

The need to flex uncertain parameters has been noted in earlier chapters. If the decisions to be made will be the same across the full range of values of the uncertain parameters, then while the value of the project may be affected, the uncertainty has no bearing on the best strategy. If the optimum decision changes within the likely range of a parameter, more information needs to be obtained so that the correct value is used and the best decision made.

Some parameters, such as future prices, will always be intrinsically unknowable. Others may be knowable but because of a lack of data, their values have not been accurately identified. In some cases, it may be impossible to access the information needed before the decision must be made. So whether the parameter is unknowable or currently uncertain, the best option is to find strategies that take account of the possible values of uncertain parameters. This chapter illustrates how the upside of making assumptions about uncertain parameters that turn out to be correct can be traded off against the risk associated with decisions based on incorrect assumptions.

IDENTIFYING UPSIDE REWARD AND DOWNSIDE RISK

Figure 12.1 shows value versus cut-off curves for two metal price predictions. What cut-off strategy should the operation adopt? The temptation is to select the cut-off that maximises the value at the higher price, since this clearly boosts value overall.

Figure 12.2 reveals that if a higher cut-off that maximises value at the lower price is selected and the higher price then occurs, most of the potential value increase associated with that price is procured anyway. The real additional value gained by using the lower

FIGURE 12.1

Value versus cut-off for two price scenarios.

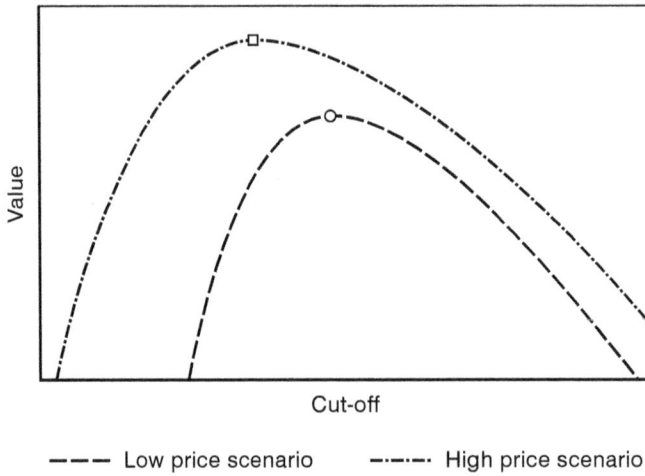

Cut-off

- - - Low price scenario   - - - - High price scenario

FIGURE 12.2

Risks and rewards of optimum cut-offs.

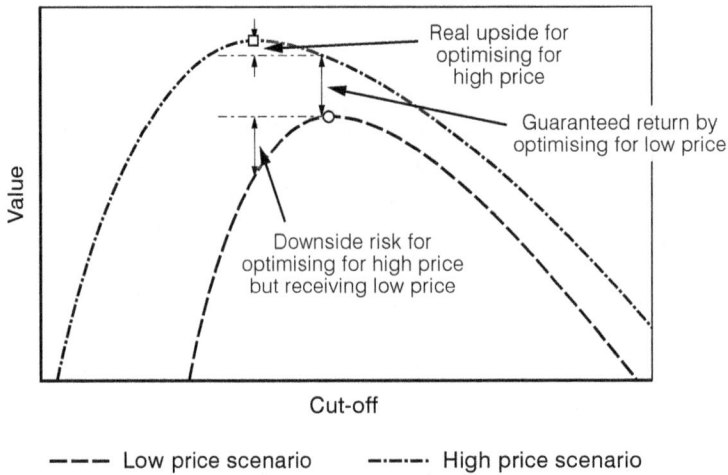

Cut-off

- - - Low price scenario   - - - - High price scenario

cut-off, which improves value with the higher price, is in fact quite small. But if the lower cut-off that maximises value at the higher price is used and the lower price occurs, the loss of value may be significant.

The key features of price behaviour of many metals are short periods of high prices and long periods of low prices. Anecdotal evidence suggests that price predictions tend to be optimistic when prices are low or falling, which is more often than not. The problems described in Chapter 3 regarding the dangers of break-even cut-offs during times of falling prices are still seen at many sites. If the reader accepts that predicted prices are often optimistic, and that cut-offs are often set at break-even values that are lower than those that maximise value, the dangers of an optimistic price prediction can be seen to

FIGURE 12.3

Risks and rewards of incorrect price predictions and suboptimal cut-offs.

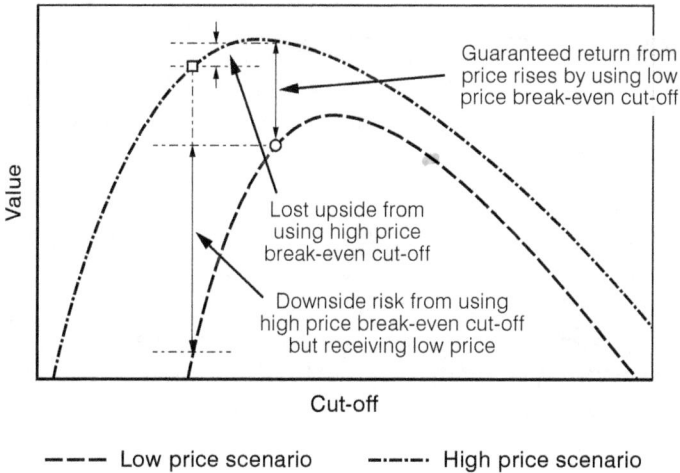

Guaranteed return from price rises by using low price break-even cut-off

Lost upside from using high price break-even cut-off

Downside risk from using high price break-even cut-off but receiving low price

Value

Cut-off

— — — Low price scenario   — · — · — High price scenario

be significant. Figure 12.3 shows how the downside risk in Figure 12.2 may be greatly magnified by operating policies using break-even grades as cut-offs.

Again, if pessimistic prices set the strategy, the higher price will automatically deliver the upside of the higher price. (Coincidentally, with these parameters, the break-even cut-off at the lower price is close to the optimum cut-off at the higher price, but this is a coincidence.) Compared with what could have been obtained by optimising cut-offs for the lower price, setting the mining plans to a break-even grade with the higher price will result in a great loss of value if the lower price eventuates. This is the practical effect of falling prices and break-even cut-offs described in Chapter 3. In many cases, the final value could well be negative.

Even if it is not accepted that the predicted prices are more likely to be optimistic than not, Figure 12.3 still shows that setting a cut-off policy based on a low or pessimistic price will be a more robust option than basing policy on a high or optimistic price. It should also be noted that the curves tend to steepen with increasing distance from the optimum, so the further the selected cut-off is from the optimum, the greater the loss incurred by any change in cut-off. These curves have been constructed to illustrate the point being made, but real cases do exhibit similar behaviour. Real curves may be more symmetrical about a vertical axis through the maximum value than shown in these figures. However, it is not uncommon to find that the gradients at cut-offs less than the optimum are steeper than those above the optimum. The author has case studies that show that setting a break-even cut-off only 20 per cent above a lower price can result in a loss of 50 per cent of the net present value (NPV) of the maximum value at the lower price, if that is actually received.

The preceding discussion has focused on managing the risk associated with unknown future prices, but the same principle can be applied to evaluate the impact of all uncertain parameters. Figure 6.14, for example, shows value versus cut-off curves for one deposit and a sequence of deposits. At the time that the decision must be made regarding the cut-off for the first deposit, it is uncertain whether other deposits will be mined. The same processes described in this section for price uncertainty could be applied to operating life or the reserve tonnage uncertainty evident in Figure 6.14.

REWARDING MANAGEMENT PERFORMANCE?

There are some other interesting effects flowing from these figures. Imagine that over a period of time the price fluctuates between the high and low price represented by the two lines in Figures 12.1, 12.2 and 12.3.

There is much kudos to be gained if one is Manager A, the manager of an operation or company when the price is increasing from the low-price line to the high-price line at the high-price break-even cut-off (at left of Figure 12.3). Dramatic improvements in profitability are gained. It is, however, most unpleasant to be Manager B when the price is doing the reverse, wiping out reserves that no longer cover variable costs, and the operation goes into panic mode to keep the mill full and generate cash. Manager B might be a better manager than Manager A, but carries the opprobrium of low profits or losses and an inability to meet production targets.

Manager C, on the other hand, operates at the low-price optimum cut-off (illustrated in the centre of Figure 12.2). The profitability fluctuates as the prices go up and down, but at a considerably higher average level and with much less volatility. But there is no glory for Manager C: if the mine is not lurching from crisis to crisis, then things aren't obviously being pushed hard enough. When the prices go up, there is nowhere near the percentage increase in profitability that Manager A generated from a close-to-zero base.

The curves for NPV in these figures are all assumed to be generated at the same discount rate. It is beyond the scope of this book to delve into the intricacies of selecting a discount rate; however, for those who have an interest in the topic, there is a secondary effect on value resulting from what has been illustrated in these curves. The low return and high volatility of results obtained by Managers A and B would increase the riskiness of the company, which would drive up the required rate of return for shareholders, thereby raising the weighted average cost of capital (WACC) and reducing the NPV. On the other hand, the higher return and lower relative volatility obtained by Manager C would reduce the riskiness of the company, which would lower the required rate of return for shareholders, thereby lowering the WACC and increasing the NPV.

Rationally, everything seems to drive the optimum strategy towards a value-maximising cut-off using conservative price forecasts. Yet market forces seem to drive companies to dangerous positions with low cut-offs that maximise reported reserves but expose the operation to major losses when prices fall, as they inevitably do.

A CASE STUDY

Figure 12.4 shows curves from a study conducted for an underground mining operation. The figure shows NPVs for a range of cut-offs for gold prices of $500 and $600/oz. The minimum value on the NPV scale is zero and the grades of interest on the plot are:

- operating cost break-even   price $600/oz   2.25 g/t
- operating cost break-even   price $500/oz   2.70 g/t
- planning cut-off   3.00 g/t
- NPV-maximising cut-off   price $600/oz   4.25 g/t
- NPV-maximising cut-off   price $500/oz   4.55 g/t.

FIGURE 12.4

Case study value versus cut-off results at different prices.

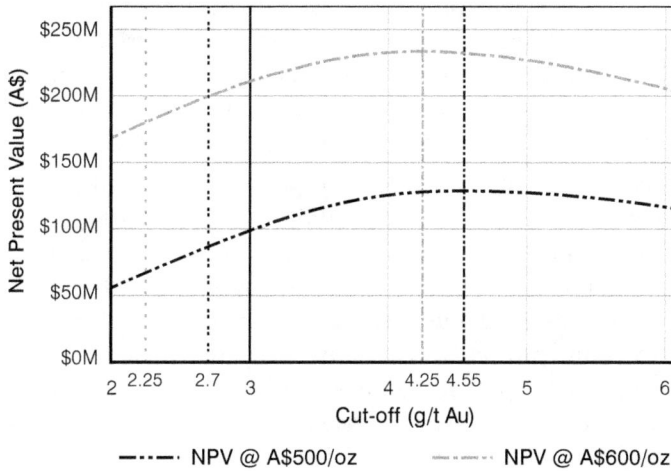

It can be seen that for a 20 per cent increase in price from $500 to $600, the break-even decreases by 17 per cent, but the optimum cut-off decreases by only seven per cent. A cut-off selected in the range of, say, 4.0 to 4.5 g/t Au is near the flat top of the hill of value (HoV), and will vary the NPV of the order of one to two per cent of the maximum value at each gold price. Incorrectly forecasting the price does not have a big impact on value if the optimum cut-off is selected.

If the $500 break-even were selected as the cut-off and a price of $600 were received, the NPV would be 12–15 per cent greater than if using the $600 break-even as the cut-off. Yet if the $600 break-even were selected as the cut-off and a price of $500 was received, the NPV would be 20–25 per cent less than if using the $500 break-even as the cut-off. At suboptimal break-even cut-offs, the higher the cut-off, the greater the value. If a break-even cut-off is to be used, the highest break-even for the most pessimistic price will always generate the best NPV (even though there will be fewer ounces in the reserve).

NPVs received using the planned break-even, $500 break-even and $600 break-even cut-offs are respectively ten per cent, 15 per cent and 25 per cent less than the NPV using an optimum cut-off of 4.0 to 4.5 g/t Au if the price received were A$600/oz. NPVs received by using the planned break-even, $500 break-even and $600 break-even cut-offs are respectively 20 per cent, 30 per cent and 45 per cent less than the NPV using an optimum cut-off of 4.0 to 4.5 g/t Au if the price received were A$500/oz.

In this case study, there is little risk in using an incorrect metal price to select the optimum cut-off – any value in the 4.0–4.5 g/t range will generate close-to-optimum values regardless of price. Technical staff can in this case make recommendations without being aware of the company's risk–reward profile; however, if lower cut-offs (such as break-evens) are to be used, there are significant risks with the selection of the metal price to be used for determining the strategic policy. The gradients of the value versus cut-off curves grow steeper the further away from the optimum cut-off, so small errors in specifying the cut-off can have bigger impacts on value.

MAKING THE STRATEGIC DECISIONS

The trade-off between risk and reward evident in the figures in this chapter will be dependent on the shapes of the HoV, and these will vary from project to project. The magnitude of the risks and rewards flowing from cut-off selection have a direct impact on the value of the company, and must be a matter for executive consideration. At the very least, decisions must involve senior executives, with the role of the technical staff being to present appropriate information, such as the curves illustrated.

It is the owners of the company, or their representatives, who should determine how risk-averse the decision-making processes should be. This may change from time to time, and may also depend on the project's size. Whether to seek to maximise the potential or minimise the risk may depend on the company's financial strength. If it is strong, decision-makers might opt for the upside, but if weak, may wish to guard against the downside. Similarly, if the project is small, it may be reasonable to accept the downside risk and target the upside reward. If the project is large and the downside could lead to financial ruin, the loss-minimising decision may be preferable. The same project outcomes, represented by the curves, could therefore lead to different decisions in different times or circumstances depending on information that is almost certainly outside the expertise of those conducting the evaluation.

Recommending the best strategy can clearly not be delegated to junior staff or external consultants. Even senior staff who might normally be tasked with this may be insufficiently aware of the corporate circumstances. Senior executives do not need to make every company decision, but should be involved in those that have the potential to incur significant losses if key assumptions (such as price) turn out to be inaccurate. Note that the loss referred to is not simply a negative cash flow or accounting loss. Indeed, the values obtained may all be more than acceptably positive. Rather, the loss is the difference between what would be obtained and what might have been gained with a better strategy if the assumptions underlying the decisions are incorrect.

Various levels of management are typically given authority to make decisions within certain ranges of cost or value. The author suggests that for the decisions required to trade off the upside rewards and downside risks of uncertain parameters, approvals authorities should go beyond the costs of an investment decision and extend it to the potential losses; that is, reduced value incurred if assumptions regarding uncertain parameters do not transpire. The potential losses may be significantly greater than the cost of the investment, which may shift the decision to a different level in the organisation from where it might otherwise be made.

The simple rule-of-thumb is this: if the desire is to maximise the upside, use the optimum cut-off derived with the higher prices. If it is to minimise the downside, use the optimum cut-off derived with the lower and more pessimistic prices.

CHAPTER SUMMARY

The need to flex uncertain parameters to determine whether they affect decisions to be made has been noted. These parameters may be intrinsically unknowable or merely currently uncertain. The best option is to find strategies that take account of the possible range of future values of the uncertain parameters. It is necessary to trade off the reward of making assumptions about uncertain parameters that turn out to be correct against the risk of incorrect assumptions.

Value measures can be plotted against decisions to be made for values of uncertain parameters. Future product prices will almost universally be uncertain, and many other input parameters may not be as well-defined as desired. Higher prices will lead to higher values and may, though not always, lead to a lower optimum cut-off. Often with price uncertainty, optimising the cut-off for a lower or pessimistic price will result in a mine plan that delivers most of the additional value gained with a higher price if the cut-off were optimised for the higher price.

Setting the mine plan for the optimum cut-off with the higher price typically generates a small true gain if the higher price eventuates, but has the potential to generate a substantial loss if the lower price occurs.

The situation is exacerbated by the common use of break-even cut-offs rather than optimum cut-offs. If a break-even cut-off for the higher price prediction is used but a lower price eventuates, the value obtained will be the lowest resulting from any of the potential cut-offs that could be applied in this analysis. The value versus cut-off curves also tend to steepen with increasing distance from the optimum, so the further the selected cut-off is from the optimum, the greater the loss incurred by any change in cut-off. For marginal operations, this can drive the mine into a loss-making situation until plans and schedules can be reworked to account for the new circumstances and the revised plans implemented.

For dealing with price uncertainty, the simple rule is: if the desire is to maximise the upside, use the optimum cut-off derived with the higher prices. If it is to minimise the downside, use the optimum cut-off derived with the lower, more pessimistic, prices. If value-destroying break-even cut-offs are decreed, resist the pressure to maximise reserves using the highest forecast price to generate the lowest cut-off. Rather, use the lowest price to generate the highest break-even and keep the cut-off as close to the optimum as possible to generate the highest possible value within the bounds of the suboptimal constraints.

CHAPTER 13

Optimisation Methods

DIFFERENT WAYS OF FINDING THE OPTIMUM STRATEGY

There are many ways in which strategy optimisations can be carried out. This chapter describes several methods, but does not purport to cover all approaches that readers may be familiar with. It does, however, provide some background to the types of methods commonly employed. Rather than being exhaustive, it aims to give the reader some idea of what might be available, describing briefly their strengths and weaknesses. With ever increasing computing capabilities, the author has deliberately refrained from discussing software in any detail due to the frequently changing nature of this technology.

The methods discussed are:

- exhaustive calculations
- genetic algorithms
- dynamic programming
- linear and mixed integer programming.

Together these methods complement each other, and a combination may provide the best solution for finding optimum strategies and communicating results to decision-makers.

Examples of products in each category are provided. These are a limited set with which the author has some familiarity, and should not be seen as an endorsement; nor should the absence of other products imply that they are inferior to those named.

EXHAUSTIVE CALCULATIONS

Simplistically, exhaustive calculation or *brute force number crunching* calculates the outcome for all potential combinations of inputs, both options and scenarios, to find the set that best delivers on corporate goals. Hills of value (HoV) are simple graphical examples of the result of exhaustive calculations, though there are usually many dimensions to the optimisation problem. The intention is to establish all the possibilities so that the best value and strategy can be identified.

Implicit in this method is that software enables the simultaneous flexing of values for any number of value drivers. Being able to report on multiple resulting values, such as ore tonnes and grades produced, components of the total cash flow and so on is also necessary. If specialist software is to be built using general purpose programming languages, these capabilities should be included in the software design specification. Models can also be constructed to provide these capabilities in Microsoft Excel™ using standard functions, without recourse to user-written programming code. However, spreadsheet software, while essentially unlimited in terms of flexibility, is not an efficient software development environment. Spreadsheet tools are widely used in the minerals industry, but the modelling techniques typically employed are quite inefficient. More efficient modelling constructs and functions can be used to significantly improve the efficiencies of spreadsheet models and will usually be required for strategy optimisation models.

Advantages

As all decisions are being flexed in the analysis, exhaustive calculations automatically analyse the response of value to decisions, not just to external parameters. It is also possible to calculate values for a number of measures to trade off between conflicting corporate goals, or to impose constraints that limit ranges of some outcomes.

Depending on the software used, there may be restrictions on modelling complex relationships within the optimisation model developed. Spreadsheet software, for example, has no limits on the types of relationships that can be handled – if something can be described, it can be modelled. The same may be possible for software built using general purpose programming languages. In these circumstances, techniques from previous studies can be quickly adapted to suit unique conditions at each site, while inapplicable issues can be ignored.

Disadvantages

Despite the benefit of having exhaustive calculations to demonstrate how the values of certain corporate goal parameters may vary with changes in operating strategies, there is still one major disadvantage: as the number of parameters increase, the number of combinations very quickly becomes impossible to compute.[1] As the complexity of the analysis increases, other techniques may be necessary to handle the number of cases that exist. Mathematical programming techniques seek to find the optimum strategy using as few iterations as possible, calculating outcomes for what will often be a small proportion of the number of feasible cases along a path from the initial starting point to the final solution found. Some of these techniques are discussed in the following sections.

---

1. The author has built relatively simple models in Microsoft Excel™ where, with a large number of option and scenario parameters that could be flexed, the number of potential combinations is of the order of 10^{90} – the number of photons believed to exist in the known universe. Clearly, it is not feasible to exhaustively calculate all those potential combinations, but the construction of the models is such that any of that number of combinations could be selected and reliably modelled.

GENETIC ALGORITHMS

Genetic algorithms (GAs) have been described by their proponents as being one of the best techniques for being reasonably sure of being reasonably close to an optimum solution when analytical methods are computationally intractable, when the number of cases is too great to permit an exhaustive evaluation to identify the best.

It is beyond the scope of this book to describe the operation of a GA in detail; however, simplistically, GA software is given control of various decision-related inputs. It is given allowable values for each, and then proceeds to evaluate randomly generated combinations of inputs. Combinations leading to poor outcomes are discarded, while those generating good results are retained, their input values mixed and the processes repeated. A GA may be thought of as a *hill-climbing* optimiser, but to avoid the problem of standard hill climbing methods (which will only climb to the top of the hill on which they start), the GA will not only climb the hill, but will also look around to see if there is a higher hill in the solution space. This is done by making random variations (mutations) in some of the good sets of input data specifications from time to time. Value drivers controlled by GAs could include:

- cut-offs applied separately to different mining stages, pits and underground areas
- sizes of both final and intermediate stages of pits being mined
- mining and treatment rates, in total and by areas, and changing over time
- sequencing of mining various deposits or mine areas
- allocation of deposits and mining areas to multiple treatment plants.

Advantages

As noted, GAs provide a process to find a strategy approaching the optimum when analytical methods are computationally intractable, and when the number of cases is too great to permit exhaustive evaluation. If a GA analysis starts with, for example, the best results from a necessarily constrained and limited exhaustive calculation analysis (that is, at the top of a multidimensional HoV), further improvements in value are common.

GA add-ins are commercially available for spreadsheet software[2] such as Microsoft Excel™. They are relatively quick and easy to implement in a spreadsheet model. Results may indicate how improvements can be made in existing strategies, or identify new strategies not previously considered. Logs of the analyses can be used to identify not only the best solution, but also factors common to high-value and low-value cases to guide further planning and investigations.

Disadvantages

There is no guarantee that a GA will find the highest value hill, nor that it will find the maximum value of any hill that it is on. Opponents of GAs say that there is no way of knowing how close the best value found by the GA is to the optimum value. Other methods, such as mixed integer programming, provide this information. While this may be true from a mathematical point-of-view, for the planning team any gain is better than nothing until a better one is found.

---

2. The author has obtained useful results in a number of studies using Palisade Corporation's Evolver™ and RiskOptimizer™ add-ins for Microsoft Excel™.

Conducting a reasonable number of calculations may take a long time with a complex model. Because the GA has no way of knowing whether it has found the optimum or not, it will continue operating indefinitely. A GA will therefore be terminated by running for a fixed time or a number of iterations; specifying the maximum number of iterations without an increase in value being found; or direct user intervention to terminate the process.

DYNAMIC PROGRAMMING

Dynamic programming (DP) is not a single algorithm, but rather an underlying philosophy to solving a problem. There is no general-purpose DP software: each class of problem requires its own formulation; however, many classes of mining strategy problem will share similar processes, and mining-specific software that can be applied across a range of sites and problems is commercially available.

As a general principle, DP reduces the number of evaluations by making sequential decisions, using the results of one decision as an input to the next decision. Determining a life-of-mine optimum cut-off policy using Lane's methodology (Lane, 1988) is a DP process. Decisions are made sequentially regarding the optimum cut-off for each mining step. The opportunity cost brings the results of other decisions – the assumed cut-offs in future periods – into the decision for the period under consideration.

A simple Lane-style analysis is limited to a single production stream with a prespecified mining sequence and it can be applied using spreadsheet software. Specialist DP software with more complex capabilities is commercially available.[3]

Advantages

As with GAs, DP finds a strategy approaching the optimum when analytical methods are computationally intractable and the number of cases is too large for exhaustive evaluation. In many ways, the DP process mimics real-world mining decision-making process in that sequential decisions are used to assess the best strategy by accounting for assumed future decisions and the future value of the operation.

Disadvantages

Because DP is not a specific algorithm but an approach to address the problem, each project must be modelled from scratch; however, software will provide a framework for a class of problems with the capacity to make modifications that fit within the framework.

The more complex the problem, the greater the likelihood that sequential decision-making will be suboptimal. Thus, as with GAs, some DP solutions have no guarantee that the true optimum will be found, and there is no way of knowing how close the best value is to the optimum value. Again, for the planning team seeking to optimise the operating strategies, any gain is better than none until a better one is found.

---

3. Comet® by Strategy Optimisation Systems extends the *simple Lane*-style optimisation of cut-offs for a specified mining sequence to the more general *complex Lane* process to optimise a wide range of strategic decisions, involving, for example, sequencing and scheduling, mining and treatment rates and multiple ore sources using the DP approach.

LINEAR AND MIXED INTEGER PROGRAMMING

Linear programming (LP) and mixed integer programming (MIP) are classical analytical techniques for maximising or minimising the value of an objective function subject to a number of constraints. *Linear* indicates that all the relationships in the formulation are both linear and continuous. *Mixed* identifies an analysis with a mixture of continuous and integer variables. Because the continuous components are linear, MIP is also referred to as mixed integer linear programming (MILP). The techniques apply to several common mining industry problems. There are a number of powerful general-purpose MILP software packages available commercially as well as some with much of their power available freely on the internet. In addition, the Solver utility in Microsoft Excel™ includes MILP capability, but while it can handle some mining-related problems, it lacks the capacity to handle a major mining strategy optimisation. There are also several mining-specific software tools available, either for purchase or via services provided by specialist consultancies.[4] These are built on state-of-the-art commercial solvers with standard modules and mining-specific user interfaces. Some customisation may be required. The interface software converts the user inputs into the strict formats required by the solver software.

Linear and mixed integer linear programming advantages and software capabilities

LP can solve problems with thousands of variables and constraints. It is particularly suited to allocating scarce resources and blending problems; however, it cannot account for discrete variables needed to optimise schedules. Problems to be solved will typically be for a single time period. Geometrically, the set of feasible solutions is bounded by planes in multidimensional space such that the surface is convex. As indicated by the name, all relationships between variables in pure LP must be linear, but non-linear relationships can be approximated by a number of linear segments, so long as the required convexity is maintained. Efficient algorithms known for decades will rapidly find the optimum point, which can be proven to be so. Increasing computer power is merely increasing the speed of calculation and the number of variables that can practically be handled, but the underlying algorithms are essentially unchanged. Some top-end solvers, however, also permit the use of curvilinear relationships, such as quadratic, but with a significant increase in the time required to find the optimum solution.

Using the concept of HoV introduced in Chapter 6, LP is applicable when there is only one hill and no valleys or discontinuities in the multidimensional surface of resulting value versus inputs, where the inputs are both the options and scenarios included in the analysis. The solution algorithms move efficiently and rapidly from a feasible starting point on the surface, along edges between plane segments of the overall surface, from vertex to vertex, to the vertex defining the top of the hill. In some cases, the continuous surface may have multiple hills but the LP algorithm will only climb to the top of the hill on which it starts. It cannot go into a valley and up the other side, so to speak, onto another hill.

As well as the types of problems handled by LP, MILP can be used for scheduling and sequencing optimisation, but is more limited in terms of problem size. In principle it

---

4.   Minemax Planner by Minemax and Whittle Consulting's 'Prober' software are two known to the author.

can solve for all time periods simultaneously, so that decisions later in the mine life are not constrained by earlier ones, as is often the case with other optimisation techniques.[5] Optimum sequencing, timing of upgrades and similar time-related options can therefore automatically be part of the solution.

While for LP the set of feasible solutions is bounded by planes in multidimensional space such that the surface is convex, MILP can be used for problems with the set of feasible solutions bounded by surfaces that are not fully convex. Rather than being a single hill as for LP, with MILP there may be multiple hills, valleys and discontinuities in the value versus inputs surface, as illustrated in Figure 6.5 in Chapter 6.

The integer variables will typically be binary (taking values of 1 or 0 only), representing, for example, whether various mining areas are producing or whether one area is finished so that another can start. Integer variables may also be used to identify each of the mutually exclusive options for relevant variables; for example, the cut-offs for each mining block in an underground operation with different tonnages, grades, development requirements and so on for each cut-off. Notionally, when there are binary variables, the number of potential combinations is 2 raised to the power of the number of variables.

Geometrically, this may also be thought of as the number of hills in the value versus inputs surface, with the shape of each hill defined by continuous variables amenable to LP. Each hill is a locally convex region in the overall multidimensional surface. Notionally, the efficient LP algorithms could be applied to each hill, rapidly finding the maximum value and associated strategy for each hill, and the best one found. This is essentially a modified exhaustive calculation style of solution, and there would typically be an intractable number of integer variable combinations. Expert users working with the software will attempt to reduce the search space by eliminating physically impossible options; for example, in an open pit, eliminating from the analysis any potential sequences that involve mining a block of rock before all overlying blocks are mined. Mining-specific interfaces constructed for non-expert users will have similar problem size reduction processes automated and built in. The MILP software also includes processes to eliminate regions of the solution space where the best solution is not to be found. As with LP solutions, there is a provably optimum solution; however, given the large number of combinations, there is no guarantee that the hill with the best value will be found in reasonable time. The MILP process will predict from solutions already found what the optimum value will be, and will continue processing until the difference between the best value found and the predicted maximum has fallen below a specified tolerance.

MILP algorithms are continually improving. Along with greater processing speeds, this is increasing the likelihood of successfully finding the optimum or close-to-optimum solution in a reasonable time frame.

---

5. Many optimisation techniques that do not have this look-ahead capability make use of what is known as a greedy algorithm – taking the apparently best available option now without any recognition of potential problems that this could create later. This is similar to decision-making with a short-term focus that occurs at many operations, and results in short-term gain for long-term pain. Although greedy algorithms may work in some circumstances, more powerful optimisation techniques such as MILP have demonstrated that they frequently generate suboptimal results.

Disadvantages

Solving complex problems can be time-consuming with MILP. Both LP and MILP require strict mathematical formulation, and necessary simplifications may introduce inaccuracies. The use of MILP in particular requires an experienced user who understands the problem as well as modelling methodologies, so that the solver settings may be altered to suit the problem and maximise the likelihood of a solution.

CONCLUDING COMMENTS

Exhaustive calculation or brute force number crunching does not automatically optimise for any goal. Instead, visualisation or search techniques must be used to inspect the results and find the set of options that maximise value or satisfy a number of corporate and external stakeholder goals. The HoVs described offer but one way of graphically assisting this process.

All the other processes are specifically designed to target an optimum and can only optimise for one measure at a time. Different value measures could be optimised separately, but there is no simple way to assess the trade-offs in selecting the best strategy. It may be possible to combine the values of two or more parameters into a single value for optimisation, or to apply penalties to the primary value measure for secondary measures that are not at their optimum values. Because these mathematical techniques are intentionally evaluating a limited subset to find the optimum solution as quickly as possible, the full set of value versus option and scenario values is not available for assessing trade-offs. This is not to say that a software product using one of these optimisation methods cannot include facilities to provide that capability, but it is not intrinsic to the methodology as it is with exhaustive calculations.

Other techniques, such as goal programming and hierarchical programming, can be used to optimise for multiple goals, but the algorithms are not as widely used as those for one value measure. These techniques require goal ranking and specifying outcomes for each measure. This results in optimising the primary goal, subject to subsidiary goals or constraints achieving the minimum or maximum outcomes. The author has no experience with these, but notes them here so that readers may be aware that more advanced, though little used (in the mining industry, at least), techniques exist.

The mathematical optimum from any of these methods may not be the best strategy, particularly when risk–reward trade-offs are required. Qualitative value judgements based on the shapes of the various value versus decisions curves and surfaces, as described in Chapter 12, may be the best technique for final decision-making.

Finally, it should be noted that the inputs to and outputs from any of these optimisation processes will usually be simplified to generate quick results with the least calculation. For presentation purposes, it will usually be necessary to post-process the results from optimisation modelling into a form or level of detail appropriate for company reports. Changes necessary will include regularising the time scales for reporting and breaking down summary cost and revenue parameters into more detailed components.

CHAPTER SUMMARY

This chapter has described briefly four major types of optimisation methods to provide some background on what is commonly used in the industry.

Exhaustive calculations attempt to calculate combinations of options and find, by inspecting the results, the set that best delivers the corporate goals. They assess the response of value to both controllable decisions and uncontrollable external parameters. The value versus decisions responses for a number of measures can be compared to trade off between conflicting corporate and stakeholder goals; however, as the number of parameters increases, the combination of scenarios to be evaluated and inspected very quickly becomes impractical to compute.

Genetic algorithms (GAs) are often used when analytical methods are computationally intractable. The GA software is given control of various decision-related inputs and evaluates randomly generated combinations of values for inputs. Combinations leading to poor outcomes are discarded, while those generating good results are retained, their input values mixed and the processes repeated. A GA is a hill-climbing heuristic optimiser, but to avoid the problem of standard hill-climbing methods (that only climb to the top of the hill on which they start), the GA will not only climb the hill it is on, but will also look to see if there is a higher hill elsewhere in the solution space. There is no guarantee that it will find a higher hill if one exists, nor even that it will reach the top of any hill it climbs. Though lacking a certain mathematical rigour, GAs are powerful and useful for a pragmatic evaluation team if they give a better solution than through other means.

Dynamic programming (DP) is a philosophy of approach to solving a problem rather than a specific algorithm. There is no general-purpose DP software: each class of problem requires its own formulation; however, many classes will have similar processes, and mining-specific software applicable across a range of sites and problem types is commercially available. DP reduces the number of options to be evaluated by making sequential decisions, using the results of one decision as an input to the next decision. Determining a life-of-mine optimum cut-off policy using Lane's methodology is a DP process – the opportunity cost brings results of future cut-off decisions into the decision being made for the increment of rock being considered. As with GAs, there is no guarantee that the true optimum will be found by DP, and there is no way of knowing how close the best value found is to the optimum value. Again, for the planning team seeking to optimise operating strategies, any gain is worthwhile until a better one is found.

Linear and mixed integer programming (LP/MIP) or mixed integer linear programming (MILP) are classical techniques for maximising or minimising the value of a function subject to constraints. Powerful general-purpose MILP software applications are available both commercially and on the internet. Mining-specific software tools available for purchase or via specialist consultancies. These are built on commercial solvers with standard modules and user interfaces. The interface converts the user inputs into the strict input formats required by the underlying solver software.

For LP, geometrically, the set of feasible solutions is bounded by planes in multidimensional space such that the surface is convex. All relationships between variables must be linear, but non-linear relationships can be approximated by linear segments, so long as the required convexity is maintained. Efficient algorithms will rapidly find the optimum point. Increasing computer power is increasing the speed of calculation and the number of variables that can be handled, but the underlying algorithms are essentially unchanged. LP is particularly suited to allocating scarce resources and blending problems within a single time period.

MILP can be used for scheduling and sequencing optimisation, but is more limited in terms of problem size. In principle it can solve for all time periods simultaneously, so that decisions later in the mine life are not constrained by earlier decisions, as is the case with other optimisation techniques. Geometrically, the surface bounding the set of feasible solutions may be thought of as a number of hills and valleys. Experts working directly with the software will try to reduce the search space by eliminating physically impossible options. Mining-specific interfaces constructed for non-experts to interact with the solver software will have similar processes automated. The MILP software itself also includes processes to eliminate regions of the space where the best solution is not to be found. Given the large number of cases, it cannot be guaranteed that the hill with the best value will be found in the time available. Rather, the MILP process will predict what the optimum value would be based on solutions already found, and continue processing until the difference between the best value found and the predicted maximum has fallen below a specified tolerance.

In summary, each method has strengths and weaknesses, proponents and opponents. If there were one best method, we would all be using it. Together the various methods complement each other, and a combination of methods and software may provide the solution for finding optimum strategies and communicating results to decision-makers.

CHAPTER 14

Answering Misconceptions and Objections

MISCONCEPTIONS OR VALID CONCERNS?

A number of common objections have arisen when conducting optimisation studies, especially when results indicate that the optimum strategy is to increase the cut-off, thereby reducing reserves and mine life. There are two main categories of objections – misconceptions and valid concerns. Both categories need addressing, but in essence the former simply needs to be considered logically to expose the underlying misconception.

Valid concerns, however, are the real issues that management or a strategy optimisation team may need to grapple with. Some are issues of communication, understanding and philosophy. Others, however, can be dealt with by identifying these concerns at the start of the evaluation and including them in the process, so that the optimum strategy incorporates them. If they are raised as valid concerns at the end of a study, it indicates that the scope of the investigation was poorly established when the project was initiated.

MISCONCEPTIONS

Strategy optimising high-grades the resource and sterilises valuable resources

One of the key issues that has been stressed in this book is the need to specify the corporate goal. It is true that plans focused on maximising cash-flow-related measures, such as net present value (NPV), often lead to using higher cut-offs than typical break-even cut-offs. This is often seen as *high-grading* the orebody, the term used pejoratively to suggest it is inappropriate. If it is a result of focusing on the corporate goal, it should be seen as *right-grading*, and break-even cut-offs could then be disparagingly described as *low-grading*. If material that might have been included in the reserve at a suboptimal cut-off is excluded by an NPV-maximising cut-off, it must be recognised that the material is destroying value. If it added value, the optimum cut-off would have brought it into the reserve. By definition, such material is not a valuable resource that has been sterilised, it is value-destroying mineralisation that is correctly excluded from the reserve.

Value is maximised by producing till marginal cost equals marginal revenue

This principle is taught in basic economics courses. It has been developed in the context of the manufacturing industry where the main assets of the firm are its production facilities. Simplistically, resources and markets are external to the firm, and successive time periods are essentially independent. If resources are not acquired and goods are not made and sold in the period considered, the opportunity is lost forever. Decisions made in one period about what to produce and sell do not influence the life of the firm, which is assumed to be infinite. The firm's value is therefore maximised by making independent decisions that maximise the value of each period, and producing to the point where marginal cost equals marginal revenue will achieve that.

Applied to mining operations, this principle is interpreted to imply that the cut-off should be the marginal break-even grade; however, a mine is significantly different from a factory. Although it has production facilities, its prime asset is its mineral resource, which is both finite and internal to the company. Decisions made about what to do with a portion of the resource in one time period will affect what remains of the resource for exploitation in later periods and what can be done with it, and hence its value.

Producing so that marginal cost equals marginal revenue, at least in the sense that is applied in the industry, almost guarantees that a mineral deposit won't deliver the maximum value possible. This is at variance with the expectations of many senior mining industry leaders without a mining background, and also of many who do, since the difference is not widely recognised in the industry. There is a real risk that unthinkingly applying conventional economic wisdom will reduce the value of a mining operation.

It must be recognised that the principle comes from the field of economics and is expressed in economic terms. The marginal costs are those as defined in economic theory, which are not necessarily those defined as such at a mining operation, which will typically be the variable costs extracted from the company's cost accounting reports. Rather, the marginal cost for an overall planning cut-off must include fixed costs, since the treatment of an additional tonne of material as ore will extend the life of the operation and incur further fixed (or time-variable) costs. Lane's opportunity cost must also be included in the marginal cost. The astute reader will recognise that we have now described Lane's treatment-limited cut-off, which will be significantly higher than the marginal cost as typically defined at a mine. The author suggests that most technical and financial mining staff would consider the marginal cost to be more akin to Lane's mining-limited cut-off, which is essentially the downstream variable cost of treating ore and dealing with product.

By invoking Lane, we have highlighted the other key issue that differentiates mining from manufacturing – the interrelationship between the capacity of various stages of the production process, from mine to sale of product, and the grade distribution in the material being mined. The optimum (NPV-maximising) cut-off may well be a Lane-style balancing cut-off, which depends only on the physical parameters noted and is independent of prices and costs. So even employing an economic definition of marginal cost may be inappropriate if the optimum cut-off is a Lane-style balancing cut-off rather than a Lane-style limiting, or break-even, cut-off.[1]

The best strategy can be identified by simple studies and intuition

Industry results have been inadequate for a number of years, and even if they were good, strategy optimisation analyses indicate that they could be much better. The author suggests that nobody is born with this sort of intuition; rather, it is based on experience. This experience – on the basis of which some claim to have developed an intuitive feel for what is right for an orebody – is of strategies that have led to suboptimal outcomes.

If such intuition does exist, the author suggests that it is faulty and cannot be relied upon. The risk of arriving at a suboptimal result from failure to do a rigorous study is high.

A higher cut-off won't account for the option value of unmined material

One must first ask: what is meant by option value? As indicated in Chapter 11, this is usually understood to refer to a value derived using the Black–Scholes option pricing model.

As also noted, if the mineralisation has been sterilised by mining at a higher cut-off, there is no option to mine it and in reality it has no value. If it has not been sterilised, there is the potential for a second phase lower cut-off mining plan, which could realistically be included as part of the optimum mining plan. If the initial optimisation assumes one cut-off for all areas or for the whole mine life, there may be some additional upside, though the effect of this would probably be to further increase the optimum cut-off for the first pass, followed by a lower second-pass cut-off. As noted in the discussion of strategy optimisation in Chapter 6, variation in cut-offs, both by location and over time, is potentially part of optimisation, so if a truly optimum strategy has been identified, there is again no value-adding option and hence no option value.

Effectively, this value is already accounted for in most open pit mining plans. Stockpiles are built up and drawn down as grades mined fluctuate during the mining phase of the operation. Grades reduce to a minimum grade as lower grade stocks are drawn to depletion and treated after the mining operation has finished. The predicted cash flows and values generated account for grade variations over time and, theoretically at least, no material with a grade that could generate a positive net cash flow will be left untreated.

It is in underground operations that this objection is most likely to be raised. Here, the equivalent of open pit stockpiles is effectively the material left unmined. The planning team will need to identify potential reserves at a range of cut-offs. In some areas, material marginally below the initial cut-off may be in narrow haloes around the higher-grade material that has been stoped, and may therefore be effectively sterilised. In other areas, lower-grade zones may comprise volumes of sufficient dimensions to allow productive mining, albeit at a lower cut-off, after the higher-grade portions have been extracted.

In many cases, this material may be mined opportunistically to generate an incremental value, particularly during the run-down-to-closure phase of the mine's life. It may contribute extra value at the time, but is not likely to be of high value during the main

---

1. It is interesting to note that the title of Lane's book is *The Economic Definition of Ore* (Lane, 1988). Lane applies a number of economic principles correctly to identify optimum cut-offs, not a misinterpreted catchphrase to destroy value, as is the case with this misconception.

production phase of the operation. In both cases, its mining will be short-term focused and tactical, rather than a significant part of the mining strategy.

Cut-off optimisation is not applicable to bulk commodities

Bulk commodities, such as coal, iron ore and bauxite, are often produced to meet minimum quality specifications for the valuable component and maximum levels of impurities or contaminants. Value maximisation in these circumstances will often be achieved by maximising the reserve above the contract specification. This may include judicious blending of lower-quality material with higher-quality so that the total blend remains within the limits of the quality specifications. The concept of cut-off may not be as common in such cases as it is in base and precious metal operations.

Nevertheless, some way of specifying the mining limits will be required. However it is expressed, the boundary defining what is valuable for extraction to become product may be thought of as a cut-off.

While the current sales contract specifications may be strict, contracts can change in the long term. So long as any variations in quality within the resource are correlated with variations in value, there is potential to investigate whether mining higher or lower grades will facilitate creating different product mixes and volumes that can be sold for different prices and add value. This is as much cut-off and strategy optimisation as anything else described earlier in this book. The time required to implement a change may be longer and may need more marketing effort than for a base or precious metals operation, where the times between evaluation, decision-making and practical implementation are small, but the underlying principles are similar.

VALID CONCERNS

As already indicated, the author considers some objections to be real issues that management or a strategy optimisation team need to grapple with. Some are organisational or investor relations and communications concerns, while others are related to the scope of work for evaluation. In principle, all of these matters can be overcome by identifying the issue and responding to it appropriately.

There is no time or money available for optimisation in feasibility studies

This is an organisational and perhaps investor relations issue. An optimisation study can obviously be done at any time, and there is an understandable desire to reduce the time and cost involved in preproduction studies. However, the task of a full feasibility study is to prove the technical and economic feasibility of a particular strategy for an operation, with no guarantee that it is optimal. One of the main purposes of the prefeasibility study is to find the best strategy for the feasibility study. If this is not done, a suboptimal yet still feasible strategy may be selected. Once the strategy is selected, items of plant and equipment are sized, and reserves are reported publicly. Both of these factors may limit an operation's ability to change to a plan that generates higher values, either through the cost of removing physical limitations, or through a perceived inability to report a reduced reserve to the market or admit that the best plan was not implemented in the first place.

This objection highlights the industry's tendency to focus on cost rather than value creation. A full strategy optimisation study may be more expensive than conducting, for example, a simple break-even analysis to derive a cut-off, and arbitrarily specifying a unoptimised production rate. The additional cost of such a study is usually small compared to the feasibility study cost, not to mention potential value gains identified by optimisation studies.

While a full optimisation study will identify significant value that may be available whenever it is done, the sooner this occurs, the greater the likely gain and the chance it can be realised. Figure 14.1 illustrates the potential risk with failure to conduct a rigorous analysis, particularly where there are widely different options available.

Suppose two strategies – A and B – are compared at a single cut-off, as depicted by the vertical line connecting the markers on the curves and the cut-off axis of the plot. If the values represented are for the only cases evaluated, the decision is obvious – select strategy A. If, however, a range of cut-offs were evaluated, the curves might represent the response of value to changes in cut-off for the two strategies. It might be that, co-incidentally, the cut-off value used was close to optimum for strategy A, but without the full analysis, it could not have been predicted that the maximum value of strategy B is significantly higher than strategy A. Significant shareholder value can be destroyed by inappropriate cost- and time-saving at the point in the evaluation process where value-maximising decisions can still be made.

FIGURE 14.1

Potential loss of value from not conductiong a full optimisation study.

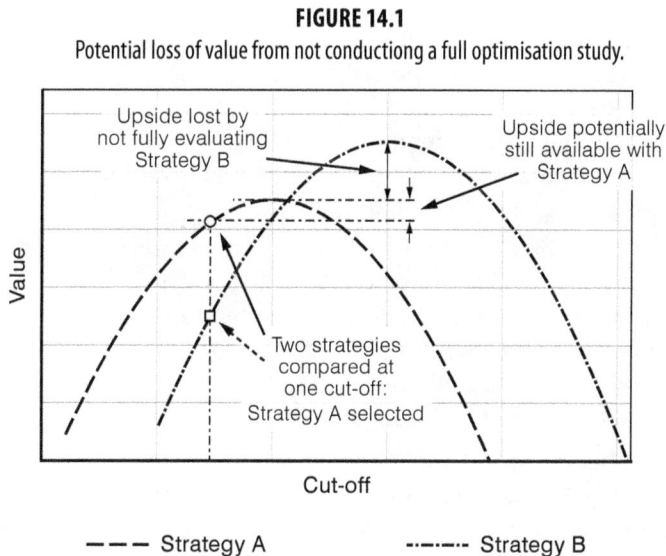

The market will react adversely if a reduced reserve is reported

Ultimately, this is an investor relations and communication issue, but the author has encountered it as a real concern for the study and resulting decision-making on a number of occasions. There is a real aversion to reporting a reserves reduction that would result from, for example, increasing the cut-off from a break-even grade to an NPV-maximising cut-off, or reducing the planned size of an open pit. The common experience of optimisation practitioners and mine planning personnel is that the pressure is always to reduce the cut-off and maximise the reported reserves, often accompanied by a fear that a reduction will lead to a fall in the company's share price. It is easy for the technical team to place additional material into the reported resource or reserve, but it can be very difficult to remove something that should not be there.

More reserves are good if they come from exploration success, but maximising reserves is not economically sensible for a given resource; rather, there is an optimum size of reserve that maximises returns. As noted earlier, it is assumed that everything in the reported reserve adds value, though this is often not the case – most mines will be working with a planned cut-off or pit size where increasing the reserve is negatively correlated with value, as described in Chapter 10. As long as the market appears to, for example, value gold mining companies by multiplying the reported ounces in reserve by some dollar value per ounce, company managers will attempt to maximise reported reserves.

The quantum of the reserve base available for depreciation and amortisation is also a concern for companies. Increasing the reserve will increase the tonnage over which capital costs can be depreciated or amortised, thereby reducing the annual charges. This focus on one particular charge against profits overlooks that increasing the cut-off, for example, will increase the head grade and annual revenue will increase. Frequently this increase will be more than enough to offset the increased depreciation and amortisation charges, resulting in an increased annual profit – not a decrease. Even if this fails to occur, the total net profit over the mine life will increase. The perverse effect of these issues is that there is a fear that announcing strategies that increase value (measured by cash generation potential, epitomised by NPV) will actually drive share prices down.

Ideally a study should be done to optimise the mine strategy, including cut-off grade and pit size, before reserves are first reported publicly, so that an optimum figure is in the public domain from the beginning. Mines that are already in operation with a suboptimal reported reserve should be able to demonstrate to analysts and financial commentators the wisdom of their plans to improve value, but it is acknowledged that this may not be as easy as might be hoped.

It may incorrect to assume that reporting reduced reserves will cause the share price to fall. If so, reporting reduced reserves should not be a concern to mining companies. There have been cases where newly reported reserves that were lower than analysts' expectations did not result in share price reductions. This might not be an problem for companies with large resources and long life, but it could pose an issue for less robust companies with smaller reserves and shorter lives.

There is a concern that those who make these comments are unaware of how optimum cut-offs and values may change as a result of, for example, price or cost changes. A real risk exists that, through lack of knowledge and understanding, shareholder value is being driven down by those whose role it is to improve it, both officially (mining company boards) and unofficially (the analysts who make pronouncements that affect share prices).

How to handle this issue is, of course, the prerogative of each company. Strategy optimisation studies merely provide decision-makers with a lot more information than they've traditionally had. A full hill of value (HoV) – a value versus decisions surface – shows how much shareholder value is being written off if the decision-makers choose, for whatever reasons, to select a strategy that fails to deliver maximum real value.

Reducing mine life limits the likelihood of more profitable discoveries

A strategy optimisation study often indicates that increasing the cut-off or reducing the pit size, and hence reducing the mine life, increases value. The impact of the reduced life on realising any exploration potential therefore may be seen as an issue to be addressed.

But if it is an issue for an optimised mine plan, it should be an issue for *any* mine plan, including the current plan developed by traditional planning practices. Frequently, exploration needs will not have been accounted for in developing the unoptimised existing plan, yet are deemed compulsory for the optimisation study. Ignoring the logical inconsistency of these different requirements, it is nevertheless a valid issue that deserves consideration in the development of the strategy optimisation study.

As indicated in Chapter 6, it is necessary to identify all the relevant issues. If the interplay between mine life and exploration success is a matter of concern, it simply has to be included in the scope of the evaluation. In principle this is just another relationship that needs to be modelled. If it is not modelled, the issue is simply with the setting of the study's scope, not with any underlying flaw in strategy optimisation methods.

One way to evaluate this effect is to conduct a probability analysis. The evaluation model can be enhanced with a stochastic simulation[2] to include, in each year of operation, a probability of exploration success and probability distributions of the tonnage and grade of material discovered, if any. These discoveries would become additional reserves in the model. While simple in concept, acquiring the data for the probability distributions may be problematic, and interpreting the results would be yet more complex.

These comments assume that annual spending and exploration activity do not change, regardless of the predicted mine life, but this is unrealistic. The alternative treatment, therefore, is to recognise the implicit commitment to spend a certain amount on exploration over the life of the mine, given the current plan and targets. If the mine life is to be shortened, the funding should still be spent before existing resources are depleted. The probability of exploration success is then identical in all scenarios, and the peak of a HoV for NPV will take into account the timing of the exploration cash flows. The essence of the objection is that the probability of discovery differs for each mining strategy; the suggested solution makes it the same for all strategies. The implicit assumption that makes comparisons between reported values for different strategies valid is *all other things being equal*. The suggested process explicitly makes that the case.

If a new resource were to be found, value would be maximised by bringing it into production at a point in time where the increase in NPV resulting from its earlier production (that is, bringing forward the receipt of the NPV of the new resource) balances the loss in NPV from reducing the life of the current resource. This reduction in life would be achieved by increasing the cut-off (assuming the production rate cannot be increased). The optimum cut-off for the known resource will therefore be higher with additional resources than without.

It may not be possible to physically carry out the exploration program required over the identified mine life for each case, in which case it may be necessary to revert to a maximum exploration rate program regardless of the cut-off or pit size and mining rate specified. For regular annual expenditure programs, companies should recognise the loss of value from a suboptimal production strategy as a real cost of the exploration program, in addition to its direct costs for drilling and the like. Shareholders should be made aware that the cost of ongoing exploration is not just what is expended on drilling, assaying and geological interpretation, but also includes the costs of keeping the mine operating longer simply to support exploration, which may be a significantly larger cost.

---

2. As described in Chapter 11, using software such as the @Risk™ or Crystal Ball™ add-ins if the model is built in Microsoft Excel™.

In some situations, such as deeper exploration below a deep mine, the operation's continuation is required to provide logistical support and a platform for the exploration program. In other cases, this may not be required and is simply assumed to be the best course to avoid the costs of closing down and restarting if more resources are discovered. The analysis should also evaluate the costs and benefits of continuing to mine at lower rates or cut-offs to support the exploration as well as mining at higher cut-offs or rates, versus putting the operation on care and maintenance and restarting if the program is successful.

To the extent that the value of the mining operation is reduced by extending the life to support an exploration program, that value reduction should be identified and an informed decision made as to whether that money should be invested in exploration, returned to shareholders as dividends or invested in other value-adding projects.

Reducing mine life increases the probability that price increases adding value won't occur until after the mine has closed

This concern is easily dealt with by evaluating different price scenarios. If there is a view that a price rise may occur at some time in the future, it is a simple process to model one or more price paths with an increase. Potentially the increase could occur at different times in the future, and a range of values for the price increase could also be modelled. The need to account for this should be part of the scope of the evaluation when the project is initiated.

Optimum strategies can then be identified for each price scenario. Where price increases at some future point result in different (lower cut-off, longer life) strategies relative to strategies for no price increases, trade-off assessments need to be made between the rewards of selecting strategies based on eventually correct assumptions and the risk of making decisions based on wrong estimations, as described in Chapter 12.

Discussions in Chapter 10 regarding maximising mine life as a measure of value suggested that, in general, mine life and cash generated will be negatively correlated; however, this should not be taken to mean that this is wrong in all cases. The criticism is of the mistaken assumption that a longer mine life is associated with higher value. Where it can be demonstrated that there are rational reasons for selecting a longer life in order to improve value, that is an entirely appropriate decision.

CHAPTER SUMMARY

Several objections are sometimes raised regarding mining strategy optimisation. Some basic misconceptions include:

- value is maximised by producing to the point where marginal cost equals marginal revenue
- strategy optimisation leads to high-grading of the resource and sterilising valuable resources
- setting a higher cut-off does not account for the option value of the additional material left unmined or untreated.

These arise from misunderstandings of the underlying economic reality and what the optimisation process is trying to achieve. The first fails to recognise that commonly understood economic principles have been derived from the manufacturing industry where the factory is the main asset. Mining is different – the depleting resource is the main asset, and the optimum cut-off may well be a Lane-style balancing cut-off that depends only on the grade distribution and the capacities of the production system, rather than a break-even cut-off grade that depends on prices and costs.

The remaining two objections fail to recognise that if the strategic goals have been established to maximise the benefit for all shareholders, the operation will be right-grading, not high-grading. It will be the best value obtainable to meet all the goals of various stakeholders. Any material not mined and treated would destroy value if it were mined, while unmined material does not have any value that needs to be recognised.

Other objections have some validity and need to be addressed in the evaluation process. Some are organisational or investor-relation concerns, while others relate to the scope of work. In principle, all of these can be overcome by identifying the issue and responding appropriately:

- there is no time or money available for optimisation in feasibility studies; it can be done when the mine is producing, and costs and performance are better known
- the market will react adversely if a reduced reserve is reported
- reducing the mine life limits the likelihood of exploration making another discovery that could profitably extend the life
- reducing the mine life increases the probability that a price increase that could add value won't occur until after the mine has closed.

The first of these is an organisational issue. The reality is that the optimum plan should be identified before such major value drivers and cut-offs, mining methods, pit sizes and mining and treatment rates have been decided upon. Indeed, an optimisation study should be the major source informing those strategic decisions. The cost of such studies is significantly less than the value gained by doing such a study.

The second is an investor relations and communications problem. There is a perception that analysts and fund managers focus on reported reserves and mine life, apparently believing that these are correlated with value or cash generation when they are not.

The final two are study scope issues. If the interplay between mine life and exploration success is a matter of concern, it simply has to be included in the evaluation. In principle this is just another relationship that needs to be modelled. The simplest way of accounting for it is to assume the same exploration program is conducted – faster or slower – to suit the mine life resulting from other strategic decisions made. This ensures that every case has the same likelihood of making a life-extending value-adding discovery. Similarly, if price increases are predicted, it is only necessary to model these as different scenarios, identifying the trade-offs between the upside of making correct decisions and the downside of wrong assumptions.

CHAPTER 15

In Conclusion

A REVIEW

We will now briefly review the topics dealt with by considering the common questions:

- Where have we come from?
- Where are we now?
- Where are we going?

The first of these is answered by summarising the main technical arguments advanced in the preceding chapters. The second is very much this author's perception of current practices. The third considers both technical and philosophical directions, and notes some hopeful signs of where the industry might be heading.

WHERE HAVE WE COME FROM?

Cut-off and strategy optimisation theory and practice have come a long way. We have seen how an increasing number of dimensions in the analysis can lead to a more comprehensive understanding.

Break-even analysis was seen to be a one-dimensional process, dealing with financial parameters – costs and prices – only. Simplistically, if the grade of mineralised material covers the costs of extracting and processing it as ore, it is classed as ore. Although widely used in the industry, break-even is a very limited cut-off model, not accounting for the geology, the nature of the mineralisation, nor the mining and processing plant capacities. Break-even cut-offs can be useful so long as they are used within the constraints of a more comprehensive model. The reality is that, in practice, a simple break-even model is all that is used by many operations, leading to mine plans that are almost assured not to deliver the company's stated goals. There is also no guarantee that an orebody delineated by a break-even cut-off will deliver a profit.

Break-even cut-offs ensure that every tonne classified as ore generates at least enough revenue to pay for costs that are included in the break-even calculation; however, there are no commonly accepted definitions of the costs to be included in those calculations.

Mortimer's Definition can be thought of as a two-dimensional model of cut-offs. As well as the financial aspects of break-evens, it brings in the need to consider the nature of the mineralisation via the tonnage and grade versus cut-off relationships. Mortimer's Definition has explicit goals of ensuring that the lowest grade of material mined pays for itself, and that the average grade of ore will deliver a specified profit target.

Lane's methodology has been state-of-the-art in cut-off optimisation for several decades, though its use has been limited. Its explicit goal is to maximise the net present value (NPV) of the operation. It can be described as a three-dimensional process – as well as the financial and geological dimensions of break-evens and Mortimer, Lane accounts for the production system's capacities to handle three classes of material: rock, ore and product. The rationale for this classification is fundamental for an understanding of all cut-off determinations. The very term *cut-off grade* indicates the need to deal with these materials – *grade* describes the amount of product within an amount of ore, and *cut-off* indicates distinguishing ore from the rock mass in which it is contained and separated.

Lane's methodology introduces the concept of a balancing cut-off, which ensures that two of the three system components – for rock, ore and product – are operating at capacity. A balancing cut-off is a function of geology and plant capacities only and not related to costs and prices. It will often be the optimum cut-off to apply. Lane also formalises the concept of opportunity cost, which has always been implicitly understood in the industry – one can mine down to the marginal break-even to fill the mill if there is a shortfall in ore supply, but this can only be done where low-grade material does not displace higher-grade feed. Opportunity cost quantifies the extent to which such displacement can occur economically.

With the development of powerful computers, full mine strategy optimisation has become the aim, even though current computer power is still only able to address parts of the problem. This is a multidimensional analysis, taking account of everything. The values of the decision variables can all be optimised while also taking account of the future-projected economic, financial, social and geological parameters. The goal must be explicitly stated in order to focus the process on a specific target. NPV maximisation is a logical goal, but in principle a number of value measures could be the target of the optimisation process. Decisions to be optimised include such things as:

- cut-offs
- production capacities in various parts of the overall production process, perhaps including multiple plants
- sequencing and timing of mining from stopes, mining blocks, pits and separate mines
- products and product mixes
- stockpiling policies.

All of these, where appropriate, can vary over time, by location, or both. In principle, if a relationship can be described, it can be modelled and included in the evaluations. The only limits are computing power, knowledge of the relationships within the system being evaluated and the ability to model them within the constraints of the optimisation techniques being used.

There are a number of cut-off models available to the mine planner. To deliver the corporate goals, the overall mine planning process must be focused on those goals. Long-term plans should be developed using the more complex higher-dimensional models of cut-off and strategy optimisation. Short-term plans that detail the earlier years may make use of simpler models to account for deviations and fluctuations from long-term average conditions on which the longer-term plans are based.

To generate robust long-term, goal-focused plans, an early consideration must be to decide upon the grade descriptor. This is the value attributed to each block of rock to describe its value relative to all other blocks – the number that tells us that this bit is more valuable than that bit. Actual metal grades, metal equivalents and money equivalents are commonly used. The best grade descriptor, however, will usually be value generated per unit of constraint in the production process. For simple styles of mineralisation, the simple grade may well be it. If there are different styles of mineralisation that impact on recoveries and product qualities, while affecting throughput rates, a measure such as dollars per operating hour may be the best grade descriptor to apply.

There are a number of measures that might be used to determine that one strategy is better than another, and alternative methodologies that might derive some of these measures. Various optimisation techniques can be applied to find the best strategies. The trade-off between maximising the reward from making correct assumptions about future conditions, such as price, and minimising the risk of making incorrect predictions, may be more important than optimising any one particular measure of value. Many techniques are useful and none can necessarily be relied upon to give the best answer. The project evaluator might well use a combination of techniques to allow the strengths of one to complement the weaknesses of another.

Given all these capabilities and considerations, corporate decision-makers can be provided with substantially more information on which to base their decisions than they have had in the past.

WHERE ARE WE NOW?

The situation describe here is very much the author's perception of current practices, based on experience and discussions with technical and management staff at many companies and operations in a number of countries. It is also informed by conference presentations, technical press reports and discussions with industry colleagues in the same field. While not rigorously and statistically valid, it comprises a large enough sample to identify both what is and is not being done in general practice around the industry.

Decisions regarding cut-off and other strategic policies cannot be delegated to technical staff on-site or in head offices. Cut-off, in particular, is one of the most misunderstood drivers of value for a mining company. It is a number that can be selected to deliver the corporate goals, but its derivation is often delegated to staff in junior positions with little understanding of its impact on production and financial targets. It should be the outcome of a strategy optimisation study, but is often a predetermined input into the study.

All mines will have some form of mine plan and cut-off policy in place, but the reality is that the planning and cut-off specification processes at many operations still leave much to be desired. It is not uncommon to find corporate long-term plans that are not set from the top down to achieve long-term goals. Rather, there is often an annual panic to generate a budget for the year ahead that reacts to current issues, with little or no consideration of how that may impact the long-term achievement of corporate goals. The practical reality is that staff at mine sites are dictating what they will produce, instead of being told by owners what is required. Of course, there are many organisations where plans are imposed from the top down, but the expectations are unrealistic and protests from the mine operators that this is the case fall on deaf ears higher up the chain of command. Unrealistic plans are agreed to because realistic ones will not be accepted, and failure to meet budgets is guaranteed. Communicating desired outcomes and practical outcomes both up and down the line is essential, but in reality it does not happen, or does not happen well, in many companies.

Many companies do not have a formal long-term planning process in place with regard to cut-offs and strategy optimisation, nor short-term plans that provide more detail for the achievement of long-term plans. Cut-offs are typically produced by break-even calculations, despite this being almost universally demonstrated to destroy wealth generation potential. Many companies do not feel the need to fully investigate what their optimum strategies might be – after all, it costs money and takes time to do so, and results may already be satisfactory. Again, traditional planning practices can be almost universally demonstrated to develop suboptimal strategies.

In companies that do have some form of long-term planning, there is often more optimism expressed than is justified by past outcomes. As Albert Einstein is reputed to have said, 'Insanity is doing the same thing over and over again and expecting different results.' Many in the industry fail to realise that mines do not have to lurch from one crisis to another, and do not understand the benefits that good planning can bring. It is easy to see the cost of a planning team in the mine's cost reports. If they have been doing their job well, it may be possible to save costs and get rid of some or all of them, but after some time, cracks will appear. It has been said that good planning is like good health – you don't realise you have it till you lose it. The benefit of a planning team is in the good plans that are made and, perhaps more importantly, in the mistakes that are avoided. How can that be quantified to offset against and justify the cost? If an increase in planning effort were to demonstrably change existing plans for the better, we'd be able to put a value on that change. If the planners are doing their jobs properly, the not-so-good plans and mistakes that might have been made don't happen – unnecessary extra costs are not incurred, revenue generation is not diminished, or is improved.

It is appropriate to have a challenging but achievable target for mine operators to inspire good performance; however, the very description indicates there is a high probability that the plan will fail to be achieved. In that case, the corporate projections and public pronouncements should be based on a more realistic plan.

The situation appears to be exacerbated by industry analysts and decision-makers focusing on measures not correlated with value creation. For example, mine life, ounces in reserve, productivity, unit costs and the like all appear to be useful numbers that can be used as surrogates for cash generation. Yet they are not necessarily correlated with long-term value creation. More reserves and mine life are good if they come from additional resources delineated by exploration success, but not if they arise merely from lowering the cut-off for existing resources. Because of the focus that these measures attract, they have in the past become drivers of the share price, and hence corporate planning directions, despite not being correlated with value creation. Strategies that increase cash-generating potential tend to come from higher cut-offs and reduced reserves, which will perversely drive the share price down, and vice versa. Much of this results from the mistaken belief that cut-off and break-even are synonymous, and that, if the price goes up, the cut-off has to go down.

It was noted in Chapter 10 that analysts and senior executives now recognise that earlier industry focus on the amount of valuable product in reported reserves and maximising production did not lead to increased cash generation for shareholders, despite metal prices being at all-time highs. It is also gratifying to see that there is no longer as much focus on reported tonnes and ounces in reserve as there has been in recent years, since these are usually negatively correlated with value creation (whether this continues, of course, is another matter). What is not so pleasing is that all the talk has swung around to focus on cost reduction, productivity and efficiency improvements. These are all good in their place, so long as they are done properly and only cut into fat and not muscle. There is little talk, at least in the public domain, of companies finding better strategies to maximise their cash generation potential; optimising mining and treatment rates, cut-off grades and optimising the size of pits, for example. Again we seem to have an industry focused on goals with the outward appearance of being correlated with value maximisation but in reality may not be.

Will we do any better in the next cycle than we did in the last one?

WHERE ARE WE GOING?

The author has deliberately painted a bleak picture of the state of mine planning and cut-off specification to make a point, though many would argue it is a realistic assessment. Of course, many companies do recognise the need for proper procedures focused on value creation rather than merely ensuring that every tonne pays for itself, but even they will admit off the record that they do not do as well as they might. Where might the industry be heading in the area of cut-off and strategy optimisation?

Technically

At the technical level, there are encouraging signs that more professionals are understanding and agreeing with the principles involved in cut-off and strategy optimisation. A number of writers of conference papers and course presenters are relaying the message around the world. While some of the objections discussed in Chapter 14 are still raised, in general these are addressed to the satisfaction of most.

For conducting studies, greater computer power will continue to aid optimisation models. The capabilities of high-end solvers, such as powerful mixed integer linear programming (MILP) packages, will continue to improve, and various advanced optimisation techniques that are not yet in widespread use will find their way into the mining strategy optimiser's toolkit.

Expanding software capabilities will also help fully integrate a number of powerful techniques into mainstream strategy optimisation modelling that are currently handled separately. Prediction is a notoriously dangerous activity, but the author suggests that, over the next few years, enhanced capabilities of optimisation software will incorporate conditional simulation of geological resource models and stochastic simulation of uncertain input parameters, such as prices, costs, construction and mining rates, and exploration outcomes. Together these will lead to probability distributions for multiple value measures, with enhanced decision-making tools to interpret this more-complex information.

Modelling of management flexibility will also improve. Many decisions will be optimised by the evaluation process, but others, particularly those that do not have to be made immediately, may be built into the modelling, so that the value of future management flexibility can be accounted for. The facilitation or preclusion of future options by immediate decisions will also be accounted for in the values generated. Improved visualisation techniques that enhance the interpretation of results and the comparison of multiple cases will be a likely addition.

The ultimate aim of strategy optimisation software is that it should simply be given the basic input data – the geological block model, geotechnical information, rules relating to several mining methods, productivity of workers and various items of mining equipment, metallurgical parameters, price forecasts, operating and capital cost data and the like – and of course the corporate goals. The optimum mine plan will then simply pop out at the end of the analysis. Eventually, the author suggests that the concept of cut-off will disappear. Plan-optimising software will have sufficient power to fully integrate the shortest- and longest-term plans, so that the short-term outputs will simply define what is to be mined on a daily or shift-by-shift basis while accounting for the long-term outlook and corporate goals. A cut-off grade that conveys rules about what is and is not to be mined from long-term planners to short-term planners will become obsolete.

We are still a long way from that capability, but given the history of rapid advances in technology, the author will avoid predicting how far away that situation might be.

Philosophically

We have suggested that industry analysts and decision-makers have been focusing on measures distinct from value creation, such as mine life, ounces in reserve, unit costs and the like, which give the appearance of being useful publicly reported numbers acting as surrogates for cash generation. As has been already been described, these are generally not associated with value. There is some indication that the situation is changing. Reports from the Mines and Money conference in London in December 2012 and December 2013 suggest that there may be the beginnings of a backlash against maximising reserves at the expense of cash generation. Analysts and fund managers want cash generation, not just increased reserves, and a number of mining executives have claimed they have changed focus from maximising reserves to cash generation, though, disappointingly, this seems to have morphed very quickly into a focus on cost cutting and productivity improvements, which are not necessarily correlated with cash generation.

Perhaps this is the opportune time for market expectations and corporate strategies to truly become aligned. If the market really wants cash generation, optimisations focused on this should be delivering the market's requirements. Most strategy optimisations of which the author is aware – his own and those conducted by other practitioners – are primarily focused on cash-related measures such as NPV. The market's verbally expressed requirements have therefore been met for a long time, but decision-makers are moving in a different direction because of their perceptions of how the market will respond.

Unfortunately, it seems that the only measure for maximising cash generation is maximising NPV (or similar cash-related measures). How do we focus on that at all levels in managing and controlling our operations? How do we report it regularly to a market that is traditionally looking at the results of the recent short period and a snapshot of where matters are at the time of reporting? Essentially we need to demonstrate, to the satisfaction of stakeholders, that we have identified the plan that generates the maximum value, and then find measures that show how we are implementing that plan. Short-term reporting of ounces in reserve, unit costs and productivity measures fail to actually do that. Worse, if they become surrogate value measures, since there apparently are no others, they have the potential to drag us away from the value-maximising plan because they do not correlate with real value-creation potential.

We need measures that are focused on long-term value generation at all levels of our operations, as well as in the public reporting area. Then we will not be rewarding value-destroying decisions at every level in the organisation, from the rock face to the CEO, as well as in the market.

Identifying the optimum plan is relatively easy. Developing the reporting metrics that help operators, managers and market analysts keep track of how well we are implementing it is the real challenge. The author does not claim to have an answer, but is flagging the need for a better measure that is yet to be identified.

How this will play out, time will tell. Everyone associated with value creation in the mining industry – senior corporate decision-makers, technical staff at all levels and industry analysts and fund managers – must recognise and demand that corporate strategies deliver real, not perceived, value maximisation. The focus must be on developing and using measures that are positively correlated with value. Otherwise, if surrogate measures are rewarded, the industry will continue to deliver unacceptable returns in the long run.

IN CONCLUSION

The author usually concludes courses and presentations on the topics in this book with a slide that states:

'The End of the Beginning'

It applies here, too. It is the end of this book. It is the beginning of readers doing something in their work area, to begin to change their company and the industry as a whole for the better.

References

Black, F and Scholes, M, 1973. The pricing of options and corporate liabilities, *Journal of Political Economy*, 81(3):637–654.

Cummins, A B and Given, I A (eds), 1973. *SME Mining Engineering Handbook* (Society of Mining Engineers of The American Institute of Mining, Metallurgical, and Petroleum Engineers, Inc: New York).

Dewhirst, R F, 2012. Using the handbook, in *Cost Estimation Handbook*, second edition, chapter 1, pp 1–20 (The Australasian Institute of Mining and Metallurgy: Melbourne).

Hall, B E and Stewart, C A, 2004. Optimising the strategic mine plan – methodologies, findings, successes and failures, in *Proceedings Orebody Modelling and Strategic Mine Planning Conference*, pp 49–58 (The Australasian Institute of Mining and Metallurgy: Melbourne).

Hawking, S and Mlodinow, L, 2010. *The Grand Design* (Bantam Books: New York).

Horsley, T P, 2005. Differential cut-off grades, in *Proceedings Ninth AusIMM Underground Operators' Conference*, pp 103–109 (The Australasian Institute of Mining and Metallurgy: Melbourne).

Kelly, J J and Bell, I F, 1992. Economically mineable resource in an underground metalliferous mine, in *Proceedings The AusIMM Annual Conference 1992*, pp 141–152 (The Australasian Institute of Mining and Metallurgy: Melbourne).

King, B M, 1999. Cashflow grades – scheduling rocks with different throughput characteristics, in *Proceedings Strategic Mine Planning Conference* (Whittle Programming Ltd: Melbourne).

Lane, K F, 1964. Choosing the optimum cut-off grade, *Colorado School of Mines Quarterly*, 59(4):811–829.

Lane, K F, 1988. *The Economic Definition of Ore: Cut-off Grades in Theory and Practice* (Mining Journal Books: London).

Lane, K F, 1997. *The Economic Definition of Ore: Cut-off Grades in Theory and Practice*, second edition (Mining Journal Books: London).

Lanz, T, Seabrook, W and McCarthy, P L, 2012. Operating cost estimation, in *Cost Estimation Handbook*, second edition, chapter 5, pp 57-82 (The Australasian Institute of Mining and Metallurgy: Melbourne).

Mortimer, G J, 1950. Grade control, *Transactions of the Institution of Mining and Metallurgy*, 59:1–43.

Peele, R (ed), 1941. *Mining Engineers' Handbook*, third edition (John Wiley & Sons Inc: New York).

Samis, M, Martinez, L, Davis, G A and Whyte, J B, 2012. Using dynamic discounted cash flow and real option methods for economic analysis in NI43-101 technical reports, in *Proceedings VALMIN Seminar Series 2011–12*, pp 149–160 (The Australasian Institute of Mining and Metallurgy: Melbourne).

Steffen, O K H, 1997. Planning of open pit mines on a risk basis, *Journal of the South African Institute of Mining and Metallurgy*, March/April, pp 47–56.

APPENDIX

Case Studies

INTRODUCTORY COMMENTS

In evaluations conducted by the author and his colleagues, increases in project net present value (NPV) ranging from ten per cent to 50 per cent are generated by assessing all the options and identifying better strategies. The case studies that follow illustrate some of the types of evaluations and strategy changes encountered. Results have been normalised to show the value of the base case or approved strategy at the start of the analysis as an index value of 100, and values for other cases relative to this. These should be seen as indicative only of the types of evaluations and outcomes that can be done, and are not exhaustive.

CASE STUDY 1

Underground gold – cut-off optimisation

The client operated a gold mine where underground mining had commenced below the exhausted open pit. The orebody extended to over 1000 m below surface, and two mining methods were used depending on its thickness and dip in various locations. The mining plan called for progressive deepening of the mine by a decline access, using underground mine trucks to haul the ore to the stockpile at the treatment plant on surface.

A number of major production options were evaluated over the course of the study:

- Shaft hoisting was proposed to replace the increasingly expensive truck haulage as the mine got deeper, but there was insufficient reserve for a reduction in ore handling operating costs to pay for the capital outlay. The analysis indicated the reserve tonnage needed to justify sinking a shaft.

- Several stages of upgrades in the treatment plant were proposed. At projected gold prices, only the first of these was justifiable. The second stage was only justifiable at higher than expected gold prices, where increased gold production and a higher price together were sufficient to justify the capital outlay.

- Several stockpiles of lower-grade material remained from the open pit mining. At the expected gold prices, fresh underground ore (using the optimum cut-off) and reclaimed stockpile ore had similar cash margins. The value of the operation was therefore relatively insensitive to the underground production rate. However, increases in the gold price or different mine head grades resulting from suboptimal cut-offs disturbed this balance and increased the impact of the underground production rate on value.

The major decision ultimately involved the cut-off policy. Like many operating mines, the existing LOMP used a single mine-wide cut-off for the whole mine life. Significant value was added by increasing the mine-wide cut-off, as illustrated in Figure A1. Further value was obtained by specifying separate cut-offs for each mine area. These were optimised using a genetic algorithm (GA) add-in with Microsoft Excel™. Cut-offs used in this analysis were at 1.0 g/t increments, which may be adequate to identify the shape of the hill of value (HoV) and an optimum mine-wide cut-off, but are probably too coarse to fully optimise individual mining area or time-based cut-offs. The additional value shown for GA-optimised area cut-offs in Figure A1 is therefore believed to be conservative.

FIGURE A1
Case study 1 – value increase from cut-off changes.

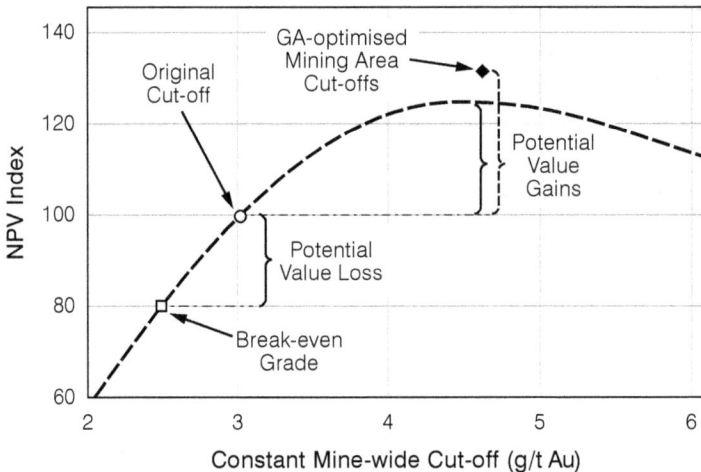

CASE STUDY 2

Underground gold – production rate and cut-off optimisation

The client had acquired a resource in a developing country that had been worked as an open pit operation for some years. There were two adjacent multilens deposits, and the title to one was less secure than the other. An underground resource had been identified, and a prefeasibility study (PFS) conducted, but only for both deposits at one production rate and one cut-off grade.

An initial high-level review of this study suggested there could be significant upside from different operating strategies, and a more detailed study was commissioned. This was a desktop study only, which involved reworking the PFS data and constructing a flexible but robust evaluation model. The investigation involved:

- definition of ore reserves at various cut-offs
- estimation of development requirements at each cut-off
- specification of ranges of the mining schedule drivers to be evaluated
  - maximum decline advance rate
  - initial stock of reserves developed before full production
  - ongoing development advance rate
  - maximum ore lens production rates
  - overall mine production target
  - sublevel interval
  - production strategy – equal priority for all exposed ore sources above cut-off, or highest grade sources targeted first (the latter requires a faster overall development rate than the former, but has the potential to deliver higher grades to more than offset the increased costs and thus add value)
- construction of a model to
  - schedule mine development and production, accounting for the controlling parameters above
  - derive other mining and production physical quantities
  - derive cash flows
    - revenue
    - operating and capital costs
    - taxes and government charges
  - derive and report NPVs, internal rate of return (IRR), returns to the government.

The modelling identified a number of characteristics of the project, including optimum sequencing strategies depending on the relativities of various mining rate parameters, such as decline advance rate and lens and total mine production capabilities, and whether the less secure deposit was mined or not. Figure A2 shows a HoV for NPV as a function of cut-off and production rate target. The value falls at high cut-offs due to a reduction in reserve, but is relatively flat at low cut-offs due to the simplified block model having low-grade regions below the resource cut-off excised, reducing the variation of tonnage and grade with cut-off.

Value generally continues to rise with increasing production rate target, as upper limits to mining rates were not well established at this stage of the evaluation. However, high level analysis suggested that the original PFS rate of 1.2 Mt/a could be increased to approximately 1.5 Mt/a. The HoV indicates that NPV could be increased by 25 to 30 per cent simply by increasing the cut-off at a production rate of 1.2 Mt/a, and by some 50 per cent if the production rate could also be increased at the higher cut-off.

The analyses also generated HoVs for the returns to government for various strategies, to indicate whether there might be any conflict between corporate and government goals and if so, what trade-offs might be appropriate.

FIGURE A2

Case study 2 – value increases from cut-off and production target changes.

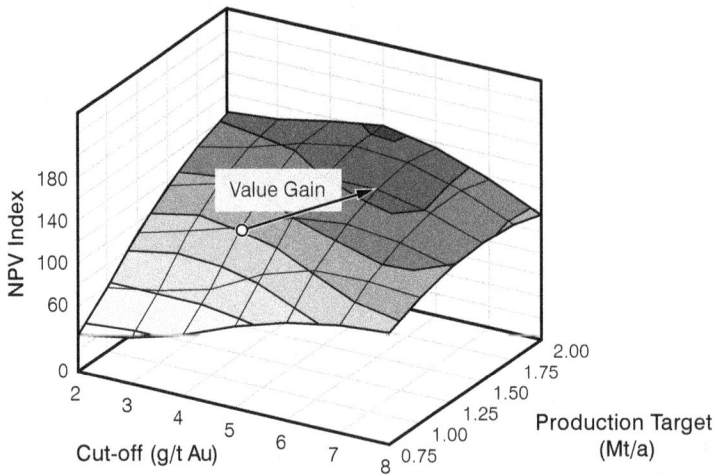

CASE STUDY 3

Underground base metals – grade descriptor, mining method, production rate and cut-off optimisation

The client operated an underground lead-zinc mine that was generating poor returns. Historically the mine had been developed as a sequence of mining blocks approximately 100 m high, with deeper blocks opened up as blocks closer to surface were depleted. There were known subvertical trends in grades, but the orebodies defined using the existing metal equivalence formula and cut-offs were such that the grades seen within the mineralised zone were relatively homogeneous along strike. The bulk of the mineralisation was mined out using open stoping with backfill to facilitate close to total extraction, with only a few lower-grade pillars being left *in situ* between the major lenses. The mining sequence was therefore predominantly along strike, with a subsidiary advance from top to bottom. Recovery of the crown pillars created between lifts presented technical challenges, with both lower productivity and reduced ore recovery in those areas.

The study first evaluated the effects of the grade descriptor used. Figure A3 shows how the true value generated by a block of rock described by the original metal equivalent value could vary significantly: the range was equivalent to the unit cost of mining. A revised equivalence formula with a much lower variation in true value was adopted.

When the mineralisation was viewed using the new grade descriptor, the vertical zonation, particularly at higher cut-offs, became more pronounced. An obvious change to the mining method was to develop immediately to the bottom of the deposit and establish stopes extending over the vertical height of the remaining orebody, with mining in a predominantly bottom-to-top sequence within lenses, and a subsidiary advance along strike. This permitted both the elimination of crown pillars and establishing extra vertical rib pillars between stopes in lower value material, allowing a significant reduction in the amount of cemented fill required. The overall value of the resource was enhanced by a reduction in the misclassification of ore and waste, which in turn resulted in a more apparent relationship between value and cut-off. Figure A4

FIGURE A3

Case study 3 – changes in spread of true values with different grade descriptors.

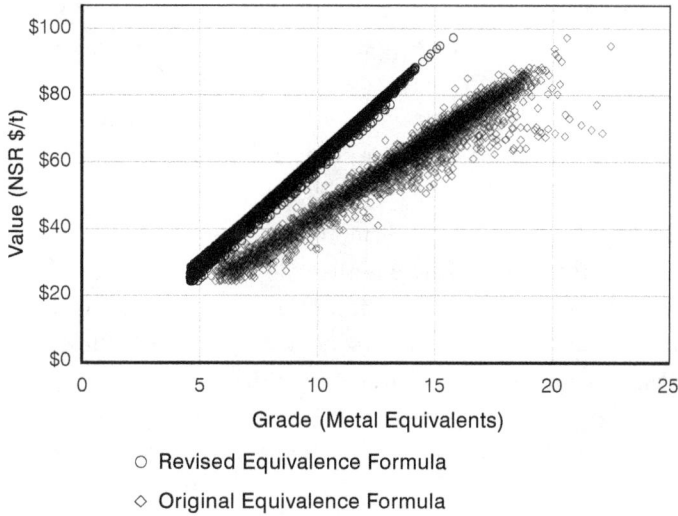

○ Revised Equivalence Formula

◇ Original Equivalence Formula

shows value versus cut-off for the original equivalence formula and mining method, and the proposed equivalence formula and mining method. The values for the latter include the effect of earlier capital development to drive immediately to the bottom of the mine to establish the full-height stopes.

The steepening of the value versus cut-off curve at high cut-offs is due to a change in the nature of the orebodies with changing cut-offs. At higher cut-offs, the orebody broke up into lenses with smaller vertical extent than at lower cut-offs, necessitating a reduction in the sublevel interval and hence a step-increase in the amount of development required.

The production rate also had a significant influence on value. Although the mine's target was nominally the same as the concentrator capacity, and the fleet size and workforce

FIGURE A4

Case study 3 – value increase from grade descriptor, cut-off and mining method changes.

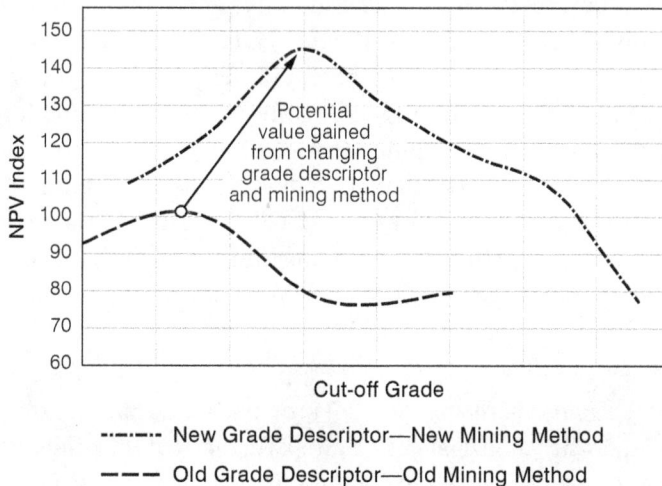

--·--·-- New Grade Descriptor—New Mining Method

－－－ Old Grade Descriptor—Old Mining Method

numbers were in place for this, the mine was in practice delivering only some 85 per cent of that target. It was identified that there were inadequate stock levels of developed and drilled ore, exacerbated by the reliance on a small number of large stopes in operation at any time. Additional value could be gained simply by producing at the target rate. This could be achieved by injecting working capital to increase ore stocks, and produce from a larger number of smaller stopes. The change in mining method also facilitated this change. Additional capital investment to upgrade the capabilities of the rock handling systems was thereby avoided.

The client elected to change the mining method and increase the cut-off substantially towards the identified optimum, so as to realise the bulk of the identified potential gain. At the same time this restricted the reduction in ore reserves to a level where the client believed the market would not react adversely and drive the value of the shares down, despite the increase in value generated for shareholders.

CASE STUDY 4

Open pit base metals – mining rate and cut-off optimisation

The client operated a large base metal open pit mine. Mine staff believed that pressure from the corporate head office to reduce costs was resulting in a reduction in waste stripping and hence overall mining rates and exposure of ore above cut-off, leading to loss of value. The study evaluated only the effects of changing the run-of-mine (ROM) cut-off and the mining rate. The existing ultimate pit limits and sequencing of mining over the life-of-mine, and the planned sequence of improvements and upgrades in the concentrator as defined in the approved plan, were retained.

Rock types had different milling rates, and combinations of mineralogies of both ore and waste minerals had complex effects on recovery in the concentrator and on product quality, which impacted on the price received. Net revenue per mill hour was identified as a better grade descriptor than a simple metal grade per tonne for ranking potential ore sources. Some two dozen rock types were identified as having potential influences on value. A number of proposed concentrator upgrade options were included in the optimisation model developed. This allowed identifying the optimum mining plans for each concentrator option and hence the real value gained from each upgrade. Figure A5 shows the results of the study, which corroborated the project sponsors' views. Faster mining of total rock allowed exposure of more mineralisation, feeding higher grades to the mill in the short term, while stockpiling the remainder for later treatment. This adds value up to a point, the additional product paying for the extra mining costs. Further mining rate increases beyond this point destroy value. Further value gains could have been obtained by fully investigating the size of the ultimate pit, sequencing and blending and alternative mill upgrade scenarios. The final decision was to save short-term costs by reducing the mining rate (and hence the ROM cut-off, to keep the mill full), which Figure A5 shows destroys more value than the costs saved.

In this case the value gains are relatively small proportionally, but the size of the project is such that substantial dollar value changes result, particularly if the difference between the values for what the company could have done and what they actually did was considered.

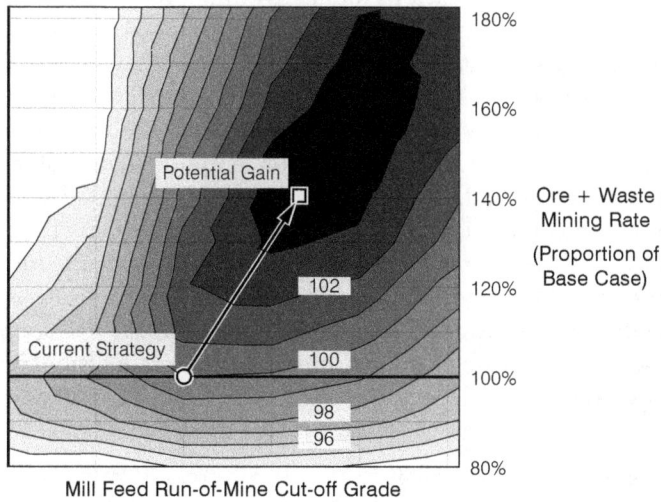

FIGURE A5

Case study 4 – effects of changing total mining rate and run-of-mine cut-off in an open pit.

CASE STUDY 5

Mineral sands – optimising deposit sequencing, mining selectivity and number of primary treatment plants

The client owned a number of mineral sands deposits containing mixtures of heavy minerals. Each deposit typically consisted of a strand line with spatial distributions of minerals differing according to the heavy mineral (HM) composition and deposition mechanics. The values of component HMs varied significantly, depending on prices and quality of the potential products.

The company's policy had been to mine and treat of all mineralised material above a specified total HM content. The company owned or planned to acquire a number of wet concentrators to separate the HM from the sand mass at the mining sites, with a further option to add magnetic separators to remove low-value ilmenite at the mine. HM concentrates, with or without ilmenite, were transported from each mine site to a single dry mill, which separated the heavy minerals from each other into various products.

The company wished to identify the optimum mining and production strategy, including such items as:

- the number of wet concentrators
- the sequence of mining the deposits, accounting for allocation of wet concentrators to deposits and the cost of moving concentrators between deposits, which also depended on the sequence
- the best plan for mining each deposit, including how much of the mineralisation to mine, how much overburden to prestrip (where applicable), and whether to separate the ilmenite and discard it as waste at the mine site.

As in Case Studies 3 and 4, the total HM grade descriptor used by the client was incapable of discriminating between higher and lower value portions of the deposits. A dollar per tonne grade descriptor was therefore used, calculated as the net revenue after deduction of wet concentration, transport, separation and realisation costs.

Each deposit was analysed individually to identify its inherent value-versus-operating-strategies relationships. Industry-standard open pit optimisation software was used to derive nested pit shells extracting various tonnages of mineralisation. These shells tended with increasing size to go rapidly to the base of the highest value parts of the strand, initially in isolated pits, then to extend along strike to form longer continuous pits, and across strike to include extensions into lower-value material. Simple HoVs were derived for each deposit, with the main value drivers being the pit shell number and the dollar value cut-off. This analysis accounted for wet concentrator moves along the length of the deposit, and identified whether isolated pits were better mined separately, or with the intervening low-grade material mined to connect them. In some deposits, value was relatively insensitive to either variable over a range of settings, but reduced significantly if either were set outside this range. In other deposits, values were relatively insensitive to one parameter while highly sensitive to the other, and vice versa. The company's existing designs were in some cases close to the optimum, and in others, significantly different. While the company's ranking of deposits by traditional planning methods were generally similar to the study rankings, there were some differences.

An evaluation model was constructed to specify several strategic decision parameters for all deposits. The number of wet concentrators was varied. Each deposit, as well as having its pit size and cut-off variable, was able to be assigned to any concentrator and in any sequence, or not mined at all. A genetic algorithm controlled the settings of all these strategic decision variables, and was run to maximise overall NPV for all deposits.

It was found that deposits mined earlier in the sequence tended to be worked at smaller pit sizes and ore tonnages than their stand-alone optimum sizes. This brings forward the value of later deposits, so that the loss of value for an individual early deposit is more than compensated for by the increase in NPV from bringing forward the values of later deposits in the sequence. However, shorter mine lives in the early deposits meant that the company had to commit to bringing other deposits into production earlier, with associated permitting and infrastructure development issues to be addressed sooner than anticipated.

Significant increases in value were obtained by the strategy optimisation study, as shown in the following descriptions of how strategies were developed. As in other case studies, the value numbers are NPV indexed to the NPV of the company's base case plan before the study commenced, which has a value of 100:

- original wet and dry plant capacities, mining sequences and reserves: 100
- original wet and dry plant capacities and mining sequences, best value stand-alone reserves: 104
- original wet and dry plant capacities, GA-optimised mining sequences, allocation of concentrators, and reserves: 110
- GA-optimised wet and dry plant capacities, mining sequences, allocation of concentrators, and reserves: 118

It was also found that an extra wet concentrator and earlier upgrades of the dry plant were justified if product prices were higher, the extra capital cost being repaid by bringing forward the net revenue from the operating mines. Again, this would have implications for permitting and infrastructure development, as well as exploration programs. It would be up to the company to decide whether the additional value justified the extra effort to bring about these changes of plan.

www.ingramcontent.com/pod-product-compliance
Lightning Source LLC
Chambersburg PA
CBHW082003190326
41458CB00010B/3054